Petrology - New Perspectives and Applications

Petrology - New Perspectives and Applications

Editor

Jason Silver

scitus academics

Petrology - New Perspectives and Applications
Edited by **Jason Silver**

ISBN: 978-1-68117-594-2
Library of Congress Control Number: 2016946449

SCITUS Academics LLC,
616, Corporate Way, Suite 2, 4766,
Valley Cottage, NY 10989

www.scitusacademics.com

Notice

Printed in United States of America on Acid Free Paper

Preface

Petrology is a geological field that focuses on the study of rocks and their conditions. There are a variety of applications for petrology, including classroom, field and laboratory opportunities for people with petrology degrees. In this area, researchers are enthusiastic about how rocks form, what they contain, how changes in conditions during formation can lead to structural and compositional changes, how rocks weather and how rocks can be used. In the field, petrology can include the identification of several types of rock, as well as the study of rock strata and different geological structures. Petrologists can use a number of technologies to take core samples and carry out imaging studies that enable them to see in the crust of the earth. Petrology is used in industry by many mining companies and the oil industry.

Petrology - New Perspectives and Applications is intended for advanced petrology graduates and professionals. The book focuses on modern petrological and geochemical methods of understanding igneous, metamorphic and sedimentary rocks.

Table of Contents

Chapter 1

Sulphide Sinking in Magma Conduits: Evidence from Mafic–Ultramafic Plugs on Rum and the Wider North Atlantic Igneous Province

Hannah S. R. Hughes[1,*], Iain McDonald[1], Adrian J. Boyce[2], David A. Holwell[3] and Andrew C. Kerr[1]

[1]School of Earth and Ocean Sciences, Cardiff University, Park Place, Cardiff CF10 3AT, UK

[2]Scottish Universities Environment Research Centre, Rankine Avenue, East Kilbride, Glasgow G75 0QF, UK

[3]Department of Geology, University of Leicester, University Road, Leicester LE1 7RH, UK

ABSTRACT

Ni–Cu–PGE (platinum group element) sulphide mineralization is commonly found in magmatic conduit systems. In many cases the trigger for formation of an immiscible sulphide liquid involves assimilation of S-bearing crustal rocks. Conceptually, the fluid dynamics of sulphide liquid droplets within such conduits is essentially a balance between gravitational sinking and upwards entrainment. Thus, crustal contamination signatures may be present in sulphides preserved both up- and down-flow from the point of interaction with the contaminant. We examine a suite of ultramafic volcanic plugs on the Isle of Rum, Scotland, to decipher controls on sulphide accumulation in near-surface magma conduits intruded into a variable sedimentary stratigraphy. The whole-rock compositions of the plugs broadly overlap with the compositions of ultramafic units within the Rum Layered Complex, although subtle differences between each plug highlight their individuality. Interstitial base metal sulphide minerals occur in all ultramafic plugs on Rum. Sulphide minerals have magmatic $\delta^{34}S$ (ranging from −1·3 to +2·1‰) and S/Se ratios (mean = 2299), and demonstrate that the conduit magmas were already S-saturated. However, two plugs in NW Rum contain substantially coarser (sometimes net-textured) sulphides with unusually light $\delta^{34}S$ (−14·7 to +0·3‰) and elevated S/Se ratios (mean = 4457), not represented by the immediate host-rocks. Based on the Hebrides Basin sedimentary stratigraphy, it is likely that the volcanic con duits would have intruded through a package of Jurassic mudrocks with characteristically light $\delta^{34}S$ (−33·8 to −14·7‰). We propose that a secondary crustal S contamination event took place at a level above that currently exposed, and that these sulphides sank back to their present position. Modelling suggests that upon the cessation of active magma transport, sulphide liquids could have sunk back through the conduit over a distance of several hundreds of metres, over a period of a few days. This sulphide 'withdrawal' process may be observed in other vertical or steeply inclined magma conduits globally; for example, in the macrodykes of East Greenland. Sulphide liquid sinking within a non-active conduit or during magma 'suck-back' may help to explain crustal S-isotopic compositions in magma conduits that appear to lack appropriate lithologies to support this contamination, either locally or deeper in the system.

KEYWORDS

sulphur isotopes, North Atlantic Igneous Province, sulphide sinking, S/Se crustal contamination

1. INTRODUCTION

Upper crustal conduits (pipes, dykes and sills) within mafic–ultramafic magmatic systems provide favourable settings for magmatic sulphide genesis, promoted by crustal contamination, resulting in some of the world's largest orthomagmatic Ni–Cu and platinum group element (PGE) deposits. Contamination can trigger S-saturation of the magma, particularly if the country rocks contain significant amounts of S in the form of sulphides or sulphates; this can be recognized by S isotope signatures outside the typical mantle range of 0 ± 2‰ (e.g. Ohmoto & Rye, 1979; versus Li *et al.*, 2002; Ripley *et al.*, 2003; Keays & Lightfoot, 2010;Hayes *et al.*, 2015; Smith *et al.*, 2016). In addition, sulphide liquid that ponds within conduits can become enriched in chalcophile elements, possibly to economic levels, as a result of interaction with magma continuing to pass through the conduit (e.g. Kerr & Leitch, 2005; Holwell*et al.*, 2014).

Sulphides within conduits commonly occur as massive to semi-massive accumulations and are present as rounded globules or droplets of varying sizes; for example, at Norilsk–Talnakh, Russia (Czamanske *et al.*, 1992), the Insizwa Complex, South Africa (Lightfoot *et al.*, 1984), Voisey's Bay, Canada (Huminicki *et al.*, 2008), a mafic dyke in Uruguay (Prichard *et al.*, 2004) and in East Greenland macrodykes (Holwell *et al.*, 2012). Droplets are generally interpreted to represent sulphide liquid that has been transported upwards, downwards or laterally along the magmatic plumbing system. The fluid dynamics involved in this transport, such as settling rate versus entrainment rate, has been the subject of many recent studies based on both empirical and experimental evidence (e.g. de Bremond d'Ars *et al.*, 2001; Mungall & Su, 2005; Godel *et al.*, 2006, 2013;Barnes *et al.*, 2008; Chung & Mungall, 2009; Arndt, 2011).

Sulphide- and chromitite-bearing ultramafic units hosting platinum group minerals (PGM) are documented within the British Paleogene Igneous Province (BPIP) on the islands of Skye, Mull and Rum (Hulbert *et al.*, 1992;Butcher *et al.*, 1999; Pirrie *et al.*, 2000; Power *et al.*, 2000; Prout *et al.*, 2002) and highlight the potential for PGE mineralization in the Scottish portion of the North Atlantic Igneous Province (NAIP; Andersen *et al.*, 2002). This is especially the case considering the potential for contamination offered by the thick crustal sequences of Mesozoic S-bearing rocks in the region (e.g. Hesselbo & Coe, 2000; Yallup *et al.*, 2013; Hughes *et al.*, 2015). Recent studies on Skye demonstrate the involvement of sediments, with light S isotope tenors, in triggering S-saturation in ascending BPIP magmas, particularly in vertical dyke swarms (Hughes *et al.*, 2015). On Rum, ultramafic, vertical or steeply inclined volcanic plugs (pipes) intrude a variety of country rocks, including sediments at the base of the Mesozoic Hebrides Basin. Power *et al.* (2003)identified a crustal $\delta^{34}S$ signature for intercumulus (net-textured) sulphides in one volcanic plug in NW Rum, but could not establish whether the sulphide liquids that formed these had been entrained upwards or had sunk through the conduit.

In this study, we use the whole-rock S-isotope composition, and sulphide-specific (*in situ*) S-isotope and trace element compositions, of ultramafic plugs on Rum to constrain if crustal contamination took place. The size range and textural associations, trace element composition of sulphide minerals and S-isotopic composition are established for 12 plugs on the island to understand the provenance of sulphide liquids in these conduits. We use these data to decipher controls on sulphide accumulation and compare plug sulphide compositions with those from peridotitic cyclic units of the Eastern Layered Series, part of the Rum Layered Suite. Where crustal contamination is evident [e.g. in NW Rum, as first discovered byPower *et al.* (2003)], these data are combined with numerical modelling of settling and entrainment rates to investigate the roles of magma entrainment and gravitational settling and sinking of sulphide liquids within conduit systems and to assess the distances that sulphide liquids could sink and over what timescale. We provide new constraints on the development of orthomagmatic Ni–Cu–PGE mineralization in small BPIP intrusions and discuss these in relation to wider processes of sulphide migration within magmatic systems.

2. REGIONAL GEOLOGICAL SETTING

The North Atlantic Igneous Province (NAIP) formed after the impingement of the Icelandic mantle plume under thick continental lithosphere, during a period of continental flood basalt magmatism (e.g. Saunders *et al.*, 1997). Magmatism initiated *c.* 62 Ma within what is now the UK, Greenland and Baffin Island, and ultimately led to the opening of the Atlantic Ocean (Saunders *et al.*, 1997). The main products of this prolonged period of magmatism were tholeiitic and alkali basalts. The British Palaeogene Igneous Province (BPIP) is part of the earliest magmatic series of the NAIP, which includes Palaeogene rocks in the Hebrides and along the west coast of Scotland and Northern Ireland (Saunders *et al.*, 1997; Fig. 1).

In Scotland, the BPIP includes magmatic rocks on the Isles of Mull, Skye and Arran, the Small Isles (Rum, Eigg, Muck, Canna and Sanday), and on the mainland at Ardnamurchan and Morvern (Emeleus & Bell, 2005). The Scottish BPIP extends across a number of Archaean and Proterozoic basement terranes (e.g. Emeleus & Bell, 2005), covered by Neoproterozoic (Dalradian and Moine) metasediments (Kinnaird *et al.*, 2007), which themselves are overlain by Mesozoic sediments from the Hebrides Basin (Hesselbo & Coe, 2000). Volcanic plugs occur within or close to lava fields and central intrusive complexes, and are primarily exposed on the Isles of Mull and Rum (Emeleus & Bell, 2005). Emeleus & Bell (2005) provided a detailed review of radiometric and palynological age determinations of various intrusions and lavas across the BPIP.

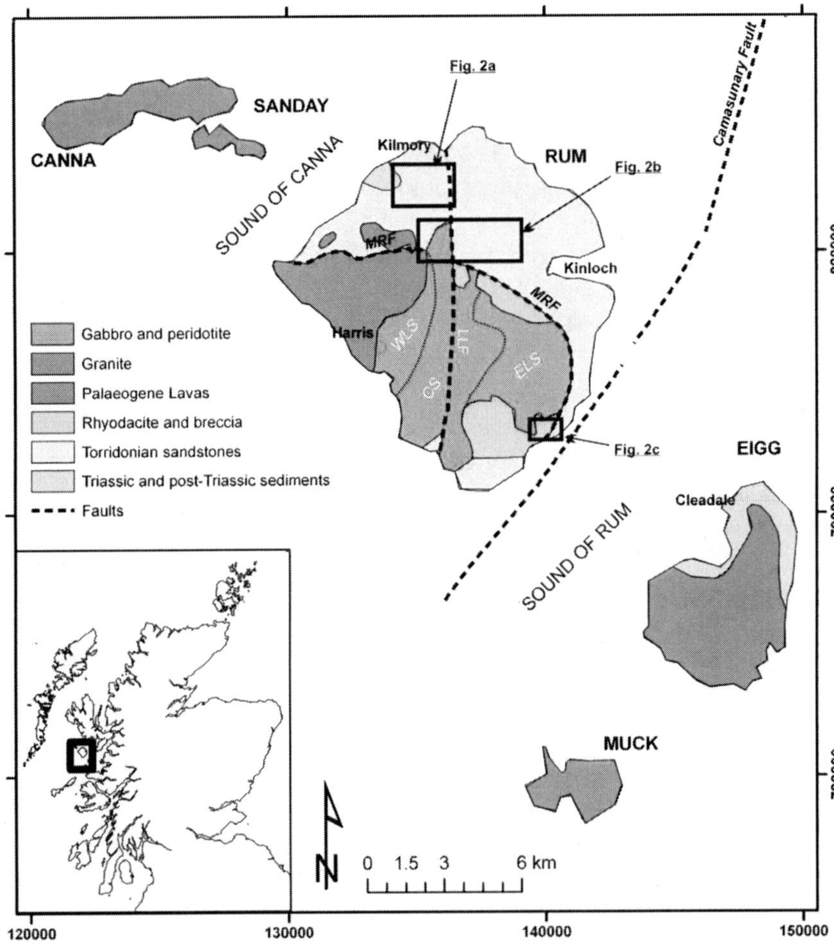

Figure 1. Simplified geology of the Isle of Rum and surrounding Small Isles. Inset shows outline map of Scotland. Rectangles show map locations in Fig. 2. LLF, Long Loch Fault; MRF, Main Ring Fault; WLS, Western Layered Series; CS, Central Series; ELS, Eastern Layered Series.

2.1 Geology of the Isle of Rum

The Rum Central Complex straddles the Long Loch Fault (Fig. 1), which probably acted as a crustal lineament that facilitated magma ascent (e.g.Upton *et al.*, 2002). The complex developed during at least three magmatic phases, forming in < 500 kyr (Emeleus & Bell, 2005; Troll *et al.*, 2008).

In Phase 1, silicic magmatism occurred along an arcuate fault system (the Main Ring Fault) at 60·33 ± 0·21 Ma (Troll *et al.*, 2008) followed rapidly by Phase 2, which formed the mafic–ultramafic Rum Layered Suite (60·53 ± 0·04 Ma; Hamilton *et al.*, 1998). Ultramafic magmas were intruded into the complex as multiple intrusions and replenishment of the chamber with picritic magma is thought to have disrupted previously formed cumulate layers (e.g. Bédard *et al.*, 1988). The mafic–ultramafic Rum Layered Suite can be divided into the Western, Central and Eastern Layered Series, bounded by the Main Ring Fault (Fig. 1). Rare late-stage picrite dykes [described by Upton *et al.* (2002)] may represent feeders to the layered suites. However, the small melt fraction (6–7%) inferred to have formed these mildly alkalic dykes and two olivine phenocryst populations observed within single picrite dykes, such as M9 (e.g. Upton *et al.*, 2002), preclude these feeders from representing a parental magma composition. During Phase 3, deep subaerial erosion and unroofing of the complex followed caldera collapse and intermittent burial by basalt flows from the Skye lava field (Chambers *et al.*, 2005).

2.2 Ultramafic plugs of the Isle of Rum

In excess of 40 volcanic plugs occur on Rum, comprising gabbros and feldspathic peridotite intrusions (*senso stricto* olivine melagabbros) probably formed during Phase 2. Plugs predominantly intrude through Torridonian and Triassic sediments, as well as portions of the Rum Layered Suite (McClurg, 1982; Wadsworth, 1994; Holness, 1999). The sedimentary host-rocks of the volcanic plugs (Triassic, Fig. 2a; and Torridonian, Fig. 2a–c) typically appear bleached, fissile and metasomatized. Locality details of the plugs sampled during this study are given in Table 1 (also see Fig. 2a–c).

Table 1: Isle of Rum sample information for volcanic plugs and associated sediments

Sample number	Location	Feature type	Plug number (Fig. 1)	Layering
Isle of Rum volcanic plug and associated dyke samples				
RM81	Dibidil	peridotite plug centre	12	none observed
RM58	Kinloch Glen	peridotite plug margin	7	olivine alignment
RM59	Kinloch Glen	peridotite plug centre	7	none observed
RM60	Kinloch Glen	gabbro plug margin	9	none
RM61	Kinloch Glen	peridotite plug centre, homogeneous but heavily weathered with some fresher nodules	8	none observed
RM63	Kinloch Glen	peridotite plug centre	6	none observed, fissile, hydrothermally altered
RM64	Kinloch Glen	peridotite plug centre	6	none observed
RM17	Monadh Mhiltich	peridotite plug margin	11	none observed
RM18	Salisbury's Dam	peridotite plug margin	10	
RM35	Loch Sgaorishal	peridotite plug centre	2	weak layering—olivine alignment
RM70	Loch Sgaorishal	peridotite plug NW margin (at contact)	2	weak layering—olivine alignment
RM71	Loch Sgaorishal	peridotite plug NW margin (5 m inwards from contact)	2	weak layering—olivine alignment
RM73	Loch Sgaorishal	peridotite plug centre	2	weak layering—olivine alignment
RM74	Loch Sgaorishal	peridotite plug NE margin	2	weak layering—olivine alignment
RM75	Loch Sgaorishal	peridotite plug NE margin (~20 m inwards from inferred contact)	2	weak layering—olivine alignment
RM85	Southern Sgaorishal	peridotite plug margin	5	weak cm-scale layering
RM86	Southern Sgaorishal	peridotite plug centre	5	weak cm-scale layering
RM87	Southern Sgaorishal	peridotite plug margin	4	none observed
RM88	Southern Sgaorishal	peridotite plug centre	4	none observed
RM89	Southern Sgaorishal	gabbro plug S margin (~3 m inwards from inferred contact)	3	none
RM29	West Sgaorishal	peridotite plug N margin (~30 m inwards from inferred contact), banded (dipping inwards)	1	indentation layering
RM30	West Sgaorishal	peridotite plug centre, banded and colloform	1	indentation layering
RM32	West Sgaorishal	peridotite plug NW margin (~35 m inwards from inferred contact)	1	indentation layering
RM33	West Sgaorishal	peridotite plug N margin, banded (dipping inwards)	1	indentation layering
RM91	West Sgaorishal	peridotite plug S margin, banded	1	indentation layering
RM95	West Sgaorishal	peridotite plug NW margin, banded (dipping inwards)	1	indentation layering

Sample number	Location	Period: Group, Formation	Sediment type	Sediment sample relationship to volcanic plugs or other proximal magmatic rocks
Isle of Rum sedimentary samples as possible contaminants of plugs				
RM21	Monadh Dubh	Triassic: New Red Sandstone Supergroup, Monadh Dubh Sandstone Formation	conglomerate with calcrete matrix	fresh, but cross-cut by series of cm-scale basaltic dykes and sills
RM22	Monadh Dubh	Triassic: New Red Sandstone Supergroup, Monadh Dubh Sandstone Formation	coarse red sandstone	fresh, but cross-cut by series of cm-scale basaltic dykes and sills
RM23	Monadh Dubh	Triassic: New Red Sandstone Supergroup, Monadh Dubh Sandstone Formation	white limestone	fresh, but cross-cut by series of cm-scale basaltic dykes and sills
RM24	Monadh Dubh	Triassic: New Red Sandstone Supergroup, Monadh Dubh Sandstone Formation	micaceous red arkose	fresh, but cross-cut by series of cm-scale basaltic dykes and sills
RM26	Monadh Dubh	Triassic: New Red Sandstone Supergroup, Monadh Dubh Sandstone Formation	limestone with calcretes	fresh, but cross-cut by series of cm-scale basaltic dykes and sills

Sample number	Location	Period: Group, Formation	Sediment type	Sediment sample relationship to volcanic plugs or other proximal magmatic rocks
RM28	Monadh Dubh	Triassic: New Red Sandstone Supergroup, Monadh Dubh Sandstone Formation	limestone	fresh, but cross-cut by series of cm-scale basaltic dykes and sills
RM94	West Sgaorishal	Triassic: New Red Sandstone Supergroup, Monadh Dubh Sandstone Formation	limestone	~60 m from plug margin, limestone is baked
RM3	Guirdil, Isle of Rum	Mesoproterozoic: Torridon Group, Applecross Formation, Scresort Sandstone Member	medium–coarse sandstone	no plugs in vicinity, some small-scale basaltic dykes
RM5	Guirdil, Isle of Rum	Mesoproterozoic: Torridon Group, Applecross Formation, Scresort Sandstone Member	medium–coarse sandstone	no plugs in vicinity, some small-scale basaltic dykes
RM69	Kinloch Glen	Mesoproterozoic: Torridon Group, Applecross Formation, Scresort Sandstone Member	medium–coarse sandstone	unbaked, but ~10 m distance from dolerite sill
RM76	West Sgaorishal	Mesoproterozoic: Torridon Group, Applecross Formation, Scresort Sandstone Member	medium–coarse sandstone	<10 m from Loch Sgaorishal plug margin, baked appearance (grey, fissile)
RM78	Papadil	Mesoproterozoic: Torridon Group, Aultbea Formation, Sgor Mhor Member	sandstone (with heavy mineral bands)	unbaked, fresh-looking sample
RM82	Dibidil	Mesoproterozoic: Torridon Group, Diabaig Shale Formation, Laimhrig Shale Member	shale	Sediment appears partially baked as proximal to S margin of layered intrusion, but still with minor pyrite
RM90	West Sgaorishal	Mesoproterozoic: Torridon Group, Applecross Formation, Scresort Sandstone Member	medium–coarse sandstone	At plug margin, baked, fissile and with evidence of in situ melting (spherules)
RM97	South shore of Loch Scresort	Mesoproterozoic: Torridon Group, Applecross Formation, Allt Mhor na h-Uamha Member	fine sandstone	Unaltered sample, no proximal magmatic features

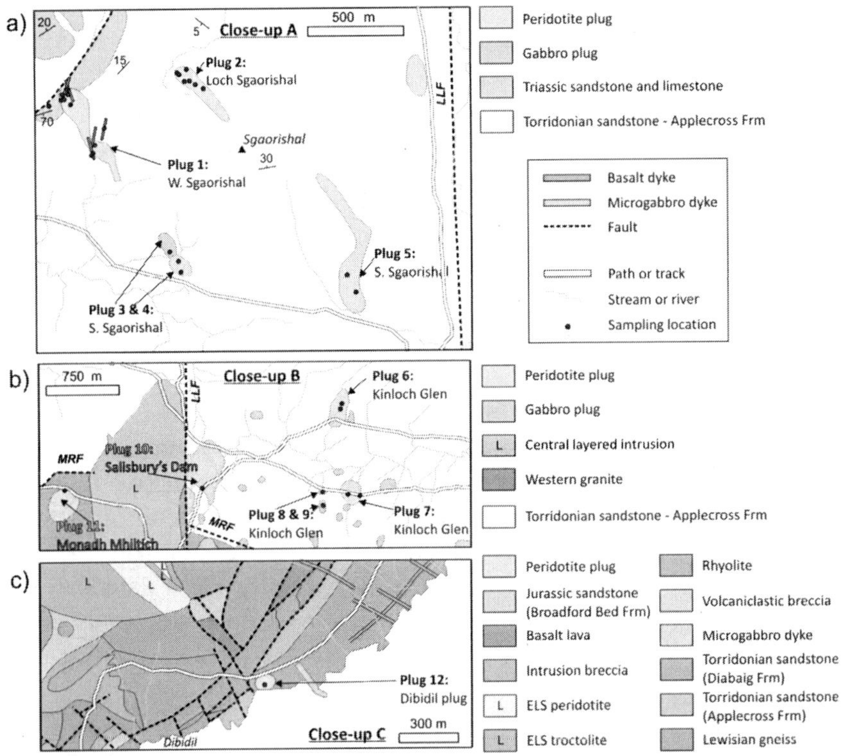

Figure 2. Geology of areas surrounding the peridotite and gabbro plugs sampled during this study. (a) NW Rum; (b) central Rum across the Long Loch Fault; (c) SE Rum near Dibidil. Sample locations are shown as dots. Plugs are numbered and correspond to Table 1.

The plugs have rounded or elongate- to teardrop-shaped outlines (e.g.Fig. 2a). Their plan-view shapes probably result from varying inclinations of conduits intersecting the present-day topography. Plugs sampled during this investigation are distinct from the small 'apophyses' of peridotite within, or straddling, the Main Ring Fault that may predate or be contiguous with the Eastern Layered Series (e.g. Fig. 2b; Butcher *et al.*, 1985; Wadsworth, 1994). None of the plugs have been radiometrically dated, and the timing of their intrusion is based upon cross-cutting field relationships alone. Most of the plugs lie within 1–2 km of the Long Loch Fault (Fig. 1) and range in size from ~100 to 600 m diameter (McClurg, 1982; Volker & Upton, 1990). The plugs may have fed lava flows, but no associated lavas have been proven (Emeleus & Bell, 2005).

The structural setting and silicate mineralogy of volcanic plugs on Rum has been studied in detail by Volker & Upton (1990), Wadsworth (1994),Holness (1999) and Holness *et al.* (2012). Ultramafic plugs are thought to represent the conduits of olivine-rich melts or 'crystal mushes', fed from an underlying magma chamber along faults such as the Long Loch Fault. These crystal mushes were lubricated by interstitial basaltic magma, hence the contemporaneous series of gabbroic plugs on the island could represent similar conduits without the entrained olivine mush. Rare plugs in the south of Rum contain rounded troctolite, dunite and chromitite xenoliths, probably derived from part of the Rum Layered Suite (Volker & Upton, 1990).

Plagioclase (10–15 modal %) and clinopyroxene (up to 8%, sometimes poikilitic) are interstitial to cumulus olivine and euhedral and rounded Cr-spinel; accessory amphibole and biotite are also observed (Wadsworth, 1994). Layering (defined by olivine cumulates) is recorded in the ultramafic peridotite plugs and typically dips shallowly at the margins (striking parallel to the margins) with the central portions of plugs taking on a colloform texture. Changes in modal mineralogy can also define layering, now weathered to 'indentation layering'. At the West Sgaorishal plug (Fig. 2a, Plug 1) indentation layering is particularly well developed and is observed to be closely spaced at the margins (5–10 cm), dipping inwards towards the plug centre at 30–50°, with layer spacing increasing towards, and ultimately disappearing, at the centre. In this plug, olivine accounts for 55–60 vol. % of the rock, but olivine abundance increases to 88 vol. % at the centre. In detail, there is a marked asymmetry in the olivine content from west to east, such that olivine is more concentrated on the western edge (see Holness *et al.*, 2012). In contrast, layering can be poorly developed, laterally discontinuous (e.g. Loch Sgaorishal plug; Fig. 2a, Plug 2) or completely absent in other plugs. Overall, layering developed after magma flow had ceased and/or during magma withdrawal back into the feeder system [compare 'withdrawal' of magmas in Hawaiian conduits, as suggested by Upton (2004)].

Olivine compositions in plugs are equivalent to those from feldspathic peridotites in the Eastern Layered Series (Fo_{89-82}) and Western Layered Series (Wadsworth, 1994; Emeleus, 1997, and references therein; Holness*et al.*, 2012) and distinct from olivine compositions in the troctolite portions of the Rum Layered Suite (Fo_{84-70}; Emeleus, 1997, and references therein).

The sedimentary host-rocks of the volcanic plugs (Triassic, Fig 2a; Torridonian, Fig. 2a–c) typically appear bleached, fissile and metasomatized. Studies of *in situ* anatexis of country rocks (e.g. Holness, 1999; Holness *et al.*, 2012) and thermal modelling of the West Sgaorishal plug (Plug 1; Fig. 2a) suggest that the plugs were active magma conduit for only a few months and cooled over a period of *c.* 30 years (Holness*et al.*, 2012).

2.3 The Hebridean sedimentary succession and country rocks to the Rum plugs

The crustal sequence on the Isle of Rum includes Archaean basement rocks of the Lewisian Gneiss Complex, Mesoproterozoic Torridonian sediments, and Triassic sediments in NW Rum. A thick sequence of Jurassic and Cretaceous sediments probably also covered the island (at least in part), as suggested by extensive outcrops of these rocks preserved on the neighbouring Isle of Skye and 'wedges' of Jurassic mudrocks preserved in the Main Ring Fault of the Rum Central Complex (Emeleus, 1997; for stratigraphic log see Hughes *et al.*, 2015).

2.4 S isotope signatures in western Scotland

Lewisian gneisses generally have variable but low sulphur concentrations [median 789 ppm according to Cameron (1994)] and $\delta^{34}S$ ranging from −1·4 to +5·5‰ on the NW Scottish mainland (Lowry *et al.*, 2005; Hughes*et al.*, 2015). The Neoproterozoic Dalradian and Moine metasedimentary successions (Grampian and Northern Highland Terranes, respectively) are mostly dominated by ^{34}S-enriched metasedimentary sequences (Lowry*et al.*, 2005). Moine metapelites have low or very low concentrations of sulphur, and where measurable have $\delta^{34}S$ = +2·5‰ (Hughes *et al.*, 2015) and +3·4 to +4·6‰ in some psammites (Lowry, 1991). Dalradian metasediments (with disseminated diagenetic and metamorphic sulphides, mostly pyrite and pyrrhotite) have a mean $\delta^{34}S$ of +8‰ (±14‰; Lowry*et al.*, 2005).

The Mesoproterozoic Torridonian formations of NW Scotland include the Stoer, Sleat and Torridon Groups (Stewart, 2002; Kinnaird *et al.*, 2007); however, the Stoer Group is thought to be absent on the Isles of Skye and Rum (Figs 1 and 2). Mudrocks are present within the Sleat (and, to a lesser extent, Torridon) Group; however, they are less extensive and thinner than the Jurassic mudrocks of the Hebrides Basin. Sulphur isotope studies of Torridonian sulphates and sulphides in Stoer Group mudrocks have identified the light isotopic shift of $\delta^{34}S$ associated with bacterial sulphate reduction in small freshwater pools (Parnell *et al.*, 2010, 2012). Hugheset *al.* (2015) established a whole-rock $\delta^{34}S$ range of +1·4 to +4·7‰ for Mesoproterozoic Torridonian sediments on Skye and Rum. They also found that these rocks contained low concentrations of sulphur (<0·018 wt %).

The stratigraphy of the Triassic to Cretaceous rocks of the Hebrides Basin is well correlated (e.g. Morton & Hudson, 1995; Hesselbo & Coe, 2000, and references therein) and the S concentration of this Mesozoic sequence is generally much greater than that of the older Torridonian sediments, Moinian metasediments, or crystalline basement (Hughes *et al.*, 2015). Therefore the Mesozoic Hebrides Basin represents the most effective source of crustal S available for magmatic contamination. Sulphur isotopes for Mesozoic (specifically Jurassic) sediments have been reported by Raiswell *et al.* (1993) and Hudson *et al.* (2001) from elsewhere in the British Isles, but Hughes *et al.* (2015), together with results from Yallupet *al.* (2013), have provided the first detailed Hebridean S-isotopic framework.

In summary, Jurassic mudrocks and ironstones have the highest S abundance in the Hebrides Basin, with $\delta^{34}S$ ranging from –33·8 to –14·7‰ (Hughes *et al.*, 2015). $\delta^{34}S$ generally becomes lighter from the Lower Jurassic (–18·4 to –14·7‰) to the Upper Jurassic (–33·8 to –29·2‰), although the ironstones in the Lower Jurassic (e.g. Raasay Ironstone Formation) generally have the highest bulk S content (up to *c.*28 wt %, in comparison with some Upper Jurassic shales with 0·1 wt %). In contrast, sediments of the Triassic and Cretaceous succession have volumetrically very few mudrocks and no ironstones, and are therefore less likely to act as major crustal S sources (Hughes *et al.*, 2015). Triassic sediments from Rum (sandstones, conglomerates, breccia and cornstone of the Monadh Dubh Formation) and the wider Hebrides Basin contain calcretes—thus if S were to be present in these rocks, it would most probably be present as sulphate with a heavy $\delta^{34}S$ composition (e.g. Poweret *al.*, 2003).

2.5 Magmatic sulphide mineralization and S contamination in the NAIP

On the Isle of Rum, disseminated and net-textured coarse sulphides (up to 5% modal abundance; Power *et al.*, 2003) occur in the West Sgaorishal peridotite plug (Plug 1; Fig. 2a) within a few metres of the margin. Sulphides are most abundant at the plug margin, and become increasingly rare and fine grained towards the centre (Power *et al.*, 2003). Sulphide-rich zones have not been identified in the other plugs on Rum. PGM are associated with sulphides in Plug 1 and include Pt- and Pd-tellurides, bismuth-tellurides and electrum (Power *et al.*, 2003). Previously reported $\delta^{34}S$ values of the coarse sulphides in Plug 1 range from −18·3 to −9·2‰, and suggest that the magmas forming the plug experienced crustal contamination by Jurassic sedimentary rocks (Power *et al.*, 2003). However, given the absence of crustal rocks with low $\delta^{34}S$ currently surrounding the plug, Power *et al.* (2003) suggested that either S contamination took place in a now-eroded parental magma chamber, or Jurassic sediments originally surrounded the plug locally (and have also since been eroded).

Hughes *et al.* (2015) found that for steeply dipping intrusions, such as basaltic dykes, incorporation of bacterially reduced sulphur from wall-rock shales produced $\delta^{34}S$ signatures in magmatic sulphides ranging from −30·7 to −2·3‰, depending on the degree of contamination and the $\delta^{34}S$ signature of the local crustal contaminant. In contrast, horizontal picritic sills on Skye do not record such crustal contamination and lack the anomalous $\delta^{34}S$ (Hughes *et al.*, 2015), indicating that the mechanism by which a conduit intruded the surrounding country rock (vertical or horizontal) and the ambient flow regime within it (i.e. turbulent or non-turbulent) were important factors in the contamination process; for example, a turbulent regime is more likely to result in brecciation of the conduit wall rocks, allowing for more efficient S incorporation into the magma (Robertson *et al.*, 2015).

Magmatic Cu–Ni–PGE–Au sulphide mineralization in the wider NAIP is recorded in a number of settings. The largest deposit is the Au- and Pd-dominant Platinova Reef in the Skaergaard Intrusion, formed through prolonged fractional crystallization that eventually triggered S-saturation (e.g. Bird *et al.*, 1991; Andersen *et al.*, 1998; Holwell & Keays, 2014). In this instance, crustal contamination played no role in the formation of an immiscible sulphide liquid. Mineralization is also found as sulphide globules in small hyperbyssal intrusions, such as the Miki Fjord and Togeda Macrodykes of East Greenland (Holwell *et al.*, 2012). Sulphur isotope analyses identified a strong crustal contribution attributed to pyritic Cretaceous black shales, and therefore in these instances crustal S contamination was the trigger for S-saturation (Holwell *et al.*, 2012).

Overall, if crustal contamination is suspected to be the main cause for S-saturation, the isotopic composition of disseminated sulphide mineralization in magma conduits, such as sills, dykes and plugs, can be used as a vectoring tool towards areas of more massive sulphide mineralization (e.g. Lightfoot *et al.*, 1984; Czamanske *et al.*, 1992;Huminicki *et al.*, 2008; Barnes *et al.*, 2016). Hence understanding the source and location of crustal contamination, and the potential crustal horizon(s) that triggered S-saturation in relation to the prevailing fluid dynamic regime in the magmatic conduit, is a valuable method for targeting sulphide-hosted mineralization up- or downstream within the magmatic plumbing systems.

3. METHODS

Twenty-six whole-rock samples from 12 volcanic plugs from across the Isle of Rum were sampled for major and trace elements, S isotopes (bulk and *in situ*), and mineralogical investigations (sample details are given inTable 1; localities are shown in Fig. 2a–c; bulk-rock geochemical data are presented in the Supplementary Material, available for downloading athttp://www.petrology.oxfordjournals.org). Additionally, 15 samples of various Torridonian and Triassic sediments were collected from areas adjacent to the plugs, and elsewhere across the island, to assess the crustal S isotopic framework of the island (Table 1).

3.1 Bulk-rock and *in situ* S-isotope analysis

For samples with visible sulphide minerals >500 μm^2, polished blocks were subjected to sulphide *in situ* laser combustion, following Wagner*et al.* (2002). Raw *in situ* $\delta^{34}S$ data were corrected by the factor $\delta^{34}S_{pyrite} = \delta^{34}S_{SO2laser} + 0{\cdot}8‰$, as laser combustion causes a small and predictable fractionation of sulphur isotope compositions for the SO_2 gas produced, compared with the actual $\delta^{34}S$ of the sulphide mineral (Wagner *et al.*, 2002). At the Scottish Universities Environment Research Centre (SUERC), SO_2 gases released by combustion were analysed on a ThermoFisher Scientific MAT 253 dual inlet mass spectrometer (conventional samples) and VG Isotech SIRA II mass spectrometer (for laser samples).

Samples without sufficiently large sulphides underwent whole-rock chemical S extraction by the chromium reducible sulphur method (CRS) followed by conventional sulphur analysis: a detailed description and evaluation of this method has been given by Hughes *et al.* (2015). In summary, CRS is based on and adapted from the techniques of Zhabina & Volkov (1978), Canfield *et al.* (1986), Tuttle *et al.* (1986), Hall *et al.* (1988), Newton *et al.* (1995), Nielsen & Hanken (2002) and Labidi *et al.* (2012). Powdered rock samples were loaded into a flat-bottomed glass reaction vessel under a nitrogen atmosphere and connected to a gas bell jar containing 0·1 M $AgNO_3$ solution. A 1M solution of chromous chloride ($CrCl_2$) was injected through a septum to cover the powdered sample and was heated to 150°C for 3 h to react as follows:

2Cr2++S0+2H+→2Cr3++H2S.2Cr2++S0+2H+→2Cr3++H2S.

This process leads to the complete breakdown of elemental and reduced sulphur species into H_2S, even in the presence of Fe^{3+}. H_2S produced is streamed to the gas bell jar to react with $AgNO_3$ solution, precipitating Ag_2S. At the end of the reaction, this black precipitate was filtered under vacuum, freeze dried and collected ready for conventional S analysis. During conventional analysis, SO_2 gas samples were run at SUERC using a ThermoFisher Scientific MAT 253 dual inlet mass spectrometer.

Reproducibility for both the laser combustion and conventional analysis is ±0·4‰ (see Hughes *et al.*, 2015) with 2σ of standards <±0·2‰. Standards used throughout all analyses were IAEA-S-3 (–31·5‰) and NBS-123 (sphalerite, +17·1‰) international standards, alongside an SUERC laboratory chalcopyrite standard, CP-1 (–4·6‰).

Bulk-rock S concentrations were determined using a LECO CS230 Carbon/Sulphur Determinator at the University of Leicester. Sample mass ranged from 0·1 to 1·0 g depending on relative bulk S content. Each sample was run in triplicate to monitor precision. Accuracy was monitored by regular analysis of the reference material BAS ECRM 877-1. The limit of minimum detection for this method is 0·018 wt % S (3 × standard deviation of the mean blank).

3.2 Mineralogy and mineral composition

Quantitative microanalysis was carried out on a Cambridge Instruments S360 scanning election microscope (SEM), using an Oxford Instruments INCA Energy energy-dispersive X-ray spectroscopy (EDX) analyser, with operating conditions set at 20 kV and specimen calibration current of ~2 nA at a fixed working distance of 25 mm. Analytical drift checks were carried out every 2 h using a Co reference standard and a comprehensive suite of standards from MicroAnalysis Consultants Ltd were used to calibrate the EDX analyser.

Sulphide trace element concentrations were determined by laser ablation inductively coupled plasma mass spectrometry (LA-ICP-MS) using polished samples and a New Wave Research UP213 UV laser system attached to a Thermo X Series 2 ICP-MS system, following procedures outlined byPrichard *et al.* (2013) and Smith *et al.* (2014). Further details are available in the Supplementary Material.

As part of our study, we also analysed the sulphide compositions of a selection of peridotites from each of the major cyclic units in the Eastern Layered Series (ELS), to place plug compositions into context. The results are presented in Table C of the Supplementary Material, alongside whole-rock $\delta^{34}S$ for corresponding ELS samples (Tables D and E, Supplementary Material).

4. RESULTS

4.1 Sulphide mineralogy and trace element compositions

Details of the mineralogy, including size, texture and mineral association of base metal sulphides in the Rum plugs are presented in Table 2together with a record of any PGM present. SEM-based PGM searches were carried out only on samples from Plugs 1, 2 and 10, as previous work had identified a correspondence between sulphide and PGM occurrences (e.g.Power *et al.*, 2003). As our study focuses on the S-isotopic composition of the plugs and trace element compositions of base metal sulphides within the plugs, we did not undertake a comprehensive search for PGM across all plug samples. Selected back-scattered electron (BSE) images of sulphide mineral textures are shown in Fig. 3.

Table 2: Sulphide mineralogy, texture and grain size.

Sample number	Sulphide minerals	Sulphide size range	Sulphide textures and associations	Silicate and oxides	PGM categories (if detected)	PGM sizes and associations	Other accessory minerals noted
RM18	Cp + Pn + Po	< 20 µm	Very rare rounded sulphides are interstitial to (but adjacent/in contact with) cumulus olivine crystals	Olivine, clinopyroxene, plagioclase, chromite (also as rare discontinuous veins)	none detected	-	Ni-arsenides, Ni-sulphide (nickeline?), Cu-sulphide (chalcocite?), monazite, apatite, magnetite
RM30	Po + Pn + Cp	< 200 µm	Sulphides are generally interstitial to cumulus olivine, but much more rare than in RM 32, RM33, RM91. Sulphides usually in contact with olivine at the crystal margins.	Olivine, clinopyroxene, plagioclase, chromite, ilmenite	not investigated	-	White vein cross-cuts sample, filled with serpentine and chlorite, minor quartz
RM32	Cp + Po + Pn ± Bo, Cc	Two size groups: < 200 µm, and 0.5 to 4.0 mm	Abundant coarse (0.5 - 4 mm) sulphides as net-textured phases, interstitial to olivine and chromite cumulates. Irregular-shaped to rounded sulphides within intercumulus plagioclase in vicinity of large net-textured sulphides. Additionally, rounded sulphides (< 200µm) within cumulus olivine marginal areas (not serpentised cracks) or embayed at edges of cumulus olivine crystals, and not in vicinity of net-textured coarse sulphides. Also, small sulphides (also < 200 µm) within serpentine-filled alteration cracks throughout slide.	Olivine, clinopyroxene, plagioclase, chromite, ilmenite, serpentinite	Pd-telluride (merenskyite), Pd bismuthide-tellurides	0.5 - 4 µm wide PGM, mostly within sulphides (Pn, Po or Cp) and some at silicate-sulphide mineral boundaries	Co-arsenide
RM33	Cp + Po + Pn ± Bo, Cc	0.5 to 2.0 mm	Abundant coarse sulphides as net-textured phase, interstitial to olivine cumulates. Margins of some of these sulphides have 'variagated' solution textures suggestive of alteration or recrystallisaton. Rarely, elongate sulphides (up to 0.5mm long) can be seen penetrating cumulus olivine crystals along fractures. Chromite crystals both as cumulate phase and sometimes as	Olivine, clinopyroxene, plagioclase, chromite, ilmenite, serpentinite	Pd bismuthide-telluride, Pt-arsenide (sperrylite), Pt-Sb arsenide	0.5 - 7 µm wide PGM, mostly within sulphides (Po, Pn and Cp) and some at silicate-sulphide mineral boundaries	electrum, Pb-telluride

Sample number	Sulphide minerals	Sulphide size range	Sulphide textures and associations	Silicate and oxides	PGM categories (if detected)	PGM sizes and associations	Other accessory minerals noted
			fine inclusions within granular cumulate olivines.			(e.g., Po or Pn - silicate)	
RM35	Cp + Po + Pn + Ga	50 to 500 μm	Sulphides within interstitial silicate minerals (e.g., pyroxene and plagioclase). Smaller rounded sulphides (typically < 100 μm) in contact with margins of cumulus olivine crystals, and associated with serpentine-filled cracks. Rare rounded sulphides (~50 μm) as inclusions in olivine (although crystals are highly fractured).	Olivine, clinopyroxene, plagioclase, chromite, ilmenite, serpentinite	Ir-Pt arsenic sulphide (irarsite-platarsite), Ir-arsenic sulphide (irarsite)	Very rare PGM typically < 1μm wide and found within Pn.	electrum, zircon, chlorite (as replacement mineral overprinting serpentine)
RM58	Cp ± Po ± Pn	< 50 μm	Rounded rare sulphides, as discrete minerals interstitial to cumulate olivine and chromite, and within interstitial pyroxene and plagioclase. Also within serpentinite-filled cracks at margins of cumulus olivines	Olivine, clinopyroxene, plagioclase, chromite	not investigated	-	-
RM61	Cp + Po + Pn ± Bo	< 50 μm	Rounded rare sulphides, as discrete minerals interstitial to cumulate olivine and chromite, and within interstitial pyroxene and plagioclase. Also within serpentinite-filled cracks through cumulate olivines	Olivine, clinopyroxene, plagioclase, chromite	not investigated	-	-
RM64	Cp + Po + Pn ± Bo	< 150 μm	Sulphides are discrete and interstitial to coarse granular cumulate olivine crystals, and sometimes within olivines themselves.	Olivine, clinopyroxene, plagioclase, chromite	not investigated	-	-
RM70	Cp + Pn ± Po	< 90 μm	Sulphides are discrete interstitial to olivine and chromite cumulate crystals, and within intercumulate pyroxenes and plagioclase	Olivine, clinopyroxene, plagioclase, chromite	not investigated	-	-
RM74	Po + Pn ± Cp	< 80 μm	Sulphides are discrete interstitial to olivine and chromite cumulate crystals, and within intercumulate pyroxenes (and variable chlorite-filled zones in OPX ± apatite)	Olivine, clinopyroxene, plagioclase, chromite, ilmenite, serpentinite	not investigated	-	apatite, Fe-oxide
RM75	Cp + Pn ± Po	< 110 μm	Discrete, rounded sulphides, interstitial to granular cumulate olivines, and within intercumulate pyroxenes and	Olivine, clinopyroxene, plagioclase, chromite	not investigated	-	-

Sample number	Sulphide minerals	Sulphide size range	Sulphide textures and associations	Silicate and oxides	PGM categories (if detected)	PGM sizes and associations	Other accessory minerals noted
			plagioclase. Sometimes sulphides at triple junction between cumulate olivine crystals.				
RM81	Cp ± Po ± Pn	< 50 μm	Sulphides are discrete and both interstitial to granular cumulate olivines, and within olivines (close to olivine crystal margin)	Olivine, clinopyroxene, plagioclase, chromite, serpentine	not investigated	-	Fe-oxide, serpentine, and replacement of serpentine by chlorite
RM85	Pn + Cc + Bo	< 40 μm	Sulphides are discrete, rounded, and interstitial to granular cumulate olivines, and within intercumulate pyroxenes and plagioclase	Olivine, clinopyroxene, plagioclase, chromite, serpentine	not investigated	-	Fe-oxide, serpentine, and replacement of serpentine by chlorite
RM91	Cp + Po + Pn	0.2 to 4.0 mm	Abundant coarse sulphides as net-textured, interstitial to olivine cumulates. Also more rounded sulphide phases (up to 0.2 mm diameter) in contact with cumulus olivine margins or as embayments at olivine margins.	Olivine, clinopyroxene, plagioclase, chromite	not investigated	-	-
RM95	Cp + Po + Pn	Two size groups: < 50 μm, and 0.25 to 2.0 mm	Abundant coarse sulphides as net-textured phase, interstitial to olivine cumulates. Rare sulphides filling or partially filling cracks through olivine (± serpentine and Fe-oxides) or as discrete rounded/irregular shaped sulphide crystals (< 50 μm) within cracks.	Olivine, clinopyroxene, plagioclase, chromite, serpentine	not investigated	-	-

Mineral abbreviations: Cp, chalcopyrite; Pn, pentlandite; Po, pyrrhotite; Bo, bornite; Cc, chalcocite; Ga, galena; Ol, olivine; Cpx, clinopyroxene; Px, pyroxene, Pl, plagioclase; Ilm, ilmenite; Chr, Cr-spinel; Serp, serpentine; Fe-ox, iron oxide.

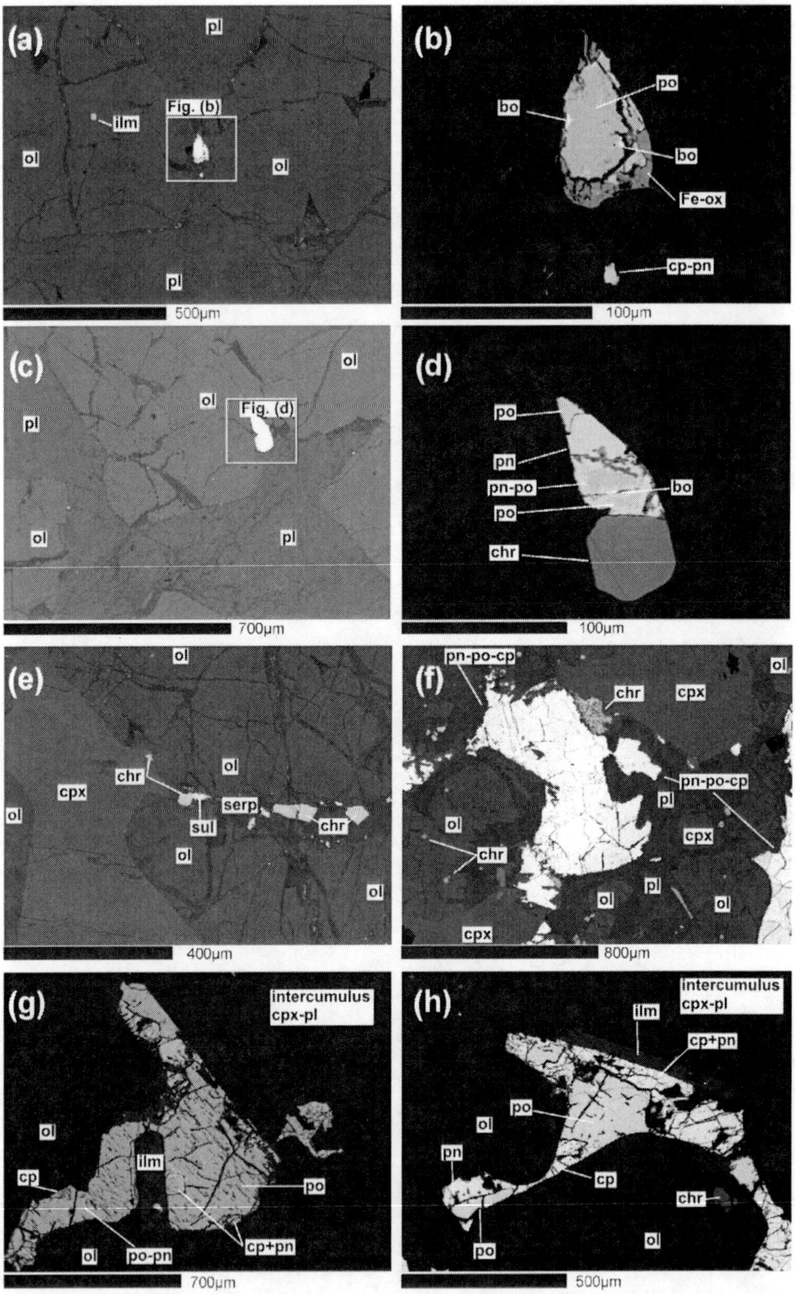

Figure 3. Back-scattered electron images of sulphides in peridotite plugs from scanning electron microscope investigations. (a) Small sulphide nucleating onto and between two cumulus olivine crystals, within

intercumulus plagioclase, plug 6. (b) Close-up of sulphide in (a) with contrast adjusted to show sulphide mineralogy. (c) Sulphide grain nucleating on small Cr-spinel at the margin of granular cumulus olivine, plug 6. (d) Close-up of sulphide minerals and Cr-spinel in (c). (e) Sulphides and Cr-spinel within serpentinized fracture at margin of cumulus olivine, with intercumulus clinopyroxene, plug 5. (f) Coarse sulphides as a poikilitic phase with clinopyroxene and plagioclase, plug 1 margin. (g) Poikilitic sulphides surrounding cumulus olivine and elongate ilmenite, with intercumulus plagioclase, plug 1 margin. (h) Similar to (g), demonstrating poikilitic, sometimes 'net-textured' sulphides surrounding granular cumulus olivine phenocrysts. In both (g) and (h) the contrast has been adjusted to display the sulphide internal texture and mineralogy. A summary of sulphide characteristics and PGM is presented in Table 2. ol, olivine; cpx, clinopyroxene; pl, plagioclase; ilm, ilmenite; chr, Cr-spinel; serp, serpentine; Fe-ox, iron oxide; pn, pentlandite; po, pyrrhotite; cp, chalcopyrite; bo, bornite.

With the exception of plugs 1 and 2, all the peridotite plugs contain small sulphide blebs (e.g. Fig. 3a), typically 20–150 μm in length, rounded and equant, composed of chalcopyrite, pyrrhotite and pentlandite, with rare bornite and chalcocite (Table 2). Occasionally a rim of iron oxide is present around the sulphide blebs in more altered samples (e.g. Fig. 3b). Sulphides normally occur at the margins of cumulus olivine (Fig. 3a and c), or within intercumulus plagioclase and clinopyroxene, and sulphides are locally in contact with granular Cr-spinel (e.g. Fig. 3d). In rare cases, sulphides occur within the fractured and serpentine-filled margins of cumulus olivine (e.g. Fig. 3e). In contrast, plug 1 at West Sgaorishal contains coarser sulphide minerals, up to 4 mm in length, which occur interstitially to pyroxene and plagioclase, or as net-textured assemblages (Fig. 3f–h; Table 2). Sulphides have been observed to almost enclose cumulus olivine, as well as ilmenite and Cr-spinel (e.g. Fig. 3g and h), and include pyrrhotite and pentlandite (and pyrrhotite–pentlandite flames) as well as chalcopyrite. Minor galena is also observed; however, unlike the other plugs, no bornite or chalcocite was encountered in plug 1. Finer sulphide minerals are of the same species as the course sulphides occurring in the centre of plug 1.

PGM (plugs 1 and 2 only) are found within sulphide minerals or at silicate–sulphide mineral contacts. These range in size from 0·5 to 7 μm wide and include Pd- and Pt-tellurides, bismuth-tellurides, Pt-arsenides and Pt–Sb-arsenides (Table 2). In plug 2 (Loch Sgaorishal) sulphides range from 50 to

500 µm and are interstitial to cumulus olivine, with rare rounded sulphides (~50 µm) occurring as inclusions within cumulus olivine, albeit in serpentine-filled fractures. Pentlandite in plug 2 contains small (<1 mm wide) Ir and Pt sulpharsenides of the irarsite–platarsite series. In both plugs 1 and 2, small grains of electrum were observed in pyrrhotite (normally <5 µm) as well as one crystal of Pb-telluride (Table 2).

Representative trace element compositions, including S/Se ratios of sulphide minerals and PGE abundances in plug sulphides (grouped per region of Rum), are presented in Table 3 (full data are available inSupplementary Material Tables C and D). Overall, the variation of PGE concentrations in sulphide minerals from plugs in various areas of Rum (NW, SE and central) overlaps significantly (Fig. 4a–d), such that there is no systematic difference in PGE tenor between plugs. Pentlandite is the principal carrier of Pd but there is no correlation between Pd and Rh for any of the plugs (Fig. 4a), nor are Rh concentrations higher in pentlandite or correlated with Ni content (Fig. 4b). By contrast, Rh and IPGE (Os, Ir, Ru) abundances in pentlandite are not distinct from other base metal sulphides or 'mixed sulphide' analyses (Figs. 5b and d).

The chondrite-normalized PGE composition of coarse sulphides at the margins of plug 1 (Fig. 5a–c) is subtly different from the composition of fine sulphides at the centre (Fig. 5d). This difference is particularly noticeable for Os, Ir and Ru, with the concentration of all PGE in marginal sulphides being lower than in finer sulphides at the centre. Overall, for each type of base metal sulphide (chalcopyrite, pentlandite, pyrrhotite), the sulphide PGE patterns of plug 1 (particularly the margins) are fairly consistent and parallel to one another. However, chalcopyrite (Fig. 5a) generally has lower PGE abundances than pentlandite and pyrrhotite (Fig. 5b and c, respectively), but the overall normalized PGE spectra are similar or parallel between Ni–Fe mineral species (e.g. Fig. 5b and c). Sulphides in plug 2 (Fig. 5e) also have two distinctive PGE patterns, with sulphides at the margin being comparatively homogeneous, and of a similar composition to those from the margin of plug 1. However, two sulphides analysed from the centre of plug 2 have differing PGE compositions—one that overlaps the sulphide compositions of the margins, and one entirely different sulphide composition with notably elevated IPGE. In the instance of plug 2, this may be due to a fundamental difference in sulphide mineralogy, as the sulphide analysed in sample RM35 has a notably lower Cu content (Table 3).

Table 3: Representative LA-ICP-MS data for plug sulphides

Sample number	Mineral	*n*	Plug number	Spot or line	S* (wt %)	Se (ppm)	S/Se	Co (ppm)	Ni (wt %)	Cu (wt %)	Zn (ppm)	Os (ppm)	Ir (ppm)	Ru (ppm)
RM30	Sulph (mean)	4	1	spot	36	137	2911	3451	28·09	1·45	1252	3·90	0·35	2·90
RM32	Cp (mean)	2	1	line	35·27	74	4961	290	1·61	23·54	8491	0·08	0·04	<0·09
RM32	Pn	1	1	line	33·38	111	3015	4540	27·59	1·48	466	1·06	0·37	0·84
RM32	Pn–Po (mean)	2	1	line	39·61	79	5067	2200	14·21	2·63	889	0·91	0·19	0·24
RM32	Po (mean)	4	1	line	41·62	104	4054	311	2·51	0·06	99	1·30	0·48	1·02
RM33	Cp	1	1	line	40·78	38	10812	7	0·19	13·36	1804	<0·02	<0·02	<0·09
RM33	Pn (mean)	2	1	line	32·34	53	6120	3440	28·21	0·12	38	1·16	0·25	0·50
RM33	Pn–Po	1	1	line	37·40	31	12128	472	1·96	0·49	36	0·53	0·13	0·13
RM33	Po (mean)	3	1	line	36·39	46	8195	72	0·81	0·16	34	0·52	0·16	0·35
RM91	Cp (mean)	3	1	line	32·03	76	4248	617	5·88	15·41	3778	0·18	0·71	0·54
RM91	Pn (mean)	3	1	line	35·00	113	3094	2493	30·31	0·50	239	2·39	1·10	2·00
RM91	Po (mean)	3	1	line	41·89	111	3827	51	0·61	0·22	1200	1·33	0·61	1·35
RM95	Cp (mean)	2	1	line	33·62	44	7694	298	3·29	28·59	6725	0·02	0·02	<0·09
RM95	Pn (mean)	6	1	line	35·49	76	4878	6121	27·78	1·28	1295	0·49	0·20	0·23
RM95	Po (mean)	2	1	line	39·36	65	6055	191	1·69	0·52	115	0·33	0·07	0·10
RM35	Sulph (mean)	2	2	spot	36	359	1030	3571	25·61	0·23	1529	72·78	21·86	13·84
RM70	Sulph (mean)	4	2	spot	36	237	1721	2260	13·91	6·07	8199	0·68	0·24	1·07
RM75	Sulph (mean)	3	2	spot	36	301	1339	3135	20·12	1·57	907	1·71	0·80	5·32
RM87	Sulph (mean)	3	4	spot	36	137	2885	1064	3·46	7·18	10617	4·25	1·58	3·90
RM85	Sulph (mean)	2	5	spot	36	630	647	6451	36·24	1·11	1203	1·73	0·10	0·10
RM64	Sulph (mean)	4	6	spot	36	292	1450	1621	12·03	8·83	7667	1·52	1·12	2·04
RM58	Sulph	1	7	spot	36	146	2463	1015	8·65	9·19	9124	0·16	0·10	0·14
RM61	Sulph (mean)	3	8	spot	36	240	1539	1876	7·30	13·20	13940	0·40	0·17	0·12
RM61	Po	1	8	line	36·19	34	10530	400	0·76	3·58	6767	0·09	0·02	<0·09
RM18	Sulph	1	10	spot	36	328	1099	805	5·60	8·45	3966	0·50	0·35	0·61
RM81	Sulph (mean)	4	12	spot	36	159	2304	1699	12·00	3·96	4295	1·20	0·13	1·79

Sample number	Rh (ppm)	Pt (ppm)	Pd (ppm)	Au (ppm)	Cd (ppm)	As (ppm)	Bi (ppm)	Te (ppm)
RM30	0·18	0·40	8·92	0·23	6·99	143·58	5·57	43·62
RM32	<0·22	<0·02	0·34	0·02	51·51	<15	0·69	16·63
RM32	<0·22	<0·02	0·59	0·02	10·64	<15	1·29	15·71
RM32	<0·22	<0·02	3·49	0·22	9·64	<15	2·33	4·32
RM32	0·51	<0·02	1·29	0·10	3·88	<15	1·79	10·15
RM33	<0·22	<0·02	0·43	0·03	45·53	<15	0·36	17·35
RM33	0·40	<0·02	22·50	1·87	<1·1	<15	0·31	7·90
RM33	<0·22	<0·02	2·96	<0·01	<1·1	<15	0·60	1·60
RM33	0·29	<0·02	0·25	<0·01	<1·1	<15	0·34	2·96
RM91	<0·22	0·02	6·25	0·02	25·78	20·03	0·61	9·06
RM91	1·16	0·04	40·11	<0·01	1·35	20·42	1·52	21·87
RM91	0·71	0·02	0·98	<0·01	7·42	<15	1·41	11·18
RM95	0·24	<0·02	<0·2	0·04	66·06	<15	2·17	12·42
RM95	0·24	0·06	28·11	0·02	6·01	46·11	1·39	12·17
RM95	<0·22	<0·02	1·02	<0·01	<1·1	<15	1·61	11·15
RM35	1·49	4·79	9·03	0·01	1·25	38·58	5·43	27·39
RM70	0·23	1·14	6·48	<0·01	36·32	54·86	0·64	15·55
RM75	0·70	0·23	8·23	0·07	3·29	36·13	11·33	20·03
RM87	0·42	0·12	0·71	0·03	14·97	16·50	0·33	1·17
RM85	0·13	0·23	13·84	0·06	7·35	637·68	3·28	110·73
RM64	0·74	3·03	46·60	0·04	13·02	<9	1·87	<0·9
RM58	<0·11	<0·02	<0·25	<0·01	2·72	<9	3·59	5·28
RM61	0·12	<0·02	0·61	0·27	15·99	<9	0·24	3·92
RM61	<0·22	0·03	<0·2	<0·01	5·70	<15	<0·05	1·30
RM18	1·78	<0·02	1·24	0·20	8·72	<9	0·59	9·98
RM81	0·18	0·52	2·61	1·64	24·90	27·48	0·48	13·75

*S content of sulphide is from quantitative analyses of the sulphide mineral or assumed as 36 wt % for spot analyses.
(See Supplementary Data Table C for all plug sulphide analyses and Table D for all ELS sulphide data.)

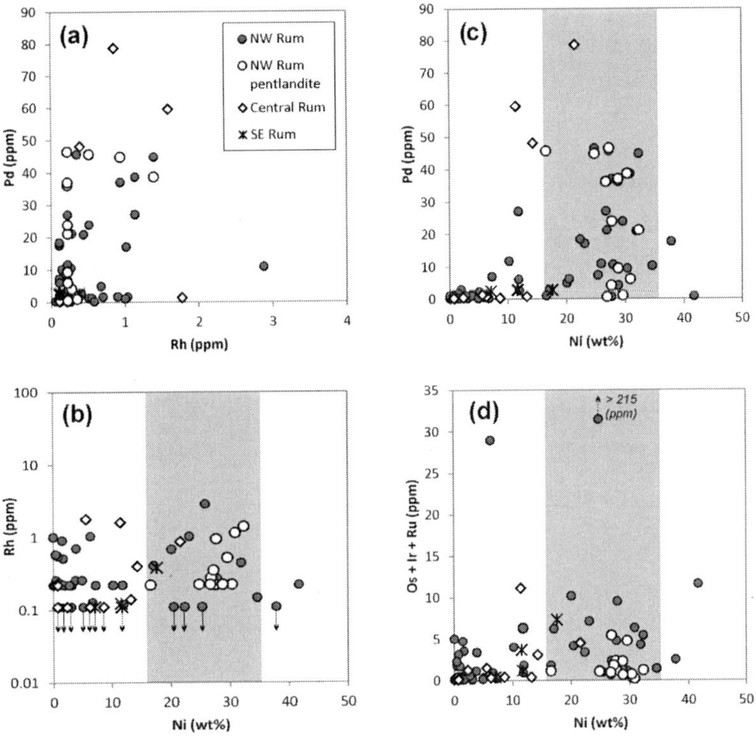

Figure 4. Trace element variation diagrams for all sulphide minerals analysed by LA-ICP-MS, grouped by area on Rum (NW, Central, and SE). Open circles represent pentlandite where it could visually be recognized and targeted by LA-ICP-MS. This serves to highlight PGE vs Ni abundance for known pentlandite, for comparison with fine sulphide grains where analysis by individual base metal sulphide was not possible, resulting in a 'mixed sulphide' analysis. Grey shaded area in (b)–(d) delineates the pentlandite field (based on a minimum of 17 wt % Ni, assuming a 50:50 mix between pentlandite and pyrrhotite). Note that where PGE abundances were less than the detection limit, results are calculated as half the detection limit and delineated by a small downward arrow.

Overall, sulphide compositions from plugs 4–12 are heterogeneous (Fig. 6). This may in part be explained by their being analyses of 'mixed sulphide' species (owing to their small size) but this cannot account for all the observed variation, as even sulphides with comparable Fe, Cu and/or Ni contents fluctuate between plugs. We note that whole-rock geochemistry is similarly variable between these plugs (albeit subtly) (seeSupplementary

Material). Broadly, peridotite plug sulphide compositions resemble those from peridotite units in the Eastern Layered Series (Fig. 6).

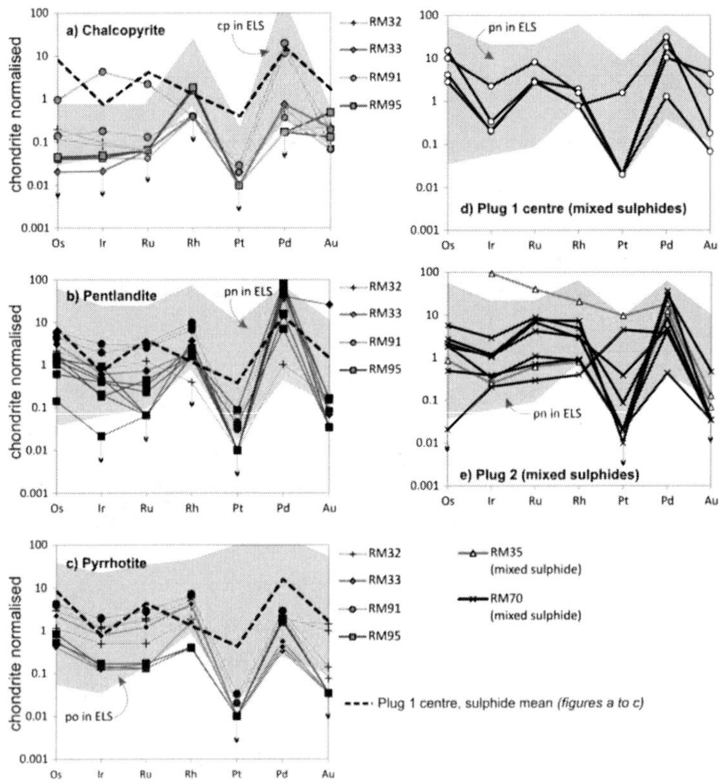

Figure 5. Chondrite-normalized PGE spectra for sulphide minerals, measured by LA-ICP-MS, for plugs 1 (centre and margin) and 2. Diagrams for analyses from the margin of plug 1 are divided according to sulphide phase: chalcopyrite (a), pentlandite (b), and pyrrhotite (c). Also shown are 'mixed sulphides' in plug 1 [centre; (d)] and plug 2 (e). Normalizing values are from McDonough & Sun (1995). (SeeTable 3 for summary of data.) It should be noted that where PGE abundances were less than the detection limit, results are calculated as half the detection limit and delineated by a small downward arrow. Grey shaded areas show the range of compositions for chalcopyrite (cp), pentlandite (pn) and pyrrhotite (po) measured in the Eastern Layered Series (ELS; seeSupplementary Data Table D) for comparison in plots (a)–(c). Dashed line shows the mean 'mixed sulphide' composition from the centre of plug 1. In plots (d) and (e) the grey shaded area delineates ELS, pn only.

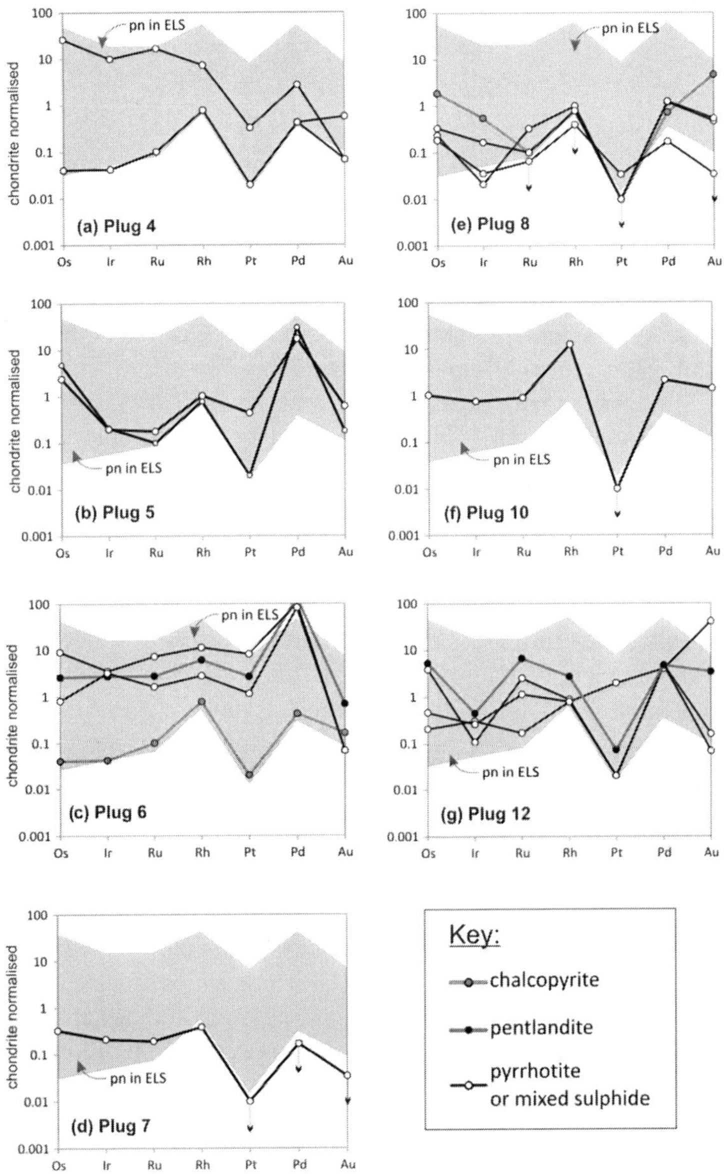

Figure 6. Chondrite-normalized PGE spectra for sulphide minerals, measured by LA-ICP-MS, for peridotite plugs 4, 5, 6, 7, 8, 10 and 12. Normalizing values fromMcDonough & Sun (1995). (See Table 3 for details.) It should be noted that where PGE abundances were less than detection limit, results are calculated as half detection limit with small downward arrow. Grey shaded areas delineate range of compositions for pentlandite (pn) in the Eastern Layered Series (ELS) for comparison.

4.2 S-isotope compositions and S/Se ratios of the Rum plugs

Figure 7a shows the average $\delta^{34}S$ value for whole-rock samples per plug (data in Table 4, along with S wt %) and for the ELS, plotted against S/Se ratio per sulphide. S/Se was calculated from LA-ICP-MS analyses of Se and calibrated by prior quantitative SEM analyses of S from the same sulphide grain (or in cases of small mixed sulphides a mean stoichiometric S abundance was assumed) and therefore quantifies S/Se on a mineral-by-mineral basis per sample. Sulphide minerals large enough to be analysed by laser combustion (i.e. only in plug 1) on a sulphide mineral-by-mineral basis (Table 5) have been averaged to the mean $\delta^{34}S$ value listed inSupplementary Material Table A, so as to be comparable with other samples analysed by whole-rock S extraction and conventional analysis. The $\delta^{34}S$ results of Hughes *et al.* (2015) for various sediments on the isles of Rum and Skye are also shown in Fig. 7b.

The S/Se ratios of ELS samples are mostly outside the mantle sulphide range (2550–3750; Lorand *et al.*, 2003) with only 8% of analyses falling within this range, 84% of analyses being below mantle values (seeSupplementary Material Table D). The full range of ELS S/Se ratios is 489–5416 (Fig. 7a) and $\delta^{34}S$ ranges from –4·4 to +1·2‰. Excluding plug 1 and one anomalous value from plug 8, the S/Se ratios of plug sulphides are up to 4000. With the exception of plugs 1 and 2, the peridotite plugs have $\delta^{34}S$ ranging from –0·6 to +2·1‰, in comparison with the ELS peridotites, which range from –1·0 to 0·9‰. Thus, most of the peridotite plug sulphides have $\delta^{34}S$ and S/Se ratios comparable with, and within range of, sulphides from ELS peridotite units. In contrast, sulphides from plug 1 have extremely light $\delta^{34}S$, ranging from –14·7 to –8·0‰, and S/Se ratios ranging from 3000 to 12 000. The centre of Plug 1 shows least variation and deviation of S/Se from the ELS; however, $\delta^{34}S$ is –10·9‰. Whereas the S/Se ratios of plug 2 are within range of the ELS, the $\delta^{34}S$ composition ranges from –6·5 to –2·9‰ (Fig. 7a and Table 4).

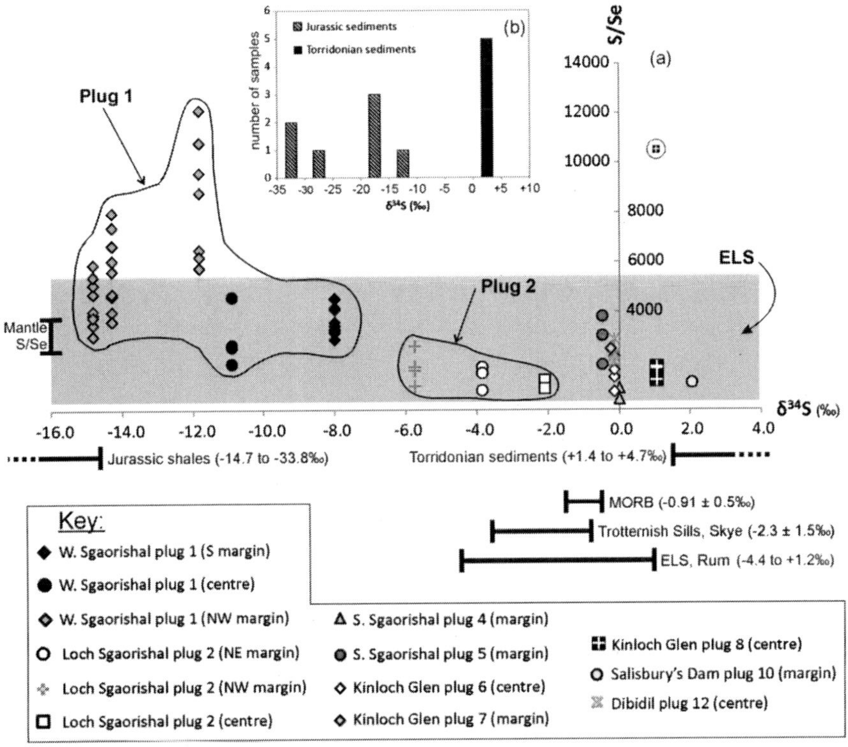

Figure 7. (a) S-isotope composition ($\delta^{34}S$) vs S/Se ratio of peridotite plug sulphides. Grey shaded area delineates the variation in S/Se ratio for the ELS, and brackets show the range of S/Se for the mantle based on Lorand*et al.* (2003), and $\delta^{34}S$ for Jurassic shales, Torridonian sediments and the Trotternish Sills (Isle of Skye) from Hughes *et al.* (2015). MORB $\delta^{34}S$ is from Labidi *et al.*(2012). $\delta^{34}S$ for the ELS is also shown (see Supplementary Material Table E). (b) Histogram of $\delta^{34}S$ for Proterozoic (Torridonian from Rum and Skye) and Mesozoic (specifically Jurassic mudrocks and ironstones from Skye and Ardnamurchan) sediments present in this area of the Hebrides; data from Hughes*et al.* (2015).

Table 4: Sulphur isotopic composition ($\delta^{34}S$, mean per sample) and whole-rock sulphur concentration

Sample number	Location	Plug no. (Fig. 1)	Plug type	Margin or centre	$\delta^{34}S$ (‰)	S (wt %)
RM81	Dibidil	12	peridotite	centre	−0·2	0·03
RM58	Kinloch Glen	7	peridotite	margin	−0·2	—
RM59	Kinloch Glen	7	peridotite	centre	—	—
RM60	Kinloch Glen	9	gabbro	margin	—	—
RM61	Kinloch Glen	8	peridotite	centre	+1·1	—
RM63	Kinloch Glen	6	peridotite	centre	−0·6	—
RM64	Kinloch Glen	6	peridotite	centre	−0·1	—
RM17	Monadh Mhiltich	11	peridotite	margin	N/P	—
RM18	Salisbury's Dam	10	peridotite	margin	+2·1	—
RM35	Loch Sgaorishal	2	peridotite	centre	−2·9	—
RM70	Loch Sgaorishal	2	peridotite	NW margin (at contact)	−5·7	0·05
RM71	Loch Sgaorishal	2	peridotite	NW margin (5 m inwards from contact)	−3·7	—
RM73	Loch Sgaorishal	2	peridotite	centre	−4·2	—
RM74	Loch Sgaorishal	2	peridotite	NE margin	−6·5	0·06
RM75	Loch Sgaorishal	2	peridotite	NE margin (~20 m inwards from inferred contact)	−3·9	0·03
RM85	Southern Sgaorishal	5	peridotite	margin	—	—
RM86	Southern Sgaorishal	5	peridotite	centre	+0·3	0·02
RM87	Southern Sgaorishal	4	peridotite	margin	−0·5	—
RM88	Southern Sgaorishal	4	peridotite	centre	−1·3	0·04
RM89	Southern Sgaorishal	3	gabbro	S margin (~3 m inwards from inferred contact)	—	—
RM29*	West Sgaorishal	1	peridotite	N margin (~30 m inwards from inferred contact)	−11·0	—
RM30*	West Sgaorishal	1	peridotite	centre	−10·9	0·02
RM32*	West Sgaorishal	1	peridotite	NW margin (~35 m inwards from inferred contact)	−14·7	0·08
RM33*	West Sgaorishal	1	peridotite	N margin	−11·8	—
RM91*	West Sgaorishal	1	peridotite	S margin	−8·0	0·18
RM95*	West Sgaorishal	1	peridotite	NW margin	−14·3	—

Sample number	Location	Period: Group, Formation	Sediment type	$\delta^{34}S$ (‰)	S (wt %)

RM21	Monadh Dubh	Triassic: New Red Sandstone Supergroup, Monadh Dubh Sandstone Formation	conglomerate with calcrete matrix	N/P	—
RM22	Monadh Dubh	Triassic: New Red Sandstone Supergroup, Monadh Dubh Sandstone Formation	coarse red sandstone	N/P	—
RM23	Monadh Dubh	Triassic: New Red Sandstone Supergroup, Monadh Dubh Sandstone Formation	white limestone	N/P	—
RM24	Monadh Dubh	Triassic: New Red Sandstone Supergroup, Monadh Dubh Sandstone Formation	micaceous red arkose	—	—
RM26	Monadh Dubh	Triassic: New Red Sandstone Supergroup, Monadh Dubh Sandstone Formation	limestone with calcretes	—	—
RM28	Monadh Dubh	Triassic: New Red Sandstone Supergroup, Monadh Dubh Sandstone Formation	limestone	N/P	—
RM94	West Sgaorishal	Triassic: New Red Sandstone Supergroup, Monadh Dubh Sandstone Formation	limestone	—	—
RM31	Guirdil, Isle of Rum	Mesoproterozoic: Torridon Group, Applecross Formation, Scresort Sandstone Member	medium–coarse sandstone	+1·4	—
RM51	Guirdil, Isle of Rum	Mesoproterozoic: Torridon Group, Applecross Formation, Scresort Sandstone Member	medium–coarse sandstone	+3·0	—
RM69	Kinloch Glen	Mesoproterozoic: Torridon Group, Applecross Formation, Scresort Sandstone Member	medium–coarse sandstone	N/P	—
RM76	West Sgaorishal	Mesoproterozoic: Torridon Group, Applecross Formation, Scresort Sandstone Member	medium–coarse sandstone	N/P	—
RM78	Papadil	Mesoproterozoic: Torridon Group, Aultbea Formation, Sgor Mhor Member	sandstone (with heavy mineral bands)	N/P	—
RM82	Dibidil	Mesoproterozoic: Torridon Group, Diabaig Shale Formation, Laimhrig Shale Member	shale	N/P	—
RM90	West Sgaorishal	Mesoproterozoic: Torridon Group, Applecross Formation, Scresort Sandstone Member	medium–coarse sandstone	N/P	—
RM97	South shore of Loch Scresort	Mesoproterozoic: Torridon Group, Applecross Formation, Allt Mhor na h-Uamha Member	fine sandstone	N/P	—

Sulphur isotopic composition was measured via conventional methods (following whole-rock sulphur extraction), except for samples with asterisk.
*Sulphur isotopic composition measured via laser combustion methods (for plug 1 only).
—, analysis was not run for that sample. Detection limit of bulk S is 0·018 wt %. N/P, no Ag_2S precipitate was recovered following bulk-rock sulphur extraction.

5. DISCUSSION

5.1 Magmatic context of the Rum plugs in relation to the Rum Central Complex

The peridotite plugs included in this study have whole-rock major and trace element and PGE signatures that are subtly distinctive from one another (Supplementary Material Table B) but, broadly, the range of major element compositions overlaps with those recorded for the various portions of the Rum Layered Suite, including peridotite units from the Eastern, Western and Central Layered Series (see Emeleus, 1997; O'Driscoll*et al.*, 2009, and references therein) and the M9 picritic dyke (Upton *et al.*, 2002).

The Layered Suites on Rum are the result of multiple injections of ultramafic magma, in an open magmatic system (e.g. Wager *et al.*, 1960). Current consensus postulates that layering resulted from interaction between cumulates and later magmatic pulses derived from subsequent intruding picrites (e.g. Bédard *et al.*, 1988; Volker & Upton, 1990; Emeleus, 1997; Holness, 2005; O'Driscoll *et al.*, 2007; Leuthold *et al.*, 2014). The S/Se data for the ELS display a broad range from 489 to 5416 (Fig. 7 andSupplementary Material Table D). Mantle S/Se ratios typically range from 2550 to 3750 (Lorand *et al.*, 2003) and although most of the ELS data are within this range there are significant deviations. Loss of S via oxidation of sulphide minerals will lower the primary S/Se ratio. This is also consistent with the common observation of partially oxidized sulphide minerals in thin sections of the ELS; we suggest that this is the cause of ELS S/Se ratios <2000. In contrast, the causes of an increase in S/Se ratio (i.e. >3750) are two-fold: this can be the result of either hydrothermal mobilization of S, forming a series of hydrothermal sulphide minerals, or contamination of the magmas by crustal S (which is devoid of Se, and thus has a high S/Se ratio). Although $\delta^{34}S$ is commonly used as a proxy for crustal S contamination, recent studies suggest that with enough magma throughput in the magmatic (conduit) system, the 'S signal' can be swamped by magmatic $\delta^{34}S$ (−2·0 to +0·4‰ for the Iceland plume;Torssander, 1989; and −2·3 ± 1·5‰ for the Isle of Skye picrites; Hughes*et al.*, 2015), whereas the S/Se ratio remains a more reliable proxy for crustal contamination (Ihlenfeld & Keays, 2011).

We record a minimum in $\delta^{34}S$ locally in unit 1 (ELS) of −4·4‰, but earlier studies have reported $\delta^{34}S$ as low as −14·8‰ in ELS unit 1 rocks with sulphide concentrations locally up to 3 modal % (Faithfull, 1985; Hulbert*et* *al.*, 1992). These lighter $\delta^{34}S$ values occur strictly within unit 1 rocks in the vicinity of metre-scale rafts of Jurassic mudrocks (Faithfull, 1985). In this study, we did not obtain a $\delta^{34}S$ composition for the Jurassic mudrock rafts, but we see no reason to disagree with the inference made by Hulbert*et al.* (1992). In general, Hebridean Jurassic mudrocks, including those in proximity to dolerite and picrite sills on Skye, have a similarly light S-isotopic signature, although those adjacent to intrusions have been significantly degassed owing to contact metamorphism and devolatilization, thereby reducing their bulk S content (Yallup *et al.*, 2013;Hughes *et al.*, 2015). In higher cyclic units in the ELS, $\delta^{34}S$ falls within the typical magmatic range; however, sporadic increases in S/Se ratio (>4350) may indicate localized crustal S contamination. This probably took place during intrusion of magmas into the layered suite, and may not be recorded in the $\delta^{34}S$ either because of the contaminant having an S-isotopic composition similar to the magma or because high volumes of magma throughput have obliterated the signal. Ultimately, the complexities of sulphide composition, S/Se ratio and S-isotopic composition throughout the ELS and Rum Layered Suite as a whole require further investigation, which is beyond the scope of this study.

The diversity of S/Se ratios and PGE compositions measured in ELS sulphides from our study overlaps with sulphide compositions from the plugs. Therefore we suggest that the plugs (conduits) record various crystal mush injection events broadly coincident with magma (or mush) injections into the Rum Layered Suite. As such, the magma conduits represented by the peridotite plugs were not formed from a single eruption event, and probably represent multiple eruptions.

5.2 Feeders and magma staging chamber for the plugs and Rum Layered Suite

The presence of dunite, chromitite, and troctolite xenoliths within peridotite plugs situated inside or close to the Main Ring Fault (Volker & Upton, 1990) suggests that ejection of crystal mushes that fed these plugs occurred whilst crystallization of the layered series was taking place. This demands a physical brecciation of wall-rock material, and does not provide evidence that the conduit crystal mushes were derived from the layered suites directly, but instead indicates that they erupted through them.

One striking observation regarding the Rum plugs is the lack of troctolite plugs. In contrast, feldspathic peridotites (strictly olivine melagabbros) described in our study exist alongside gabbroic plugs that have no phenocryst assemblage or other evidence for their having been intruded as a crystal mush, unlike the peridotite plugs. It is possible that the gabbroic plugs represent the basaltic magmas ejected from a staging magma chamber, and there is no evidence to suggest that these gabbroic plugs correlate strictly with the troctolite portions of units in the Rum Layered Suite. Indeed, per cyclic unit in the Rum Central Complex, the peridotite base and troctolite tops are typically paired in the Eastern Layered Series (Emeleus & Troll, 2011, and references therein; O'Driscoll*et al.*, 2014). Regardless of the mechanism that formed the overall layering of this suite, it is apparent that each of the troctolite layers is a fractionation feature associated and paired with each peridotite layer. Therefore because the peridotitic crystal mush that was ejected through the volcanic pipes is of a comparable composition to the peridotitic bases of the layered suite cyclic units, this suggests that the crystal mushes of both came from a common source, and not necessarily that the layered suites directly fed the plugs themselves. This is supported by the observation that small entrained sulphide minerals in both the plugs and layered suite peridotites are directly comparable with one another (according to both S/Se and $\delta^{34}S$, except for plugs 1 and 2, which are discussed below). In other words, there must have been a separate deeper staging chamber in which a peridotitic crystal mush evolved, before being tapped by the magmatic plumbing system to both the layered suites and the plugs.

5.3 The origin of sulphides within the Rum plugs

With the exception of plugs 1 and 2, the sulphide minerals observed are of low abundance and are small, often rounded phases (<150 μm in diameter); sometimes these sulphides occur as embayments within cumulus olivine (e.g. Fig. 3c and e, and Table 2). All plugs have distinctive chondrite-normalized PGE and Au patterns and S/Se ratios, some of which closely resemble the compositions of sulphides in peridotite units of the ELS (e.g. Figs 6–8). The Fo content of cumulus olivine in the plugs suggests that previously crystallized olivine phenocrysts were entrained within a basaltic or picritic magma, such that the peridotite plugs were essentially crystal mush conduits (e.g. Wadsworth, 1994; Holness *et al.*, 2012) sometimes with entrained chromitite, dunite or troctolite xenoliths (e.g. Volker &

Upton, 1990). We propose a similar scenario for the entrainment of sulphide liquids.

The magmatic temperature of the peridotite plugs on Rum has previously been estimated at 1400°C, based on picritic magma temperatures (Holness*et al.*, 2012). The solidus temperature of an Fe-bearing sulphide liquid is*c.* 1100°C (Bowles *et al.*, 2011; after Kullerud & Yoder, 1959; Arnold, 1971) suggesting that sulphides would be liquid at the estimated temperature of emplacement for the plugs. Given that the 'lubricating' magmas were energetic enough to entrain up to 50% olivine crystals (e.g. Holness *et al.*, 2012) and sizeable xenoliths (e.g. Volker & Upton, 1990), it is highly likely that these were also capable of entraining immiscible sulphide droplets. A similar scenario of upward sulphide liquid entrainment has been suggested in the vicinity of Grasvally in the Northern Limb of the Bushveld Complex by Smith *et al.* (2016), suggesting that S-saturation occurred in a deeper master chamber (albeit via deep crustal contamination with a distinctive S-isotopic and S/Se composition). Therefore, we suggest that at least some sulphides (with mantle S/Se ratios) were formed prior to intrusion of the plugs, and were transported as sulphide liquid droplets into these conduits.

Sulphur content at sulphide saturation (SCSS), as defined by Li & Ripley (2005), is dependent upon the bulk composition of the magma (MgO, FeO^T, SiO_2, Na_2O and K_2O), pressure and temperature. Assuming the magmatic temperature of 1400°C from Holness *et al.* (2012) and using the bulk compositions of gabbroic plugs (e.g. plugs 3 and 9) and ultramafic plugs (e.g. plug 12 at Dibidil, and plug 1 at West Sgaorishal) reported in the Supplementary Material, we find that SCSS would be 2985 ppm (plug 12), 3510 ppm (plug 1) and 1830 ppm for the two gabbroic plugs (plugs 3 and 9) at 1 km depth [using the equation of Li & Ripley (2005)]. For the ultramafic plugs, this is apparently well in excess of the measured bulk S concentration (Table 4). However, we highlight that this bulk concentration potentially incorporates up to 50% cumulus olivine, and therefore the actual S concentration of the interstitial magma carrying the crystal mush will have been significantly higher if this is taken into account (i.e. up to double that measured). Further, a late picrite dyke (M9) on Rum is calculated to reach an SCSS at 1400°C and 1 km depth of 3000 ppm S.Upton *et al.* (2002) observed up to 60 vol. % olivine phenocrysts and we observe two sets of olivine phenocrysts in the M9 dyke, thus highlighting that the bulk composition of this dyke is not suitable for use in modelling the parental magma.

According to the equation of Li & Ripley (2005), pressure (i.e. depth) has little effect on SCSS in comparison with temperature. Therefore if we reduce the ambient temperature to 1200°C (i.e. still 100°C above the liquidus temperature of sulphide) we see a significant drop in SCSS, to ~1600 ppm for the ultramafic plugs and ~1000 ppm for the gabbroic plugs. Accommodating for the bulk geochemical analyses incorporating up to 50% cumulus olivine for the ultramafic plugs, and using this lower temperature, we suggest that SCSS is within the range of the S content now observed in plugs across Rum. Coupled with observations of sulphide mineral embayments in cumulus olivine (as noted above), and Li & Ripley's calculations of decreasing SCSS with fractional crystallization of a magma, it is plausible to assume that sulphide saturation could be achieved deeper in the magmatic system of Rum, prior to plug emplacement.

Several examples of sulphide liquid entrainment in magma conduits have been recorded worldwide; for example, the Norilsk–Talnakh magma conduit setting for orthomagmatic Ni–Cu–PGE mineralization, in which sulphides were initially exsolved in a deeper staging chamber before being flushed out to higher magmatic plumbing system levels (Brügmann *et al.*, 1993; Lightfoot & Keays, 2005). Sulphide entrainment is also suggested for the Platreef (Bushveld Complex) where sulphides were transported away from a deeper zone of sulphide liquid formation (McDonald & Holwell, 2007; McDonald *et al.*, 2009; Ihlenfeld & Keays, 2011; Sharman*et al.*, 2013). In the Platreef, evidence for this sulphide transportation was based on Cu/Pd, S/Se ratios, S-isotope compositions and trapped sulphide melt inclusions (Holwell *et al.*, 2011). Clearly, the process of entrainment of pre-formed sulphides, other silicate minerals and xenoliths is a common feature of many magmatic systems. However, in many of these cases only one part of the full system is exposed, with earlier processes at depth being inferred. The volcanic plugs and the Layered Suite of Rum provide a rare opportunity to compare layered intrusion and volcanic pipe compositions and mechanisms of sulphide transport at the same present-day erosional level.

The $\delta^{34}S$ composition of the peridotite plugs (+0·1 ± 1·0‰, excluding plugs 1 and 2) and peridotite cyclic units of the ELS (–0·5 ± 1·4‰;Supplementary Material Table E) overlap with the typical Icelandic mantle plume range of –2·0‰ to +0·4‰ (Torssander, 1989), mid-ocean ridge basalt (MORB) –0·91 ± 0·5‰ (Labidi *et al.*, 2012) and the picritic Trotternish Sills on the Isle of Skye with average $\delta^{34}S$ of –2·3 ± 1·5‰ (Hughes *et al.*, 2015). Thus we suggest that, for most plugs and the ELS, S-saturation was reached prior to emplacement, in the deeper magmatic plumbing system where olivine crystallization and fractionation was taking place, and did not involve the addition of crustal sulphur with $\delta^{34}S$ different from magmatic compositions. We note that the $\delta^{34}S$ values observed in the ELS and plugs (excluding plugs 1 and 2), are slightly heavier (–0·5 to +0·1 ± 1‰) than the Icelandic plume $\delta^{34}S$ signature of –0·8‰ to –2·3‰ [Torssander (1989) and Hughes *et al.* (2015), respectively] and that this may be due to small changes in the magmatic source signature. With ejection of crystal mushes from a staging chamber, sulphide liquid was entrained and emplaced into the peridotite plug magma conduits and Rum Layered Suite alike. However, in the case of plugs 1 and 2 significant deviation of $\delta^{34}S$ and elevation of S/Se ratio indicates that localized crustal contamination also took place.

5.4 Potential crustal S-sources for the West and Loch Sgaorishal plugs (plugs 1 and 2)

The $\delta^{34}S$ of plug 1 (West Sgaorishal) ranges from –14·7 to –8·0‰ with a S/Se ratio of sulphide minerals ranging from 3000 to 12 000, whereas the $\delta^{34}S$ of plug 2 (Loch Sgaorishal) varies from –6·5 to –2·9‰ (Fig. 7a). Further, the sulphide S/Se ratios of plug 1 are elevated above the range of most ELS sulphides, and often exceed 10 000 (a critical value for crustal sedimentary sulphide compositions; Ihlenfeld & Keays, 2011, and references therein). These S-isotope compositions are significantly lighter than typical mantle or Icelandic plume values, and, together with S/Se, demonstrate that crustal S must have been added to the plug 1 and 2 magmas. There is also a change in textural character of the sulphides in plug 1, which became more abundant and coarser grained. Hence the following questions arise: what was this sulphur source, and why is this light $\delta^{34}S$ recorded only in two plugs from NW Rum?

The $\delta^{34}S$ of sedimentary S (typically pyrite) can be highly variable owing to bacterially mediated sulphate reduction processes during diagenesis. The range of $\delta^{34}S$ can be of the order of tens of per mil (‰) with both negative and positive values. Palaeozoic and Mesozoic sedimentary sulphides generally have more negative values than those observed in Precambrian rocks (Canfield & Teske, 1996; Parnell *et al.*, 2010).

Most plugs on Rum have intruded through Torridonian (Applecross Formation) sandstones at the current erosion level, but the NW tip of plug 1 also intrudes the unconformable base of a sequence of Triassic limestones and calcretes (Binns *et al.*, 1974; Fyfe *et al.*, 1993). Hugheset *al.* (2015) established that the $\delta^{34}S$ of these Triassic sediments is isotopically heavy (with very low S concentrations) and thus cannot account for the $\delta^{34}S$ of plugs 1 and 2 via assimilation. This leaves three possible explanations: (1) Torridonian lacustrine sediments with a light $\delta^{34}S$ (e.g. Parnell *et al.*, 2010) are present below plugs 1 and 2; (2) a segment of Jurassic mudrock with suitably light $\delta^{34}S$ occurs in a downthrown faulted block of a major fault zone (e.g. Main Ring Fault) through which plugs 1 and 2 have intruded; or (3) the $\delta^{34}S$ signature of plugs 1 and 2 is derived from above the current erosion level and involves the Lower Jurassic Broadford Group that probably covered this area at the time of conduit activity.

Despite some Torridonian lacustrine sediments from the mainland Stoer Group having $\delta^{34}S$ as low as −30·1 ± 17·3‰ (Parnell *et al.*, 2010), all the Sleat and Torridon Group Mesoproterozoic sediments analysed by Hughes*et al.* (2015) had values > 0·0‰. Further, the Stoer Group is not recorded in the sedimentary pile exposed on either Skye or Rum, suggesting that scenario (1) is extremely unlikely.

Hebridean Jurassic mudrock and ironstone $\delta^{34}S$ values range from −10 to −35‰, and associated basaltic dykes intruded through this sequence record $\delta^{34}S$ values as light as −30·7‰ (Hughes *et al.*, 2015). Blocks of Jurassic mudrock are recorded as being present within the Main Ring Fault on Rum at Allt nam Bà and near Dibidil (Emeleus & Troll, 2008; Figs 1 and 2c). In contrast, no such sedimentary blocks are reported from the Long Loch Fault, not least owing to this being a strike-slip fault with no vertical motion that would have downthrown these sedimentary units. It is possible, however, that the subsurface route of the plug magma conduits could have intersected a downthrown block of Jurassic mudrocks, assuming that these conduits were fed from somewhere near the centre of the Rum volcanic edifice. However, only plugs 1 and 2 from the NW quadrant of the island record this light $\delta^{34}S$ and the absence of such a light $\delta^{34}S$ in plugs 3–5 renders this scenario highly improbable; and necessarily very localized, if true.

By a process of elimination, to account for the light $\delta^{34}S$, we are left with a scenario whereby crustal sulphur contamination demonstrably took place only in plugs 1 and 2. If contamination were taking place above the current erosion level, in Lower Jurassic sediments that very probably overlay this region, there remains a problem as to how these crustally contaminated sulphides are now present below the level of the potential contaminant. One possibility may be that the sulphides moved down the plug or conduit. Below we consider whether this is feasible given the upward injection of magma in a conduit, and whether the observed characteristics of the sulphides in plugs 1 and 2 support this model. What were the physical constraints on sulphides sinking and could this effect have been so localized that the other plugs (3–5) further 'downhill' on the NW quadrant of Rum, be too low in altitude to record this process?

5.5 Dynamic conduit modelling and sulphide sinking

5.5.1 Hadamard–Rybczynski conduit modelling

The settling of sulphide liquid droplets through a magma conduit can be modelled either by using Stoke's Law or by the Hadamard–Rybczynski equation (e.g. Lightfoot *et al.*, 1984) as established by the experiments ofde Bremond d'Ars *et al.* (2001). The latter takes into account the non-Newtonian fluid properties in the conduit, as well as the differences in both density and viscosity of the silicate and sulphide liquids present in the system, and is expressed as

$$V_s = \frac{1}{3}\left[\frac{\Delta\rho g r^2}{\mu f}\right]\frac{[\mu f + \mu d]}{[\mu f + \frac{3}{2}\mu d]} \tag{1}$$

where V_s is settling velocity (m s^{-1}), r is sulphide droplet radius (m), $\Delta\rho$ is the difference in density between sulphide liquid and silicate magma (= 1400 kg m^{-3}), μf and μd are the viscosities of silicate magma (0·1 Pa s;Paterson *et al.*, 1998) and sulphide liquid (0·01 Pa s) respectively, and g is the acceleration due to gravity (9·8 m s^{-2}). Sulphide droplets will sink (with velocity V_d) through the conduit pipe once the settling velocity (V_s) exceeds the upward velocity of magma intruding into and through the pipe (V_z), such that

$$V_d = V_z - V_s. \tag{2}$$

Sulphide droplet settling occurs when $V_d < 0$ m s^{-1}. The analogue experiments of de Bremond d'Ars *et al.* (2001) indicated that sulphide droplets can be transported in suspension by flowing magmas, and that owing to the large surface tension of the droplets, they do not coalesce during active transport. In part, this could explain the disparity and size of sulphide minerals in peridotite plugs 4–12. These plugs generally do not record matrix banding or indentation layering, suggesting that prolonged quiescent periods were not prevalent during cooling, hence perturbing the ability of the conduit to form crystal layering (e.g. Wadsworth, 1994) and thus arresting sulphide droplet settling and coalescence. Further, the low abundance of sulphides in plugs 4–12 would also hamper coalescence. The size and distribution of fine sulphides in these plugs probably reflects the mode of transportation and entrainment that occurred during conduit activity.

For scenarios involving higher volumes of sulphide liquid, plots of V_s and V_d according to sulphide droplet radius are presented in Fig. 8a and b. Sulphides in the margins of plug 1 range in diameter from 0·5 to 4 mm (Table 2); most intercumulus sulphides are 1–3 mm in diameter. V_z was varied in equation (2) until sulphides of the minimum average radius observed in the margin of plug 1 ($r \sim 0{\cdot}5$ mm) had $V_d < 0$ m s^{-1}. This required an upward magma velocity of 0·01 m s^{-1} (Fig. 8a and b), such that sulphide droplets would sink through the conduit at a rate of 0·001 m s^{-1} (V_d). By lowering V_z further to 0·001 m s^{-1}, sulphides with $r = 0{\cdot}2$ mm would also begin to sink at a similar rate. Thus, the upward flow of magma through the conduit would essentially have to have ceased before sulphide droplets of the sizes observed in the margins of plug 1 would settle. The viscosity of the silicate magma was assumed to be 0·1 Pa s (Paterson *et al.*, 1998); however, if the conduit was instead filled with olivine phenocrysts entrained in basaltic magma then the overall effective viscosity of the 'silicate magma' would be higher. Holness *et al.* (2012)suggested a crystal load in the range of 10–40 vol. % for the peridotite plugs, based on the volume of phenocrysts entrained in picrite dykes on Rum. Even by increasing viscosity to 1 Pa s [i.e. an assumed 50% crystal mush entrained in a mafic magma, with viscosity = 1 Pa s based onPaterson *et al.* (1998)] at $V_z = 0{\cdot}001$ m s^{-1}, sulphides of $r \sim 0{\cdot}5$ mm would still be capable of sinking through the conduit. In summary, it is possible that sulphide liquids forming through crustal contamination above the present-day level of erosion and exposure of plug 1 (and plug 2) would have sunk through the conduit once active upwards magma injection had ceased. Sulphide liquid sinking may have taken place either in a passively cooling (stationary) environment and/or sulphide liquid was 'sucked' downwards if magma withdrawal were taking place.

By assuming that the cooling rate of a stationary magma is controlled by Newton's Law of Cooling,

$$T = T_a + (T_o - T_a)e^{-\kappa t} \quad (3)$$

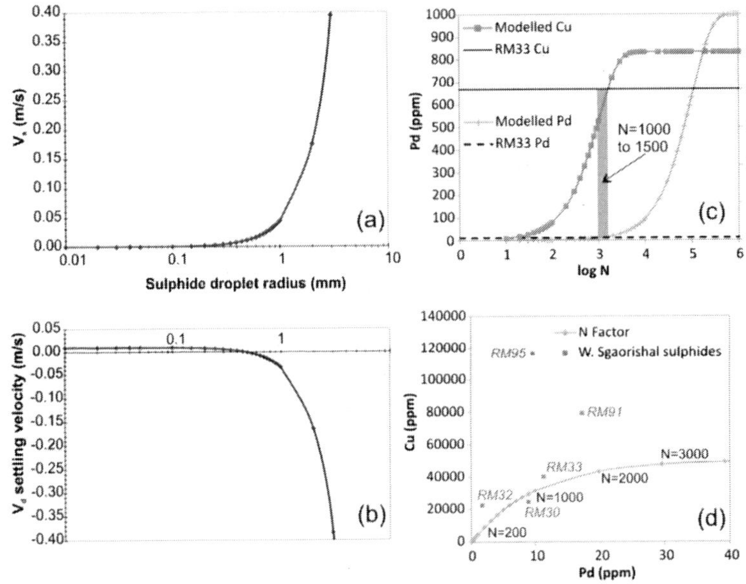

Figuer 8. Dynamic conduit modelling using the Hadamard–Rybczynski equation for sulphide settling: (a) sulphide velocity (V_s) vs size of a sulphide droplets (radius in mm); (b) sulphide settling velocity through a conduit (V_d) vs sulphide droplet radius. Sulphides will sink if $V_d < 0$. (c, d) *N*-factor modelling: (c) *N*-factor (plotted as log *N*) vs Pd and Cu concentration in the sulphide liquid. Continuous horizontal line delineates the Cu concentration in the sulphide liquid from the margin of plug 1 (RM33); the dashed horizontal line is for Pd. Both Cu and Pd were calculated from the bulk-rock geochemistry normalized to 100% sulphide [using the method outlined by Huminicki *et al.* (2005)]. Shaded rectangle shows the range of modelled log *N* values that fit Cu and Pd. (d) Modelled Cu and Pd concentrations in sulphides according to various values of *N* (labelled). Plotted for comparison are the calculated mean Cu abundances of the sulphide liquid [calculated by normalizing the bulk Cu concentration to 100% sulphide, using the method outlined by Huminicki *et al.* (2005)] and weighted mean Pd abundances (per sample weighted according to the number of analyses of chalcopyrite, pentlandite and pyrrhotite) for sulphides from plug 1, centre and margins.

it would take the margin of plug 1 *c.* 5 days to cool below 1100°C, whereT_a is the temperature of the country rock [assumed to be 600°C, based onHolness *et al.* (2012)], T_o is the initial magma temperature of the conduit [assumed to be 1400°C; Holness *et al.* (2012)], T is the target temperature of 1100°C for the sulphide liquidus (e.g. Bowles *et al.*, 2011) and κ is a thermal diffusivity constant of 10^{-6} m^2 s^{-1}. Given a downward velocity $V_d$of 0·001 m s^{-1} for sulphide droplets through the quiescent conduit, sulphides would be capable of settling through a distance of the order of 400 m, downwards from the initial level of S-saturation. As mentioned above, however, constraints on the SCSS of these plugs may suggest that their emplacement temperature was lower than 1400°C. Therefore if we use 1200°C as the emplacement temperature, we estimate that sulphides could still have settled more than 100 m down the conduit.

Ultimately, this settling distance is likely to be subject to estimation errors; for example, as sulphide droplets coalesce during settling, their settling velocity will change. In addition, as the conduit cools towards 1100°C, the effective viscosity of both the sulphide liquid and silicate magma will increase, thus slowing its descent. The 'permeability' of the host silicate mush (i.e. amount of partial melt and degree of ponding of sulphide liquid above areas of low permeability; e.g. Holwell & McDonald, 2006; Hutchinson & McDonald, 2008) may not be uniform throughout a single volcanic plug. However, the interfacial tension at silicate–sulphide liquid interfaces may preclude sinking of sulphide droplets below a critical size range (e.g. Mungall & Su, 2005). This depends on the proportions of silicate melt to solidified silicate minerals in the system, and would suggest that compaction-driven sulphide segregation is possible (Rose & Brenan, 2001).

The estimated thickness of Triassic sediments in the Hebrides Basin in the vicinity of Rum on the 'Skye High' of this Mesozoic sedimentary basin ranges from tens of metres on Ardnamurchan, to 100 m at Raasay off the Isle of Skye, and 300 m at Gruinard Bay on the Scottish mainland (Steel*et al.*, 1975). Given that the current erosion level of plug 1 occurs at the base of these Triassic sediments (Figs 2a and 9a) and the estimated minimum potential settling distance of sulphide droplets is 100 m (assuming a lower emplacement temperature of 1200°C) to 400 m [using the Holness *et al.* (2012) emplacement temperature of 1400°C], this is within reach of S-rich lower Jurassic sediment horizon(s), assumed to have been present above the current erosion level (Fig. 9a). Assimilation of wall-rock sediments, particularly light δ^{34}S crustal sulphur from Jurassic units, would induce sulphide saturation at a high level in the conduit (above the current erosion level; e.g. Fig. 9b). Thus, we can reasonably envisage that once active magma-mush injection had ceased, silicate and sulphide liquids from above sank through the conduit, forming olivine cumulates with interstitial sulphide liquid that recorded light crustal S isotope compositions and high S/Se ratios inherited from hundreds of metres above (e.g. Fig. 9c).

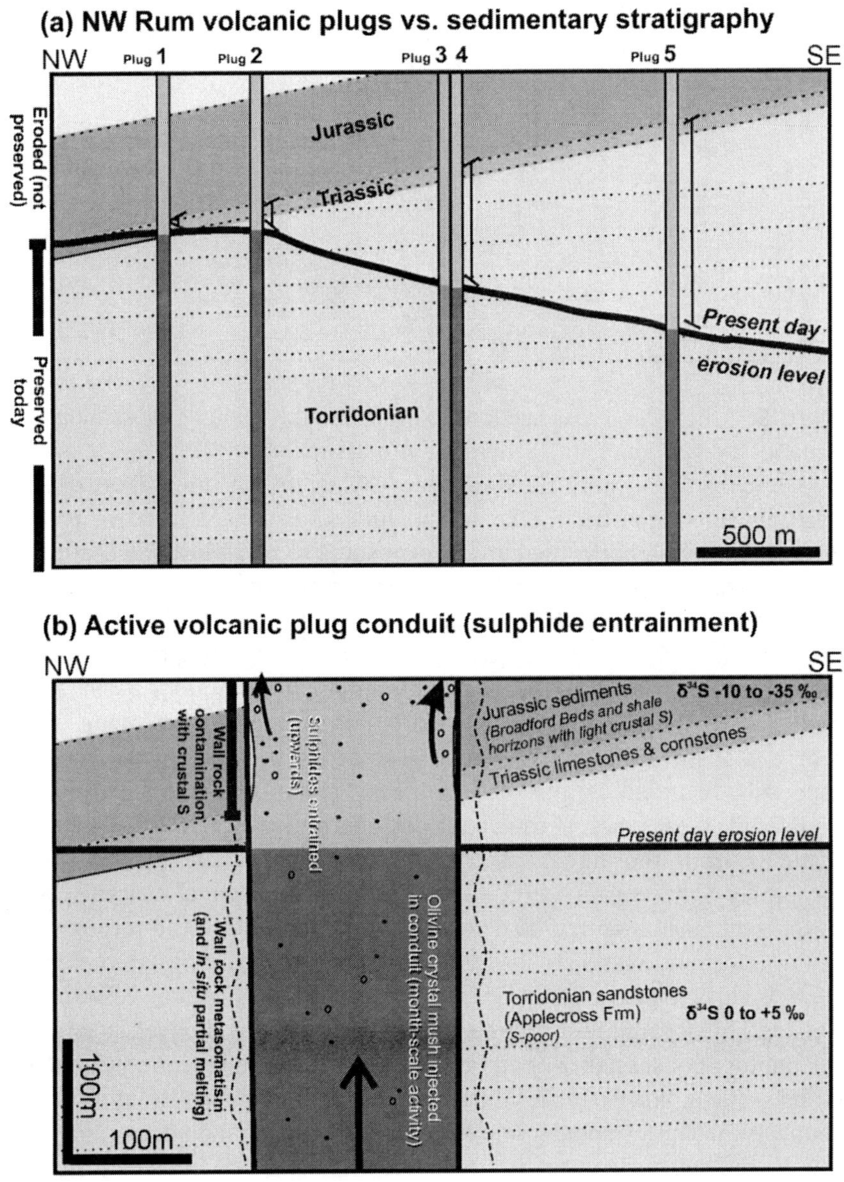

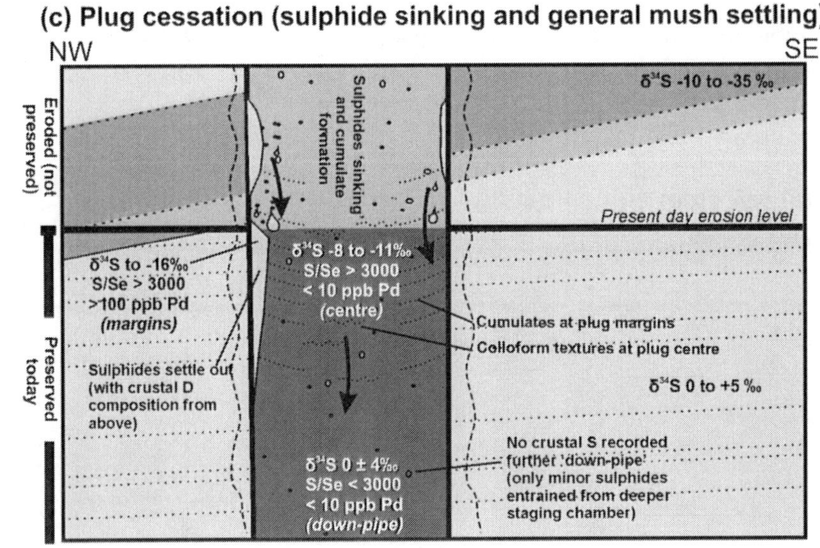

Figure 9. Schematic cross-sections of the crustal sulphur contamination scenario for NW Rum. (a) Schematic cross-section of NW Rum showing plugs 1–5 in the vicinity of Sgaorishal. Owing to the orientation of the Mesozoic Hebridean Basin Triassic and Jurassic sediments relative to the present-day topography, the current exposures of plugs 3, 4 and 5 are too great a distance below the Jurassic contamination horizon to record crustal $\delta^{34}S$ and voluminous sulphide liquids. However, plugs 1 and 2 still record light crustal $\delta^{34}S$. (b) Active volcanic plug intrudes through Torridonian sediments, and Triassic and Jurassic sediments of the Hebrides Basin (since eroded) above. The plug entrains tiny droplets of high-tenor sulphide liquid from a staging chamber. Wall-rock metasomatism and partial melting (especially recorded in the siliceous Torridonian Applecross Formation sandstones) takes place. In the upper portions of the plug, S-rich shales and siltstones from the Jurassic sediments contaminate the magmas with isotopically light crustal S. (c) Active magma intrusion of volcanic plug ceases, and magma-entrained olivine mushes are left to settle through the conduit (forming cumulate layers, particularly at the conduit margins). Sulphide droplets entrained in the plug, including more voluminous sulphide liquids from crustal S contamination horizons (above) in Jurassic sedimentary units, settle through the conduit. Sulphide liquids amalgamate during settling, and form an interstitial or intercumulus phase around the cumulate olivine. Sulphide liquid is particularly abundant around the conduit margins (owing to wall-rock S contamination) and has a higher concentration of PGE, higher S/Se ratio and lighter $\delta^{34}S$.

Although plug 2 also records $\delta^{34}S$ in sulphides significantly less than the local magmatic compositions of around 0‰, it has lower abundances of sulphide than plug 1. Projecting the dip of the base of the Triassic sedimentary units near plug 1 (approximately 20° NW; Fig. 9a) and even accounting for the current topography and elevation of the area between plugs 1 and 2, the lower sulphide abundance in plug 2 may reflect the greater distance between the potential Lower Jurassic contamination zone and the present-day level of exposure (Fig. 9a). Further, the lack of a crustal $\delta^{34}S$ signature in plugs 3, 4 and 5 is consistent with the model, in that their current level of exposure is >400 m below the projected contamination horizon, so that any sulphides that did settle downwards are likely to have been removed by erosion (i.e. Fig. 9a). The lack of a crustal $\delta^{34}S$ signature in plugs from elsewhere on the island, particularly farther east, may also suggest that the Mesozoic sedimentary sequence of the Hebrides Basin was not present, as the Isle of Rum sits on the 'Skye High' within the basin such that only the feather-edge of the basin's sediments could have been deposited over part of the island (see Hugheset *al.*, 2015).

5.5.2 Jurassic shale geochemical interaction with volcanic plugs

The sedimentary host rocks to the volcanic plugs (Triassic, Fig. 2a; Torridonian, Fig. 2a–c) typically appear bleached, fissile and metasomatized. Detailed studies of *in situ* anatexis of these rocks have provided insights into the longevity of the plugs as active magma conduits (e.g. Holness, 1999; Holness *et al.*, 2012). Static melting occurred up to 15 m from the gabbro plugs and 6 m from the larger peridotite plugs, suggesting that the peridotite plugs were a relatively short-lived magma conduits (active for *c.* 1–2 years; Holness *et al.*, 2012) whereas the gabbro plugs were longer lived features (*c.* 40 years; Holness, 1999). However,Holness (1999) concluded that anatectic melt migration did not occur, so that very limited contamination of the plug margins was likely to have taken place. However, Yallup *et al.* (2013) and Hughes *et al.* (2015) (and references therein) demonstrated that the volatile content (S, C, H_2O, etc.) of mudrocks is particularly vulnerable to mobilization by magmas during contact metamorphism of wall-rocks and/or their brecciation. In this way, we envisage a scenario whereby S is preferentially mobilized, imparting a strong crustal contamination signature onto the plug magmas, but without other non-volatile trace element contamination.

5.5.3 Comparison with Paleogene macrodykes in East Greenland and other similar intrusions

A similar set of geological circumstances exists in East Greenland, whereHolwell *et al.* (2012) described a suite of mineralized macrodykes, radiating away from the Skaergaard intrusion and intruding Archaean basement and overlying pyrite-rich Cretaceous black shales (with δ^{34}S –23 to –30‰) and Paleogene flood basalts. Magmatic Cu–PGE–Au sulphide mineralization, with crustal δ^{34}S signatures (–10 to –26‰), is present along the margins of the subvertical Miki Fjord Macrodyke, which is observed to intrude all three lithological units. Further north, the Togeda Macrodyke, which is exposed at a level where it intrudes the basement gneiss, has similar sulphides that have a light S-isotope signature (Holwell*et al.*, 2012). Holwell *et al.* (2012) concluded in both cases that Cretaceous shales were the trigger for sulphide saturation and that sulphide droplets were entrained within the macrodyke magmas during intrusion. The presence of sulphides in the Togeda Macrodyke, beneath the stratigraphic level of the contaminant, was explained by northward injection of magma, which could then incorporate S from sediments exposed deeper to the south (Holwell *et al.*, 2012).

A similar situation to that described above for the Rum plugs is also possible for the Togeda Macrodyke. The Cretaceous sedimentary units are inferred to have been present some 500 m or so above the present erosion level. Given the much larger size of the droplets (up to 10 mm) in the macrodykes compared with the Rum plugs (mostly 1–3 mm), the likelihood of large droplets settling down through the dyke conduit by a vertical distance of up to 500 m is plausible. For example, assuming a sulphide droplet with radius (r) of 0·5 mm and a silicate liquid (not crystal mush) of viscosity 0·1 Pa s, then owing to the silicate–sulphide liquid density difference, the sulphide droplets would be capable of sinking through a magma conduit with upward silicate magma velocities (V_z) of up to 1·1 m s^{-1}. However, in a dyke scenario, lateral movement of magmas along the dyke (owing to a pressure gradient from the point of intrusion along an opening fracture) must also be factored in. Hence both lateral and downward migration of sulphide droplets may be envisaged for planar features such as dykes (e.g. Todega Macrodyke), in contrast to pipe-shaped conduits.

The downward movement of sulphide liquids in a cooling magmatic system has also been suggested for ultramafic complexes in Madagascar (McDonald, 2008). In this model, sulphide liquids migrate laterally, as well as downwards, towards the centre of the intrusion during crystallization.

During their migration, the sulphide liquids fractionated such that Pt- and Pd-bearing PGM exsolved from, and detached from, the sulphide liquid (becoming trapped in olivine mushes), leaving Cu-rich sulphide liquid to continue sinking towards the centre of the intrusion (McDonald, 2008).

Our model of sulphide sinking in a cooling magmatic system (where magma flow has recently ceased) differs markedly from sulphide sinking models suggested for other mineralized conduit settings; for example, at Norilsk–Talnakh where Arndt (2011) envisaged that sulphide liquids sank (or were re-injected) through an active silicate magma conduit, with a significant upward velocity of magma injected from beneath. Sulphide pools may be re-entrained by pulses or batches of new magma (e.g.Robertson *et al.*, 2014), leading to complex scenarios of re-deposition within large, dynamic or long-lived conduit systems, with ultimate sulphide tenor upgrading (Barnes *et al.*, 2016, and references therein).

5.5.4 N-factor modelling: silicate/sulphide liquid ratios

Another parameter that can be calculated for a sulphide-bearing conduit is the proportion of sulphide liquid to silicate magma within the system. In a closed system, this is typically denoted by the '*R*-factor'; however, in the case of open-system conduits this is termed the '*N*-factor' and was modelled for the sill-hosted Norilsk–Talnakh Ni–Cu–PGE orthomagmatic mineralization in Siberia (Brügmann *et al.*, 1993; Naldrett & Fedorenko, 1995; Naldrett *et al.*, 1996). In an open system such as a magma conduit silicate magma batches are flushed through the conduit system. If this system is S-saturated, sulphide liquids may become trapped within the conduit whilst silicate magmas continue to flush through. Owing to their strong chalcophile nature, PGE and Cu will partition from each fresh silicate magma batch into the resident sulphide liquid, thereby upgrading the PGE tenor of this sulphide in a mechanism analogous to zone refining (Brügmann *et al.*, 1993; Naldrett & Fedorenko, 1995; Naldrett *et al.*, 1996). The proportion of silicate magma (as a time integrated volume) relative to the volume of the sulphide liquid will determine this degree of PGE upgrading.

The *N*-factor is typically used for conduits in which magmas move upwards past sulphide accumulations. However, in the situation we propose for Rum, sulphide droplets are formed at the top of the magma conduit and slowly settle through it (e.g. Brügmann *et al.*, 1993). The effect is the same;

in both cases silicate magma and sulphide liquids must migrate past one another to achieve upgrading of PGE tenor. Thus the *N*-factor model is appropriate in this instance to assess the volumes of ambient sulphide liquid in the conduit.

The *N*-factor equation, based on Cox *et al.* (1979) is as follows:

$$Y_i = X_i\{D - [(D - 1)e^{-(1/D)N}]\} \tag{4}$$

where Y_i is the concentration of metal *i* in the sulphide, X_i is the initial concentration of metal *i* in the silicate, and *D* is the partition coefficient for metal *i* between silicate and sulphide;

$$N = \frac{\text{volume silicate magma}}{\text{volume sulphide liquid}}. \tag{5}$$

the value of *N* increases, the concentration of strongly chalcophile metals such as Pt and Pd rapidly increases. For plug 1, the concentration of trace elements such as Pd and Cu according to the volume ratio of silicate magma to sulphide liquid (*N*-factor) can be modelled usingequation (4), assuming $D_{Cu} = 1470$ and $D_{Pd} = 190\,000$ (Mungall & Brenan, 2014), with an initial silicate magma composition of 50 ppm Cu and 10 ppb Pd (comparable with picritic compositions and Cu and Pd abundances in other low-sulphide plugs such as plug 12). By finding the intersection between the model curves for Pd and Cu and the measured concentration of these metals in the plug, we can estimate that the ratio of silicate magma to sulphide liquid (*N*-factor) ranged from 1000 to 1500 (specifically using sample RM33 from plug 1) (Fig. 8c).

To cross-check if the *N*-factor calculated is reasonable, we have used the average abundance of Pd measured in sulphides (Table 3) calculated per sample for plug 1, and plotted this against the calculated Cu abundance for a homogenized sulphide liquid, based on assigning all bulk-rock Cu to chalcopyrite, and normalizing to 100% sulphides. As sulphide assemblages comprise pyrrhotite, pentlandite and chalcopyrite (with only accessory bornite, chalcocite and galena), we have followed the method of Huminicki*et al.* (2005) using the whole-rock measured abundances of S, Cu, Ni and Fe. All of the whole-rock Cu budget was assigned to chalcopyrite, and after subtracting the mass of S required, and the mass of Ni bound within olivine [assuming a Ni content of 2400 ppm in olivine and an average of 60% olivine in the plug, based on Holness *et al.* (2012)] all of the remaining Ni was assigned to pentlandite. The proportion of pyrrhotite was then estimated from the remaining S. Following the observations and methods of Holwell & McDonald (2007), such that chalcopyrite contains negligible PGE in solid solution, we then recalculate the measured whole-rock PGE content to 100% pyrrhotite + pentlandite, as these are the principal PGE-bearing base metal sulphides. By this method, the Cu and Pd compositions of sulphides in plug 1 can be modelled by equation (4) if $N = 1000–1500$ (Fig. 8c and d). This includes the centre of the plug (e.g. RM30). However, the outlier RM95 at the southern plug margin has a substantially higher concentration of Cu (almost 120 000 ppm), which may suggest that the margin of the plug (as represented by RM95) has acquired extra Cu from the surrounding wall-rock. The Triassic Sandstone Formation in this area contains rare malachite reported with carbonate calcretions (Emeleus & Troll, 2008).

5.5.5 The significance of sulphide sinking versus upward entrainment

Comparison with examples outside the BPIP shows that a process of sulphide sinking through cooling, recently inactive conduits may well be widespread and could account for the presence of high-volume sulphides with crustal S compositions at levels stratigraphically lower than that at which they originally formed. Thus, the study of mineralized magmatic sulphide systems also needs to consider the possibility of contamination from sources above the level of mineralization, which may no longer be exposed. In terms of exploration models, this is significant in that intrusions exposed below the stratigraphic level of any potential contaminant (or where possible contaminants have been eroded) still have the potential to host magmatic sulphides. If a settling sulphide liquid had the ability to pond in a depression in the magmatic plumbing system, then massive sulphide accumulations that result from crustal contamination could potentially be found hundreds of metres below the level of the contaminant that triggered sulphide saturation. Our work provides quantifiable and traceable evidence for the processes of sulphide sinking. As such, it offers support to the growing number of studies that cite sulphide sinking, slumping and/or re-injection within conduits as a mechanism for magmatic sulphide accumulation and upgrading (Benko*et al.*, 2015; Saumur *et al.*, 2015; Barnes *et al.*, 2016, and references therein; Saumur & Cruden, 2016). This contrasts with the more traditional view of upward entrainment of sulphides into higher level intrusions (e.g.Holwell & McDonald, 2007; Naldrett *et al.*, 2009, 2011).

6. CONCLUSIONS

1. The peridotite plugs on the Isle of Rum were intruded as olivine-rich crystal mushes suspended in a basaltic or picritic magma that also entrained sulphide liquid.

2. Based on the elevation and corresponding compositions of the plugs and cyclic units, it is unlikely that the Rum Layered Suite directly fed all these volcanic conduits. Instead we suggest that a separate, deeper, staging chamber was responsible for the periodic and discrete intrusion of peridotitic crystal mushes with entrained sulphide into both the Rum Layered Suite and the volcanic conduits represented by the plugs.

3. Most plugs have a magmatic whole-rock $\delta^{34}S$ signature of +0·1 ± 1·0‰, which overlaps with the whole-rock $\delta^{34}S$ signature of the ELS (–0·5 ± 1·4‰) and proto-Icelandic plume values recorded elsewhere in the NAIP (–2·3 ± 1·5‰; Hughes *et al.*, 2015). In addition, whole-rock and sulphide-specific PGE abundances in plugs lacking crustal contamination signatures overlap compositions in the ELS.

4. Two plugs in the NW of the island have a distinctive crustal $\delta^{34}S$ signature (ranging from –14·7 to +0·3‰) and elevated S/Se ratios (up to 12 130), unlike any of the other plugs on Rum. Projecting the Hebrides Basin sediment stratigraphy above the preserved unconformable base of Triassic rocks (Fig. 9a), the volcanic conduits would probably have been emplaced through a thick package of Jurassic mudrocks with characteristically light $\delta^{34}S$ (–33·8 to –14·7‰), forcing sulphide supersaturation and the formation of voluminous immiscible sulphide liquid, locally and at a near-surface level (Fig. 9b).

5. Once active magma transport had ceased, the sulphide liquids would sink back through the conduit. Based on modelling of the conduit mush viscosity and cooling rates, this settling would take place over a distance of up to 400 m and over a period of a few days, resulting in the poikilitic and net-textures preserved today. Sulphides previously entrained in the crystal mush from the staging chamber, and this secondary immiscible sulphide liquid, became amalgamated upon sulphide supersaturation. Thus the $\delta^{34}S$ of these plugs (1 and 2) is a mixture of both early and later sulphide liquids, although the crustal isotopic contamination signature dominates owing to the production of greater volumes of sulphide.

6. Given the current topography in the NW of Rum, the other plugs probably record deeper level sulphides in the plug conduit plumbing system (i.e. with uncontaminated magmatic $\delta^{34}S$), and any crustal S influence (if originally present) has since been removed by erosion.

7. *N*-factor modelling for the crustally contaminated plugs (1 and 2) suggests a silicate magma to sulphide liquid ratio of 1000–1500.

8. We suggest that this sulphide sinking process (within a cooling non-active conduit or during magma 'suck-back') may be observed in other vertical or inclined magma conduits globally; for example, in the macrodykes of East Greenland. This model may be used to explain S-isotopic and S/Se ratios that are indicative of contamination of magmas by crustal S although no suitable lithology is present either in the immediate host rocks or deeper in the system.

FUNDING

Sulphur isotope analyses were funded by NERC Isotope Geosciences Facilities grant, IP-1356-1112. H.S.R.H. acknowledges the financial support of the Natural Environment Research Council (NERC) for her PhD studentship (NE/J50029X) and funding of open access publication. This is a contribution to the TeaSe (Te and Se Cycling and Supply) research consortium supported by NERC award NE/M011615/1 to Cardiff University and the University of Leicester.

ACKNOWLEDGEMENTS

We thank Scottish Natural Heritage, the Isle of Rum Community Trust and the residents of the Isle of Rum for granting permission to sample. Alison McDonald, Brett Davidheiser-Kroll, Anthony Oldroyd and Katherine Dobbie are thanked for their assistance in sample preparation, particularly for whole-rock S-extraction and analysis. Peter Fisher is thanked for his assistance with SEM facilities. Discussions with Kathryn Goodenough, Nick Arndt, Ben Hayes, John Faithfull, Brian Upton and Chris MacLeod helped greatly during the writing of this paper. We thank Ed Ripley, Stephen Daly and an anonymous reviewer for their constructive reviews, which greatly improved the paper, and John Gamble for his editorial handling.

REFERENCES

1. Andersen, J. C. Ø., Rasmussen, H., Nielsen, T. F. D. & Rønsbo, J. G. (1998). The Triple Group and the Platinova gold and palladium reefs in the Skaergaard Intrusion: Stratigraphic and petrographic relations. Economic Geology 93, 488–509.

2. Andersen, J. C. Ø., Power, M. R. & Momme, P. (2002). Platinumgroup elements in the Palaeogene North Atlantic Igneous Province. In: Cabri, L. J. (ed.) The geology, geochemistry, mineralogy and mineral beneficiation of platinum-group elements. Canadian Institute of Mining, Metallurgy and Petroleum (CIM) Special Volume 54, 637–668.

3. Arndt, N. T. (2011). Insights into the geologic setting and origin of Ni–Cu–PGE sulfide deposits of the Norilsk–Talnakh region, Siberia. Reviews in Economic Geology 17, 190–215.

4. Arnold, R. G. (1971). Evidence for liquid immiscibility in the system FeS–S. Economic Geology 66, 1121–1130.

5. Barnes, S. J., Fiorentini, M. L., Austin, P., Gessner, K., Hough, R. M. & Squelch, A. P. (2008). Three-dimensional morphology of magmatic sulfides sheds light on ore formation and sul- fide melt migration. Geology 36, 655–658.

6. Barnes, S. J., Cruden, A. R., Arndt, N. & Saumur, B. M. (2016). The mineral system approach applied to magmatic Ni–Cu– PGE sulphide deposits. Ore Geology Reviews (online first, in press).

7. Be´dard, J. H., Sparks, R. S. J., Renner, R., Cheadle, M. J. & Hallworth, M.A. (1988). Peridotite sills and metasomatic gabbros in the Eastern Layered Series of the Rhum complex. Journal of the Geological Society, London 145, 207–224.

8. Benko, Z., Mogessie, A., Molnar, F., Severson, M. J., Hauck, S. A. & Raic, S. (2015). Partial melting processes and Cu–Ni– PGE mineralization in the footwall of the South Kawishiwi Intrusion at the Spruce Road deposit, Duluth Complex, Minnesota, USA. Economic Geology 110, 1269–1293.

9. Binns, P. E., McQuillin, R. & Kenolty, K. (1974). The geology of the Sea of the Hebrides. Report of the Institute of Geological Sciences, p. 1269–1293.

10. Bird, D. K., Brooks, D. K., Gannicott, R. A. & Turner, P. A. (1991). A gold-bearing horizon in the Skaergaard Intrusion, East Greenland. Economic Geology 86, 1083–1092.

11. Bowles, J. F. W., Howie, R. A., Vaughan, D. J. & Zussman, J. (2011). Rock-forming Minerals: Non-silicates; Oxides, Hydroxides and Sulphides. Geological Society.

12. Bru¨ gmann, G. E., Naldrett, A. J., Asif, M., Lightfoot, P., Gorbachev, N. S. & Fedorenko, V. A. (1993). Siderophile and chalcophile metals as tracers of the evolution of the Siberian Trap in the Noril'sk region, Russia. Geochimica et Cosmochimica Acta 57, 2001–2018.

13. Butcher, A. R., Young, I. M. & Faithfull, J. W. (1985). Finger structures in the Rhum Complex. Geological Magazine 122, 491–502.

14. Butcher, A. R., Pirrie, D., Pritchard, H. M. & Fisher, P. (1999). Platinum-group mineralization in the Rum layered intrusion, Scottish Hebrides, UK. Journal of the Geological Society, London 156, 213–216.

15. Cameron, E. M. (1994). Depletion of gold and LILE in the lower crust: Lewisian Complex, Scotland. Journal of the Geological Society, London 151, 747–754.

16. Canfield, D., Raiswell, R., Westrich, J., Reaves, C. & Berner, R. (1986). The use of chromium reduction in the analysis of rediced inorganic sulphur in sediments and shales. Chemical Geology, 54, 149–155.

17. Canfield, D. E. & Teske, A. (1996). Late Proterozoic rise in atmospheric oxygen concentration inferred from phylogenetic and sulphur-isotope studies. Nature 382, 127–132.

18. Chambers, L. M., Pringle, M. S. & Parrish, R. R. (2005). Rapid formation of the Small Isles Tertiary centre constrained by precise 40Ar/39Ar and U–Pb ages. Lithos 79, 367–384.

19. Chung, H.-Y. & Mungall, J. E. (2009). Physical constraints on the migration of immiscible fluids through partially molten silicates, with

special reference to magmatic sulfide ores. Earth and Planetary Science Letters 286, 14–22.

20. Cox, K. G., Bell, J. D. & Pankhurst, R. J. (1979). The Interpretation of Igneous Rocks. Allen & Unwin.

21. Czamanske, G. K., Kunilov, V. E., Zientek, M. L., Cabri, L. C., Likhachev, A. P., Calk, L. C. & Oscarson, R. I. (1992). A proton microprobe study of magmatic sulfide ores from the Norilsk–Talnakh District, Siberia. Canadian Mineralogist 30, 249–287.

22. de Bremond d'Ars, J., Arndt, N. T. & Hallot, E. (2001). Analog experimental insights into the formation of magmatic sulphide deposits. Earth and Planetary Science Letters 186, 371–381.

23. Emeleus, C. H. (1997). Geology of Rum and the Adjacent Islands. British Geological Survey.

24. Emeleus, C. H. & Bell, B. R. (2005). British Regional Geology: The Palaeogene Volcanic Districts of Scotland. British Geological Survey.

25. Emeleus, C. H. & Troll, V. R. (2008). A geological excursion guide to Rum. Edinburgh Geological Society.

26. Emeleus, C. H. & Troll, V. R. (2011). Recent research developments on the Isle of Rum, NW Scotland. Geology Today 27, 184–193.

27. Faithfull, J. W. (1985). The lower Eastern Layered Series of Rhum. Geological Magazine 122, 459–468.

28. Fyfe, J. A., Long, D. & Evans, D. (1993). United Kingdom Offshore Regional Report: the Geology of the Malin– Hebrides Sea Area. HMSO for the British Geological Survey.

29. Godel, B., Barnes, S.-J. & Maier, W. D. (2006). 3-D distribution of sulphide minerals in the Merensky Reef (Bushveld Complex, South Africa) and the J-M Reef (Stillwater Complex, USA) and their relationship to microstructures using X-ray computed tomography. Journal of Petrology 47, 1853–1872.

30. Godel, B., Barnes, S. J. & Barnes, S.-J. (2013). Deposition mechanisms of magmatic sulphide liquids: Evidence from highresolution X-ray

computed tomography and trace element chemistry of komatiite-hosted disseminated sulphides. Journal of Petrology 54, 1455–1481.

31. Hall, G. E. M., Pelchat, J. -C. & Loop, J. (1998). Separation and recovery of various sulphur species in sedimentary rocks for stable sulphur isotopic determination. Chemical Geology, 67, 35–45.

32. Hamilton, M. A., Pearson, D. G., Thompson, R. N., Kelley, S. P. & Emeleus, C. H. (1998). Rapid eruption of Skye lavas inferred from precise U–Pb and Ar–Ar dating of the Rum and Cuillin plutonic complexes. Nature 394, 260–263.

33. Hayes, B., Be´dard, J. H., Hryciuk, M., Wing, B., Nabelek, P., McDonald, W. D. & Lissenberg, C. J. (2015). Sulfide immiscibility induced by wall-rock assimilation in a fault-guided basaltic feeder system, Franklin Large Igneous Province, Victoria Island (Arctic Canada). Economic Geology 110, 1697–1717.

34. Hesselbo, S. P. & Coe, A. L. (2000). Jurassic sequences of the Hebrides Basin, Isle of Skye, Scotland. In: Graham, J. R. & Ryan, A. (eds) Field Trip Guidebook. International Association of Sedimentologists, pp. 41–58.

35. Holness, M. B. (1999). Contact metamorphism and anatexis of Torridonian arkose by minor intrusions of the Rum Igneous Complex, Inner Hebrides, Scotland. Geological Magazine 136, 527–542.

36. Holness, M. B. (2005). Spatial constraints on magma chamber replenishment events from textural observations of cumulates: the Rum Layered Intrusion, Scotland. Journal of Petrology 46, 1585–1601.

37. Holness, M. B., Sides, R., Prior, D. J., Cheadle, M. J. & Upton, B. G. J. (2012). The peridotite plugs of Rum: Crystal settling and fabric development in magma conduits. Lithos 134–135, 23–40.

38. Holwell, D. A. & Keays, R. R. (2014). The formation of lowvolume, high-tenor magmatic PGE–Au sulfide mineralization in closed systems: Evidence from precious and base metal geochemistry of the Platinova Reef, Skaergaard Intrusion, East Greenland. Economic Geology 109, 387–406.

39. Holwell, D. A. & McDonald, I. (2006). Petrology, geochemistry and the mechanisms determining the distribution of platinum-group element and base metal sulphide mineralisation in the Platreef at Overysel,

northern Bushveld Complex, South Africa. Mineralium Deposita 41, 575–598. Holwell, D. A. & McDonald, I. (2007). Distribution of platinumgroup elements in the Platreef at Overysel, northern

40. Bushveld Complex: a combined PGM and LA-ICP-MS study. Contributions to Mineralogy and Petrology 154, 171–190.

41. Holwell, D. A., McDonald, I. & Butler, I. B. (2011). Precious metal enrichment in the Platreef, Bushveld Complex, South Africa: evidence from homogenized magmatic sulfide melt inclusions. Contributions to Mineralogy and Petrology 161, 1011–1026.

42. Holwell, D. A., Abraham-James, T., Keays, R. R. & Boyce, A. J. (2012). The nature and genesis of marginal Cu–PGE–Au sulphide mineralisation in Paleogene macrodykes of the Kangerlussuaq region, East Greenland. Mineralium Deposita 47, 3–21.

43. Holwell, D. A., Keays, R. R., Firth, E. A. & Findlay, J. (2014). Geochemistry and mineralogy of platinum group element mineralization in the River Valley Intrusion, Ontario, Canada: A model for early-stage sulfur saturation and multistage emplacement and the implications for 'contact-type' Ni–Cu–PGE sulfide mineralization. Economic Geology 109, 689–712.

44. Hudson, J. D., Coleman, M. L., Barreiro, B. A. & Hollingworth, N. T. J. (2001). Septarian concretions from the Oxford Clay (Jurassic, England, UK): involvement of original marine and multiple external pore fluids. Sedimentology. 48, 507–531.

45. Hughes, H. S. R., Boyce, A. J., McDonald, I., Davidheiser-Kroll, B., Holwell, D. A., McDonald, A. & Oldroyd, A. (2015). Contrasting mechanisms for crustal sulphur contamination in the British Palaeogene Igneous Province: evidence from dyke and sill complexes on Skye. Journal of the Geological Society, London. 172, 443–458.

46. Hulbert, L. J., Duke, J. M., Ekstrand, O. R., Scoates, R. F. J., Theriault, R. J., LeCheminant, M. J., Gunn, A. G. & Grinenko, L. N. (1992). Metallogenic and geochemical evolution of cyclic Unit 1, lower Eastern Layered Series, Rhum. In: Foster, R. P. (ed.) Mineral Deposit Modelling in Relation to Crustal Reservoirs of the Ore-forming Elements. Institution of Mining and Metallurgy.

47. Huminicki, M. A. E., Sylvester, P. J., Cabri, L. J., Lesher, C. M. & Tubrett, M. (2005). Quantitative mass balance of platinum group elements in the Kelly Lake Ni–Cu–PGE deposit, Copper Cliff offset, Sudbury. Economic Geology 100, 1631–1646.

48. Huminicki, M. A. E., Sylvester, P. J., Lastra, R., Cabri, L. J., Evans-Lamswood, D. & Wilton, D. H. C. (2008). First report of platinum-group minerals from a hornblende gabbro dyke in the vicinity of the Southeast Extension Zone of the Voisey's Bay Ni–Cu–Co deposit, Labrador. Mineralogy and Petrology 92, 129–164.

49. Hutchinson, D. & McDonald, I. (2008). Laser ablation ICP-MS study of platinum-group elements in sulphides from the Platreef at Turfspruit, Northern Limb of the Bushveld Complex, South Africa. Mineralium Deposita 43, 695–711.

50. Ihlenfeld, C. & Keays, R. R. (2011). Crustal contamination and PGE mineralization in the Platreef, Bushveld Complex, South Africa: evidence for multiple contamination events and transport of magmatic sulphides. Mineralium Deposita 46, 813–832.

51. Keays, R. R. & Lightfoot, P. C. (2010). Crustal sulfur is required to form magmatic Ni–Cu sulfide deposits: evidence from chalcophile element signatures of Siberian and Deccan Trap basalts. Mineralium Deposita 45, 241–257.

52. Kerr, A. & Leitch, A. M. (2005). Self-destructive sulfide segregation systems and the formation of high-grade magmatic ore deposits. Economic Geology 100, 311–332.

53. Kinnaird, T. C., Prave, A. R., Kirkland, C. L., Horstwood, M., Parrish, R. & Batchelor, R. A. (2007). The late Mesoproterozoic–early Neoproterozoic tectonostratigraphic evolution of NW Scotland: the Torridonian revisited. Journal of the Geological Society, London 164, 541–551.

54. Kullerud, G. & Yoder, H. S. (1959). Pyrite stability relations in the Fe–S system. Economic Geology 54, 533–572.

55. Labidi, J., Cartigny, P., Birck, J. L., Assayag, N. & Bourrand, J. J. (2012). Determination of multiple sulphur isotopes in glasses: A reappraisal of the MORB d34S. Chemical Geology 334, 189–198.

56. Leuthold, J., Blundy, J. D., Holness, M. B. & Sides, R. (2014). Successive episodes of reactive liquid flow through a layered intrusion (Unit 9, Rum Eastern Layered Intrusion, Scotland). Contributions to Mineralogy and Petrology 168, 1–27.

57. Li, C. Ripley, E. M. 2005. Empirical equations to predict the sulfur content of mafic magmas at sulfide saturation and applications to magmatic sulfide deposits. 40, 218-230. Mineralium Deposita, 40, 218–230.

58. Li, C., Ripley, E. M., Maier, W. D. & Gomwe, T. E. S. (2002). Olivine and sulfur isotopic compositions of the Uitkomst Ni– Cu sulfide ore-bearing complex, South Africa: evidence for sulfur contamination and multiple magma emplacements. Chemical Geology 188, 149–159.

59. Lightfoot, P. & Keays, R. R. (2005). Siderophile and chalcophile metal variations in flood basalts from the Siberian Trap, Noril'sk Region: Implications for the origin of the Ni–Cu– PGE sulphide ores. Economic Geology 100, 439–462.

60. Lightfoot, P. C., Naldrett, A. J. & Hawkesworth, C. J. (1984). The geology and geochemistry of the Waterfall Gorge Section of the Insizwa Complex with particular reference to the origin of the nickel sulphide deposits. Economic Geology 79, 1857–1879.

61. Lorand, J. –P., Alard, O., Luguet, A. & Keays, R. R. (2003). Sulfur and selenium systematics of the subcontinental lithospheric mantle: Inferences from the Massif Central xenolith suite (France). Geochimica et Cosmochimica Acta, 67, 4137–4151.

62. Lowry, D. (1991). The genesis of Late Caledonian granitoidrelated mineralization in Northern Britain. PhD thesis, University of St Andrews.

63. Lowry, D., Boyce, A. J., Fallick, A. E., Stephens, W. E. & Grassineau, N. V. (2005). Terrane and basement discrimination in northern Britain using sulphur isotopes and mineralogy of ore deposits. In: McDonald, I., Boyce, A. J., Butler, I. B., Herrington, R. J. & Polya, D. A. (eds) Mineral Deposits and Earth Evolution. Geological Society, London, Special Publications 248, 133–151.

64. McClurg, J. (1982). Geology and structure of the northern part of the Rhum ultrabasic complex. PhD thesis, University of Edinburgh.

65. McDonald, I. (2008). Platinum-group element and sulphide mineralogy in ultramafic complexes at western Andriamena, Madagascar. Applied Earth Science (Transactions Institute of Mining and Metallurgy B) 117(1), 1–10.

66. McDonald, I. & Holwell, D. A. (2007). Distribution of platinumgroup elements in the Platreef at Overysel, northern Bushveld Complex: a combined PGM and LA-ICP-MS study. Contributions to Mineralogy and Petrology 154, 171–190.

67. McDonald, I., Holwell, D. A. & Wesley, B. (2009). Assessing the potential involvement of an early magma staging chamber in the generation of the Platreef Ni–Cu–PGE deposit in the northern limb of the Bushveld Complex: a pilot study of the Lower Zone Complex at Zwartfontein. Transactions of the Institutions of Mining and Metallurgy, Section B 118, 5–20.

68. McDonough, W. F. & Sun, S. s. (1995). The composition of the Earth. Chemical Geology 120, 223–253.

69. Morton, N. & Hudson, J. D. (1995). Field guide to the Jurassic of the Isles of Raasay and Skye, Inner Hebrides, NW Scotland.In: Taylor, P. D. (ed.) Field Geology of the British Jurassic. Geological Society, 209–280.

70. Mungall, J. E. & Brenan, J. M. (2014). Partitioning of platinumgroup elements and Au between sulfide liquid and basalt and the origins of mantle–crust fractionation of the chalcophile elements. Geochimica et Cosmochimica Acta 125, 265–289.

71. Mungall, J. E. & Su, S. (2005). Interfacial tension between magmatic sulfide and silicate liquids: Constraints on kinetics of sulfide liquation and sulfide migration through silicate rocks. Earth and Planetary Science Letters 234, 135–149.

72. Naldrett, A. J. & Fedorenko, V. (1995). Ni–Cu–PGE deposits of the Noril'sk region, Siberia: Their formation in conduits for flood basalt volcanism. Transactions of the Institution of Mining and Metallurgy 104, B18–B36.

73. Naldrett, A. J., Fedorenko, V. A., Asif, M., Lin, S., Kunilov, V. E., Stekhin, A. I., Lightfoot, P. C. & Gorbachev, N. S. (1996). Controls on the composition of Ni–Cu sulphide deposits as illustrated by those at Noril'sk, Siberia. Economic Geology 91, 751–773.

74. Naldrett, A. J., Wilson, A., Kinnaird, J. & Chunnett, G. (2009). PGE tenor and metal ratios within and below the Merensky Reef, Bushveld Complex: Implications for its genesis. Journal of Petrology 50, 625–659.

75. Naldrett, A. J., Kinnaird, J., Wilson, A., Yudovskaya, M. & Chunnett, G. (2011). Genesis of the PGE-enriched Merensky Reef and chromitite seams of the Bushveld Complex. Reviews in Economic Geology 17, 235–296.

76. Newton, R. J., Bottrell, S. H., Dean, S. P., Hatfield, D. & Raiswell, R. (1995). An evaluation of the use of the chromous chloride reduction method for isotopic analyses of pyrite in rocks and sediment. Chemical Geology, 125, 317–320.

77. Nielsen, J. K. & Hanken, N. -M. (2002). Description of the chromium reduction method for extraction of pyrite sulphur. University of Tromso.

78. O'Driscoll, B., Hargraves, R. B., Emeleus, C. H., Troll, V. R., Donaldson, C. H. & Reavy, R. J. (2007). Magmatic lineations inferred from anisotropy of magnetic susceptibility fabrics in Units 8, 9, and 10 of the Rum Eastern Layered Series, NW Scotland. Lithos 98, 27–44.

79. O'Driscoll, B., Day, J. M. D., Daly, J. S., Walker, R. J. & McDonough, W. F. (2009). Rhenium–osmium isotopes and platinum-group elements in the Rum Layered Suite, Scotland: Implications for Cr–spinel seam formation and the composition of the Iceland mantle anomaly. Earth and Planetary Science Letters 286, 41–51.

80. O'Driscoll, B., Butcher, A. R. & Latypov, R. (2014). New insights into precious metal enrichment on the Isle of Rum, Scotland. Geology Today 30, 134–141.

81. Ohmoto, H. & Rye, R. O. (1979). Isotopes of sulphur and carbon. In: Barnes, H. L. (ed.) Geochemistry of Hydrothermal Deposits. John Wiley, pp. 509–567.

82. Parnell, J., Boyce, A. J., Mark, D., Bowden, S. & Spinks, S. (2010). Early oxygenation of the terrestrial environment during the Mesoproterozoic. Nature 468, 290–293.

83. Parnell, J., Hole, M., Boyce, A. J., Spinks, S. & Bowden, S. (2012). Heavy metal, sex and granites: Crustal differentiation and bioavailability in the mid-Proterozoic. Geology 40, 751–754.

84. Paterson, S. R., Fowler, T. K., Schmidt, K. L., Yoshinobu, A. S., Yuan, E. S. & Miller, R. B. (1998). Interpreting magmatic fabric patterns in plutons. Lithos 44, 53–82.

85. Pirrie, D., Power, M. R., Andersen, J. C. Ø. & Butcher, A. R. (2000). Platinum-group mineralization in the Tertiary Igneous Province: new data from Mull and Skye, Scottish Inner Hebrides, UK. Geological Magazine 137, 651–658.

86. Power, M., Pirrie, D., Andersen, J. C. Ø. & Butcher, A. R. (2000). Stratigraphical distribution of platinum-group minerals in the Eastern Layered Series, Rum, Scotland. Mineralium Deposita 35, 762–775.

87. Power, M. R., Pirrie, D. & Andersen, J. C. Ø. (2003). Diversity of platinum-group element mineralization styles in the North Atlantic Igneous Province: new evidence from Rum, UK. Geological Magazine 140, 499–512.

88. Prichard, H. M., Hutchinson, D. & Fisher, P. C. (2004). Petrology and crystallization history of multiphase sulfide droplets in a mafic dike from Uruguay: Implications for the origin of Cu– Ni–PGE sulfide deposits. Economic Geology 99, 365–376.

89. Prichard, H. M., Knight, R. D., Fisher, P. C., McDonald, I., Zhou, M.-F. & Wang, C. Y. (2013). Distribution of platinum-group elements in magmatic and altered ores in the Jinchuan intrusion, China: an example of selenium remobilization by postmagmatic fluids. Mineralium Deposita 48, 767–786.

90. Prout, S. J., Andersen, J. C. Ø., Barnes, S.-J. & Power, M. R. (2002). Platinum-group element mineralisation within two mafic/ultramafic intrusions of the British Palaeogene Igneous Province. In: Boudreau, A. E. (ed.) Proceedings of the 9th International Platinum Symposium, 2002, Billings, Montana, USA, pp. 383–386.

91. Raiswell, R., Bottrell, S. H., Al–Biatty, H. J. & Tan, M. M. (1993). The influence of bottom water oxygenation and reactive iron content on sulphur incorporation into bitumens from Jurassic marine shales. American Journal of Science 293, 569–596.

92. Ripley, E. M., Lightfoot, P. C. & Elswick, E. R. (2003). Sulfur isotopic studies of continental flood basalts in the Noril'sk region: implications for the association between lavas and ore-bearing intrusions. Geochimica et Cosmochimica Acta 67, 2805–2817.

93. Robertson, J., Ripley, E. M., Barnes, S. J. & Li, C. (2015). Sulfur liberation from country rocks and incorporation in mafic magmas. Economic Geology 110, 1111–1123.

94. Robertson, J. C., Barnes, S. J. & Metcalfe, G. (2014). Chaotic entrainment can drive sulfide remobilization at low magma flow rates. In: 12th International Platinum Symposium. Russian Academy of Sciences, pp. 47–48.

95. Rose, L. A. & Brenan, J. M. (2001). Wetting properties of Fe–Ni– Co–Cu–O–S melts against olivine: Implications for sulfide melt mobility. Economic Geology 96, 145–157.

96. Saumur, B. M. & Cruden, A. R. (2016). On the emplacement of the Voisey's Bay intrusion (Labrador, Canada). Geological Society of America Bulletin 128, 147–168.

97. Saumur, B. M., Cruden, A. R., Evans-Lamswood, D. M. & Lightfoot, P. C. (2015). Wall-rock structural controls on the genesis of the Voisey's Bay intrusion and its Ni–Cu–Co magmatic sulfide mineralization (Labrador, Canada). Economic Geology 110, 691–711.

98. Saunders, A. D., Fitton, J. G., Kerr, A. C., Norry, M. J. & Kent, R. W. (1997). The North Atlantic Igneous Province. In: Mahoney, J. J. & Coffin, M. F. (eds) Large Igneous Provinces: Continental, Oceanic, and Planetary Flood Volcanism. American Geophysical Union, Geophysical Monograph 100, 45–93.

99. Sharman, E. R., Penniston-Dorland, S. C., Kinnaird, J. A., Nex, P. A. M., Brown, M. & Wing, B. A. (2013). Primary origin of marginal Ni–Cu–(PGE) mineralization in layered intrusions: D33S evidence from the Platreef, Bushveld, South Africa. Economic Geology 108, 365–377.

100. Smith, J. W., Holwell, D. A. & McDonald, I. (2014). Precious and base metal geochemistry and mineralogy of the Grasvally Norite–Pyroxenite–Anothosite (GNPA) member, northern Bushveld Complex, South Africa: implications for a multistage emplacement. Mineralium Deposita 49, 667–692.

101. Smith, J. W., Holwell, D. A., McDonald, I. & Boyce, A. J. (2016). The application of S isotopes and S/Se ratio in determining ore-forming processes of magmatic Ni–Cu– PGE sulfide deposits: a cautionary case study from the northern Bushveld Complex. Ore Geology Reviews 73, 148–174.

102. Steel, R. J., Nicholson, R. & Kalander, L. (1975). Triassic sedimentation and palaeogeography in Central Skye. Scottish Journal of Geology 11, 1–13.

103. Stewart, A. D. (2002). The Later Proterozoic Torridonian Rocks of Scotland: their Sedimentology, Geochemistry and Origin. Geological Society, London, Memoirs 24.

104. Torssander, P. (1989). Sulphur isotope ratios of Icelandic rocks. Contributions to Mineralogy and Petrology 102, 18–23.

105. Troll, V. R., Nicoll, G. R., Donaldson, C. H. & Emeleus, H. C. (2008). Dating the onset of volcanism at the Rum Igneous Centre, NW Scotland. Journal of the Geological Society, London 165, 651–659.

106. Tuttle, M. L., Goldhaber, M. B. & Williamson, D. L. (1986). An analytical scheme for determining forms of sulphur in oil shales and associated rocks. Talanta 33, 953–961.

107. Upton, B. G. J. (2004). Volcanoes and the Making of Scotland. Dunedin Academic Press.

108. Upton, B. G. J., Skovgaard, A. C., McClurg, J., Kirstein, L., Cheadles, M., Emeleus, C. H., Wadsworth, W. J. & Fallick, A. E. (2002). Picritic magmas and the Rum ultramafic complex, Scotland. Geological Magazine 139, 437–452.

109. Volker, J. A. & Upton, B. G. J. (1990). The structure and petrogenesis of the Trallval and Ruinsival areas of Rhum ultrabasic complex. Transactions of the Royal Society of Edinburgh: Earth Sciences 81, 69–88.

110. Wadsworth, W. J. (1994). The peridotite plugs of northern Rum. Scottish Journal of Geology 30, 167–174.

111. Wager, L. R., Brown, G. M. & Wadsworth, W. J. (1960). Types of igneous cumulates. Journal of Petrology 1, 73–85.

112. Wagner, T., Boyce, A. & Fallick, A. E. (2002). Laser combustion analysis of d34S of sulfosalt minerals: determination of the fractionation systematics and some crystal-chemical considerations. Geochimica et Cosmochimica Acta 66, 2855–2863.

113. Yallup, C., Edwards, M. & Turchyn, A. V. (2013). Sulphur degassing due to contact metamorphism during flood basalt eruptions. Geochimica et Cosmochimica Acta 120, 263–279.

114. Zhabina, N. & Volkov, I. (1978). A method of determination of various sulphur compounds in sea sediments and rocks. 3rd International Symposium on Environmental Biogeochemistry. Ann Arbor Science Publications, 735–746

Chapter 2

Studying Petrology and Determining Tectonic Setting of Bidkhan Area Southeast of Iran

Ahmad Adib[1], Alireza Ashofteh[2]

[1]Mining Engineering Department, Islamic Azad University South Tehran Branch, Tehran, Iran

[2]Young Researchers and Elite Club, Science and Research Branch, Islamic Azad University, Tehran, Iran

ABSTRACT

Bidkhan region has an area of approximately 104 km^2. This region includes all the heights and villages that are known by the name of Bidkhan. Chemical nomenclature of igneous rocks is performed with the help of main oxide plots. Based on the analysis results of the samples with minimum alteration, position of these samples was obtained on special diagrams. The achieved data should be compatible with field evidences at a regional scale. There are some limits to the interpretation of the harvested samples on chemical nomenclature diagrams for igneous rocks that should be taken into account. Lithology of Bidkhan region was performed using the main oxides that were obtained from XRF method. 30 sample rocks, which were very fresh and without alteration, were used for drawing diagrams and determining the position of these regional rocks on the chemical nomenclature diagrams for igneous rocks. The results for the analysis of the main oxides of these 30 samples are also presented.

KEYWORDS

Bidkhan, Tectonic, Petrology, Rocks

1. INTRODUCTION

Bidkhan region is located in central Iran structural zone as a part of the Urumieh-Dokhtar magmatic arc which is evolved during the Arabia-Eurasia convergence. In this region, quaternary deformations were consistent with recent faults activities [1] - [5] . From a lithological perspective, Bidkhan region is divided into three distinct and separable parts. The exterior part, which is formed by volcaniclastic, volcanogenic, and sedimentary deposits and interior part, which is divided into two parts of central and the surrounding ring. These two internal parts are formed by volcanic and subvolcanic outcrops. They have an intermediate to acidic chemical composition. Moving from margins to the center, alteration intensity varies from low (propylitic) to severely altered (phyllic, alunite, advanced argillic and potassic) [6] . Altered rocks such as propylitics, phyllics, albites, sericites, argillics, alunites, kaolins, carbonates, silicas, advanced argillics, potassics and malachites exist in Bidkhan region and especially in the central part of this area. Alteration is a characteristic of this region. It is necessary to note that alteration is observed in this region from propylitic zone to phyllic, alunite and potassic. Several conditions and circumstances are necessary for an event to become the main phenomenon in a region. Alteration of the interior outcrops has developed geochemical as the main phenomenon. Without a doubt, this geochemical phenomenon has made Bidkhan region as one of the biggest severely altered integrated sets of Iran [7] .

2. MATERIALS AND METHODS

The position of Bidkhan rocks was studied on various diagrams Tas (Cox 1979) diagram: In which, SiO_2 is on the horizontal axis and $Na_2O + K_2O$ are on the vertical axis. 30 samples were analyzed for this diagram (Figure 1). It can be observed that the rocks of this region are generally intermediate to acidic and vary from sub-alka- line to alkaline, which is a general trend. Based on the diagram, these rocks are in three ranges of trachyandesite, andesite and dacite, which is in accordance with microscopic and field observations [8] .

Examining Tas (Le Bas 1986) diagram confirmed the results of the above diagram. Based on this diagram, these rocks are generally intermediate to acidic, sub-alkaline, and are placed within three ranges of trachyandesite, andesite and dacite (Figure 2) [9] .

Tas (Middle Most 1994) diagram is newer than the previous ones. 30 analyzed samples are presented on the diagram to validate the previous diagrams. It is also observed in this diagram that these rocks were from three types of trachyandesite, andesite and dacite (Figure 3). It seems that magmatic origin of these rocks was andesitic and other trachyandesitic and dacitic rocks were created by other processes such as differentiation and digestion of walls from andesitic magma [10] .

3. DISCUSSION

The most important tectonic structures that exist are main, reverse, and relatively long quadruplet faults that have a northern-southern trend and movement. Fracture and minor faults of the strike slip with high density exist within the interstitial space of these main faults. In fact, Bidkhan valley is one of these main faults that are connected to the southern heights from its south side. This is the main fault of Bidkhan that have converted into the main drainage of Bidkhan catchment area over time. Width of this drainage has become extended because of mechanical weathering. Other main faults in Bidkhan area are also placed and observed in a direction similar to this valley fault. Total fracture networks created by main and minor faults allows for smooth movement of magmatic fluids. There is a severely altered integrated set of rocks in Bidkhan. Despite its youth, lithology and the events causing this set are remarkable. Important, various, and different events that have geochemical-geo- logical aspects occur with high density in this area. All kinds of alterations along with sulfide stock works, leachings, and supergenes are among these events. It is also noteworthy to mention the phenomena of hot springs that make travertine and are located in the vicinity of this area (outside of Bidkhan region). They are a part of these evidence collections [11] .

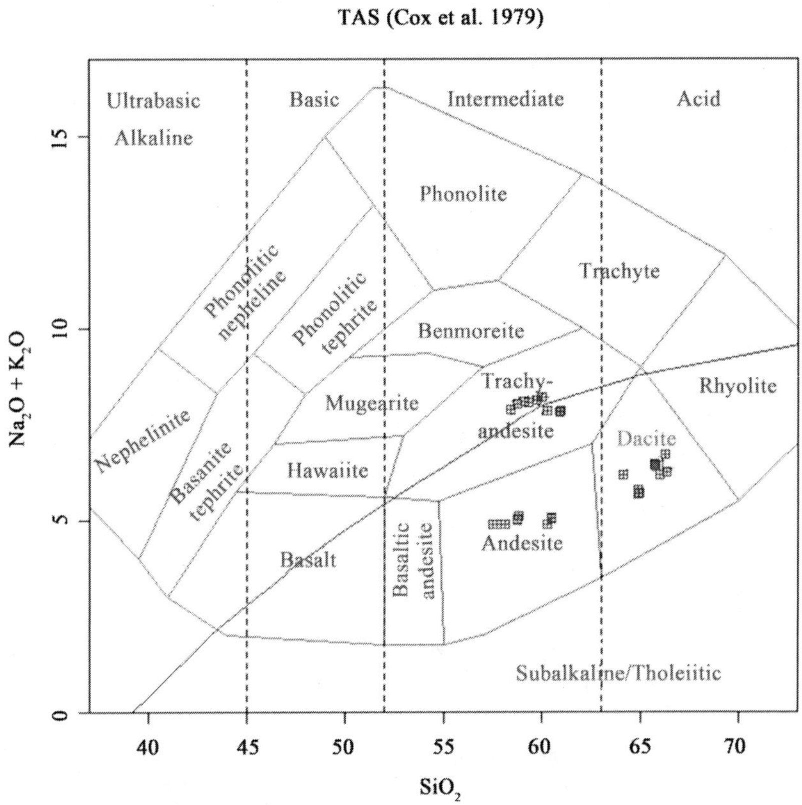

Figure 1. The position of Bidkhan rocks on Tas (Cox 1979) diagram.

4. GEOCHEMISTRY AND THE POSITION OF THESE REGIONAL ROCKS BASED ON THEIR CHEMISTRY

R_1, R_2 Plot (De La Roche 1980) diagram is adjusted according to the chemistry of these regional rocks (Figure 4). 30 analyzed samples, $R_1 = 4Si - 11(Na + K) - 2(Fe + Ti)$ in the x-axes and $R_2 = 6Ca + 2Mg + Al$ in y-axes were plotted on this diagram. These rocks were placed within three ranges of andesite, latite, dacite, and trachydacite with little differences, which is close to the observations for sections petrology.

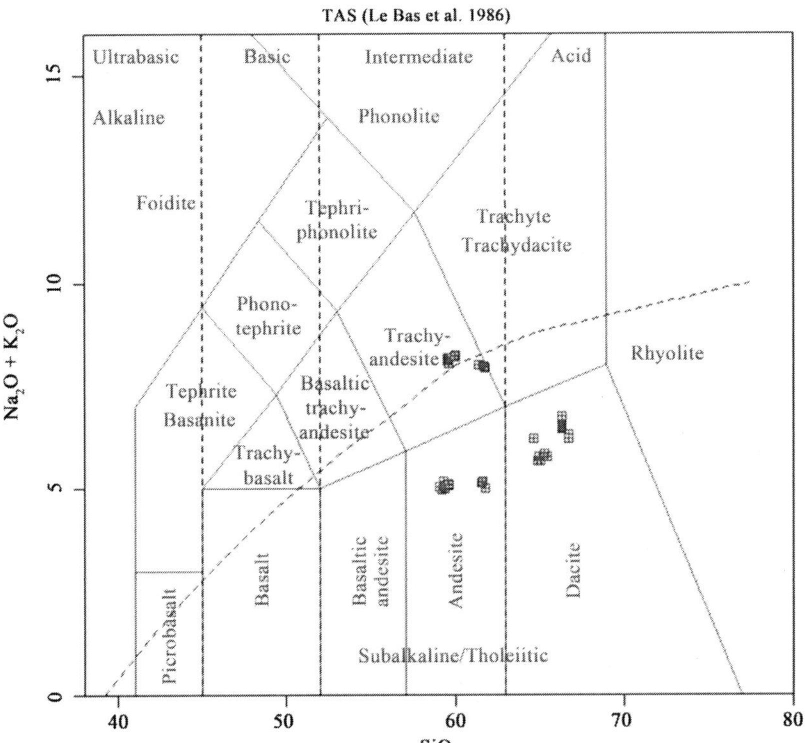

Figure 2. The position of Bidkhan rocks on Tas (Le Bas 1986) diagram.

On Harker diagram, main oxides were plotted against SiO_2. Valuable information can be gained from the changing trend of these oxides compared to SiO_2 increase. This diagram is presented in (Figure 5) and the following results were obtained:

1) Na_2O with a negative slope, decreases with the increase of SiO_2 and has a sharp (steep) slope. It was determined that generally in these three types of rocks, with the increase of silica, sodium is decreased. But more specifically, with the increase of silica, sodium is also increased with a gentle slope in each type.

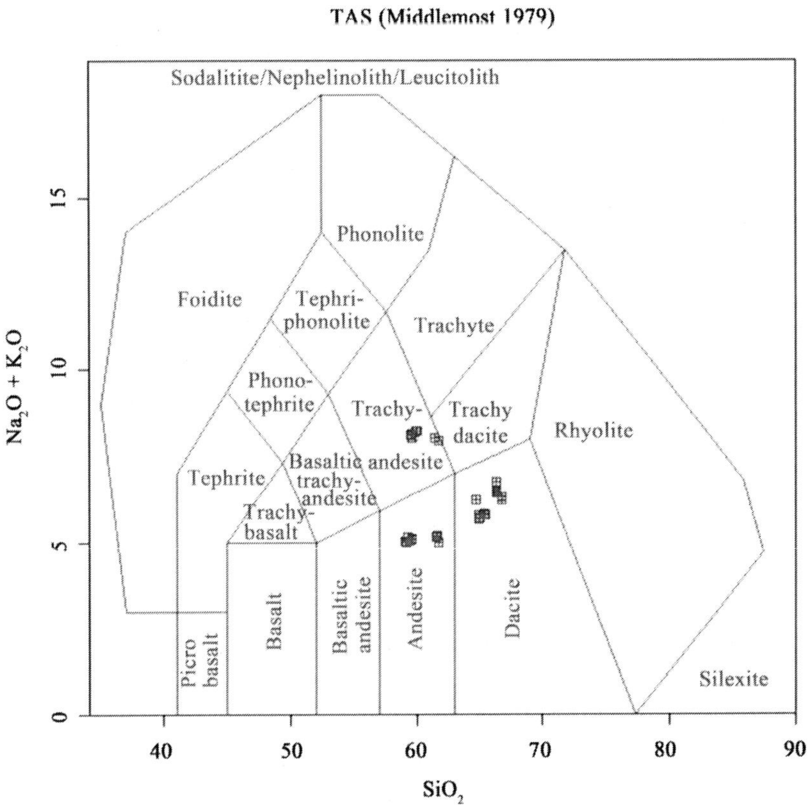

Figure 3. The position of Bidkhan rocks on Tas (Middle Most 1994) diagram.

2) K_2O can also be observed in three rock types of this diagram. This diagram demonstrates that andesitic rocks of this region do not show a significant change in the amount of potassium against the increase of SiO_2. But other rocks of this region show a decrease in potassium with a gentle slope from the increase of silica.

3) CaO decreases in relation to the increase of silica with a steep slope.

4) MgO shows the same decrease slope of calcium in respect to the increase of silica.

5) Al_2O_3 shows an overall decrease, which means aluminium decreases with the increase of silica. With a more careful examination, it is evident that

increase exists in all the separate types of rocks for this region. Andesitic rocks demonstrate positive gentle slope and dacite rocks demonstrate a positive steep slope.

6) TiO_2 decreases with the increase of silica.

7) FeO or iron oxide also decreases with the increase of silica.

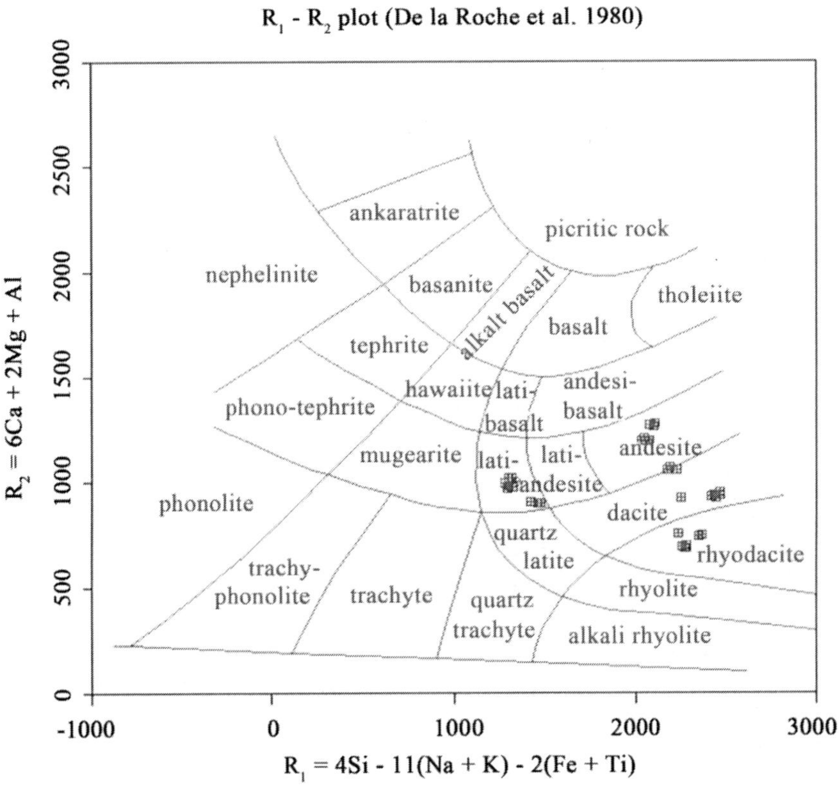

Figure 4. The position of Bidkhan rocks on R1, R2 Plot (De La Roche 1980) diagram.

Therefore, it was concluded that even though the changing trends for main oxides of igneous rocks are somewhat linear and it actually shows magmatic differentiation in these rocks, factors such as alteration and magmatic contamination affected this trend (Table 1).

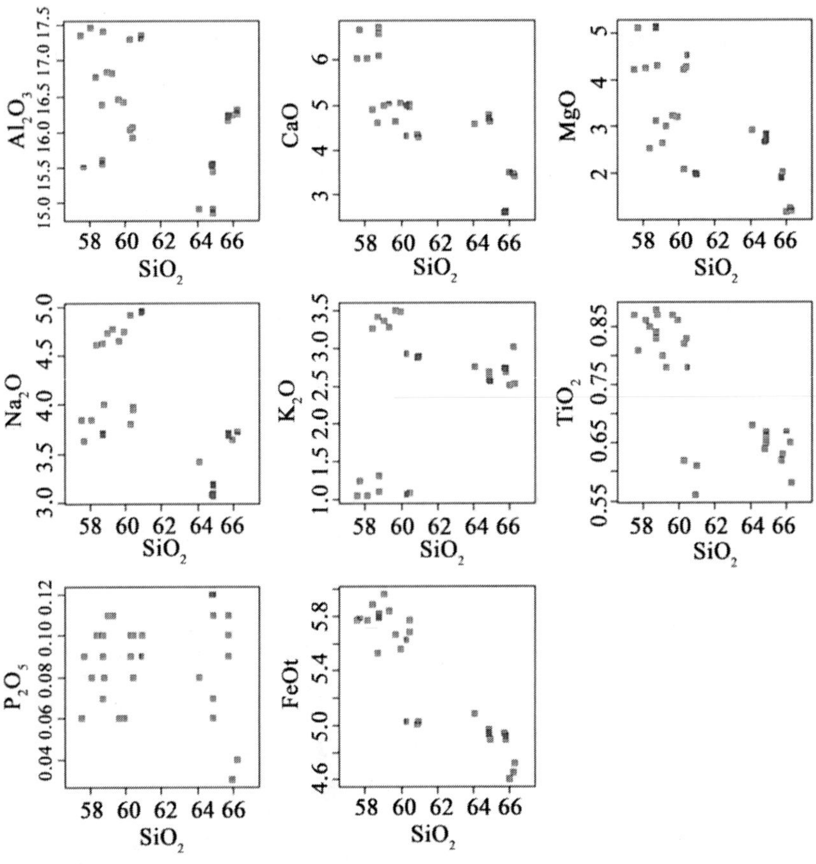

Figure 5. Harker diagram for Bidkhan rocks.

5. GEOCHEMISTRY AND THE POSITION OF THESE REGIONAL ROCKS BASED ON TYPE OF MAGMA

As it was observed in the diagrams of previous section, rocks of this region are in the range of sub-alkaline, which also tend to alkaline. After plotting the 30 samples on Miashiro 1974 diagram for SiO_2 on x and y axes, it was observed that almost all the samples are in the range of calc-alkaline series (Figure 6).

Studying AFM diagram, which is a triangular diagram, showed that Bidkhan rocks are among calc-alkaline series (Figure 7). On this diagram, A = (K_2O + Na_2O), F = FeO_t and M = MgO.

SiO_2-K_2O (Peccerillo & Taylor 1976) diagram divides the magmatic series of Bidkhan rocks into two series. Andesitic rocks are calc-alkaline and the other rocks of this region are in the range of the above potassic calc-alkaline series (Figure 8).

On A/CNK – A/NK Plot (Shand 1943) diagram, x-axes is A/NK = $AL_2O_3/(Na_2O + K_2O)$ and y-axes is A/CNK = $AL_2O_3/(CaO + Na_2O+K_2O)$. An examination of this diagram reveals that majority of these rocks are in the range of metaluminous rocks and some of them are in the range of peraluminous rocks (Figure 9). According to microscopic evidences, it can be concluded that this high level of aluminum in the rocks is due to the argillization and increase of aluminum by alteration (Table 1).

Geotectonic diagrams, which are triangular diagrams by (Pearce et al. 1977), are an attempt to study the type of magmatism in relation to tectonic processes. According to this diagram, magma source for the rocks of this region was caused by orogenry. Based on the existing information and this diagram, this magma was caused by the subduction of the oceanic crust under the continental crust (Figure 10) and (Table 1).

6. MALACHITIZATION

Even in small quantities, the existence of azurite and malachite in an exploration area has significant exploratory value. The importance of malachite is not only due to being a remarkable mineral in exploration and foundation activities, but also mostly because of the importance of a process known as malachitization. Impacts of malachitization can be observed in many areas of Bidkhan. The analogous phenomena in the malachite outcrops of these regions is feldspar crystals that are converted into a set of secondary minerals such as sericite, illite, and other clay minerals and amphibole crystals are completely altered and their framework is filled up by malachite. There is also malachite within frameworks of some feldspar crystals. Secondary minerals of the outcrops from these regions consist of sericite, illite, clay minerals, malachite, apak, oxides-hydroxides of iron, chlorite, epidote and quartz (Figure 11) [12] .

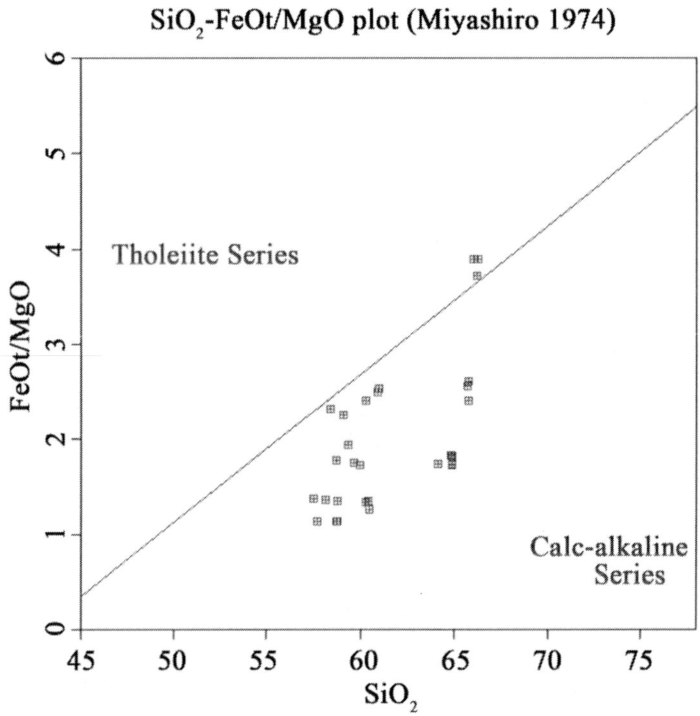

Figure 6. The position of Bidkhan rocks on Miashiro 1974 diagram.

7. GEODES AND POROUS QUARTZES CAUSED BY SUBSIDENCE

The process for the formation of porous quartzes is a kind of alteration which is observed in Bidkhan along with geodes. In this kind of alteration, the empty space between quartz grains is partially filled with all kidns of minerals. This alteration is caused by the reaction between acidic fluids and the surrounding rocks and leads to the loss of the majority of constituents except for quartz and rutile parts. Places around alteration zones are the best location for observing porous quartzes. Many hand samples and also petrographic and mineralographic samples of Bidkhan region demonstrate alteration of quartz (porous). By moving from the region of alteration for porous quartz into the alteration region, many alterations such as quartz-alunite, quartz-kaolinite, and even perhaps pyrophyllite can be reached. Geodes are also valuable for exploratory studies along with the other evidences. Such geodes are also found in Bidkhan's set. Alteration of porous quartz is usually created in the center and core of many ores and Au-Cu mineralization-with high sulphidation. In addition, these evidence can be observed in the upper parts of some porphyry ores and also in epithermal environments. Many traces of subsidence exist in Bidkhan and since faults play the main role in geological fluid dynamics, traces of subsidence are seen in fault zones. Since Bidkhan region has a significant density of main and secondary faults, almost all the central parts of Bidkhan bear traces of subsidence within themselves. These traces have different dimensions, from centimeters up to meters wide. The hand sample presented in the following figure shows pores that are caused by subsidence and are filled with silica, oxide, and iron hydroxides. Sometimes, these pores and cavities are also filled with malachite and azurite crystals (Figure 12).

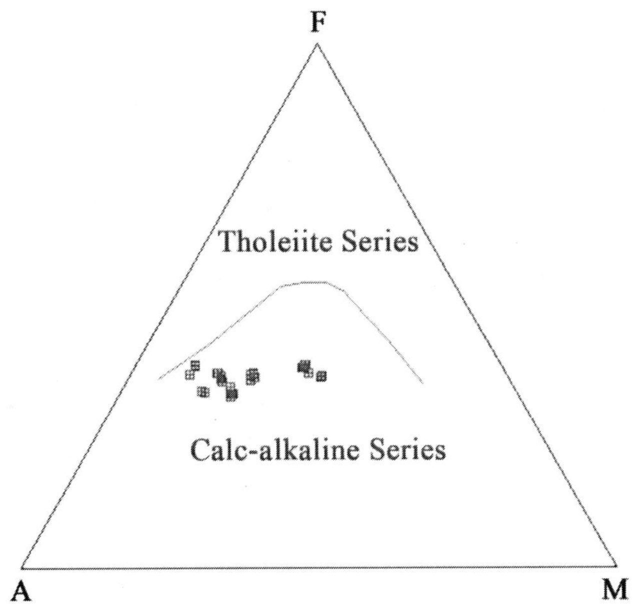

Figure 7. The position of Bidkhan rocks on AFM diagram.

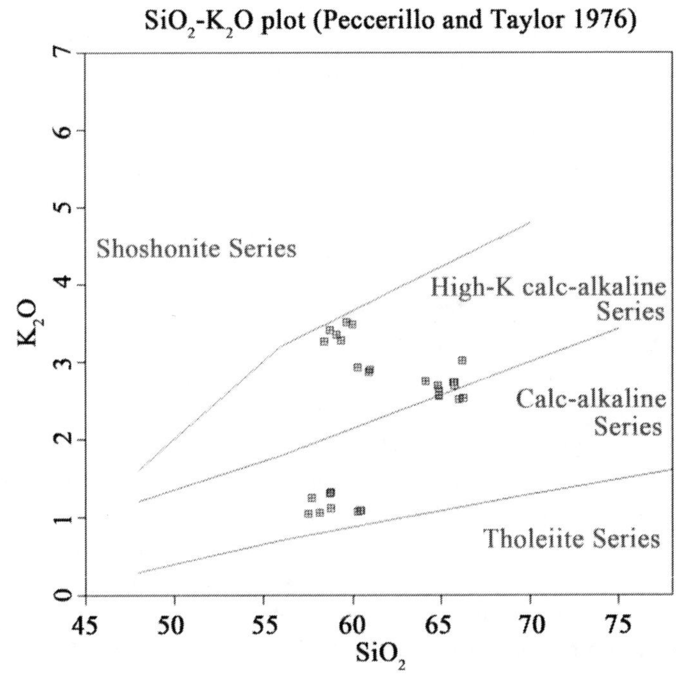

Figure 8. The position of Bidkhan rocks on SiO_2-K_2O (Peccerillo & Taylor 1976) diagram.

8. RESULTS

1) Morphology of Bidkhan resembles a ring shape or a ring structure of huge dimensions with several tens of kilometers. This huge ring in the outside edge is very lofty, which is quickly converted into a foothill towards the central direction with a negative steep slope (Figure 13).

2) Based on the Ore-Mineral findings in Bidkhan region, it can be concluded that "erosion surface" includes the central part and is shaped like a ring around the central and upper part of leaching zone. Erosion surface is for all the surfaces of the central part of Bidkhan. By performing deep excavations, some parts could be reached in which minerals and other related and expected sets exist in the supergene zone (Figure 13).

Table 1. The results of the analysis of main oxides for the retrieved petrological samples.

Major oxide analyse result of fresh or slightly altered rock samples of subvolcanic-volcanic Bidkhan outcrops														
No.	Sample No.	SiO_2	Al_2O_3	Fe_2O_3	CaO	MgO	Na_2O	K_2O	$Na_2O + K_2O$	MnO	TiO_2	P_2O_5	L.O.I	Total
1	Kb.80.ch	65.74	16.23	5.47	2.58	1.89	3.68	2.73	6.41	0.02	0.62	0.11	Nd	99.07
2	Kb.81.ch	64.86	15.56	5.52	4.79	2.71	3.06	2.62	5.68	0.01	0.67	0.12	0.02	99.92
3	Kb.82.ch	64.88	14.92	5.44	4.62	2.83	3.17	2.56	5.73	0.01	0.67	0.06	Nd	99.16
4	Kb.83.ch	60.89	17.31	5.57	4.33	2.01	4.95	2.87	7.82	0.02	0.56	0.09	Nd	98.60
5	Kb.84.ch	57.53	17.36	6.41	6.05	4.21	3.85	1.05	4.90	0.02	0.87	0.06	0.86	97.41
6	Kb.85.ch	58.72	16.39	6.15	4.59	3.12	4.63	3.41	8.04	0.01	0.88	0.07	0.88	97.97
7	Kb.86.ch	60.28	16.04	6.25	4.98	4.21	3.81	1.07	4.88	0.06	0.82	0.09	0.68	97.61
8	Kb.87.ch	65.97	16.24	5.11	3.49	1.18	3.64	2.52	6.16	0.01	0.67	0.03	Nd	98.86
9	Kb.88.ch	58.39	16.77	6.54	4.91	2.54	4.62	3.26	7.88	0.01	0.85	0.10	Nd	97.99
10	Kb.89.ch	57.71	15.51	6.43	6.68	5.09	3.63	1.25	4.88	0.04	0.81	0.09	Nd	97.24
11	Kb.90.ch	65.76	16.17	5.44	2.61	2.03	3.70	2.69	6.39	0.02	0.63	0.10	0.01	99.15
12	Kb.91.ch	64.86	15.45	5.50	4.68	2.74	3.09	2.58	5.67	0.01	0.65	0.11	0.00	99.67
13	Kb.92.ch	64.85	14.86	5.49	4.69	2.85	3.19	2.61	5.80	0.01	0.66	0.07	0.20	99.28
14	Kb.93.ch	60.95	17.35	5.59	4.29	1.99	4.97	2.89	7.86	0.02	0.61	0.10	0.75	98.76
15	Kb.94.ch	58.79	17.41	6.44	6.11	4.28	4.01	1.11	5.12	0.02	0.87	0.08	0.02	99.12
16	Kb.95.ch	59.95	16.42	6.18	5.04	3.21	4.75	3.48	8.23	0.02	0.86	0.06	0.01	99.97
17	Kb.96.ch	60.43	15.93	6.41	5.01	4.27	3.98	1.09	5.07	0.05	0.83	0.1	0.72	98.10
18	Kb.97.ch	66.21	16.31	5.17	3.47	1.25	3.72	3.01	6.73	0.01	0.65	0.04	0	99.84
19	Kb.98.ch	59.30	16.83	6.49	5.03	3.01	4.78	3.28	8.06	0.01	0.78	0.11	0.02	99.62
20	Kb.99.ch	58.75	15.55	6.45	6.74	5.11	3.69	1.32	5.01	0.04	0.84	0.09	0.53	98.58
21	Kb.100.ch	65.70	16.21	5.49	2.59	1.93	3.71	2.74	6.45	0.02	0.62	0.09	0.04	99.10
22	Kb.101.ch	64.81	15.53	5.48	4.7	2.69	3.08	2.69	5.77	0.01	0.64	0.12	0.01	99.75
23	Kb.102.ch	64.09	14.91	5.65	4.58	2.92	3.41	2.75	6.16	0.02	0.68	0.08	0.02	99.09
24	Kb.103.ch	60.28	17.3	5.59	4.32	2.09	4.93	2.93	7.86	0.02	0.62	0.1	0.78	98.18
25	Kb.104.ch	58.13	17.46	6.41	6.04	4.25	3.84	1.06	4.90	0.06	0.86	0.08	0.5	98.19
26	Kb.105.ch	59.67	16.45	6.3	4.63	3.24	4.65	3.51	8.16	0.02	0.87	0.06	0.87	99.40
27	Kb.106.ch	60.45	16.07	6.32	4.96	4.52	3.95	1.09	5.04	0.05	0.78	0.08	0.64	98.27
28	Kb.107.ch	66.25	16.25	5.24	3.39	1.21	3.73	2.53	6.26	0.02	0.58	0.04	0.06	99.24
29	Kb.108.ch	59.05	16.84	6.63	4.99	2.64	4.73	3.36	8.09	0.01	0.8	0.11	0.01	99.16
30	Kb.109.ch	58.73	15.61	6.47	6.59	5.12	3.71	1.31	5.02	0.04	0.83	0.1	0.39	98.51

3) Folds have an almost northwest-southeast trend. The impacts of these folds can mostly be seen on the parts that are outside the studied region of Bidkhan. Considering that this region is located in a way that its surrounding units are all folded symmetrically, it can be concluded that this symmetry is in accordance with positioning of an intrusive body (rocks) in the central section of Bidkhan. Due to the homogenous tension caused by this symmetrical uplift, folded surrounding units became symmetrical and

they were folded in a dome shaped pattern. This dome shaped pattern folding can be observed across the heights surrounding the central part of Bidkhan (Figure 14).

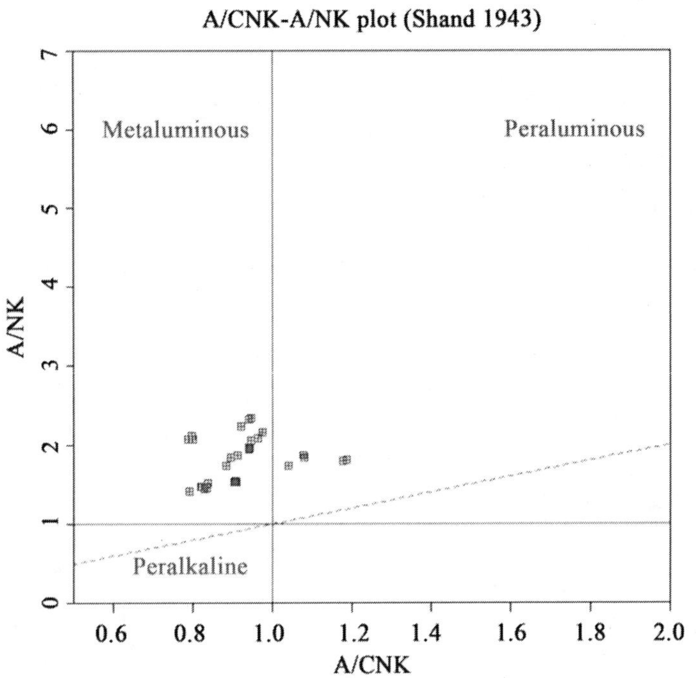

Figure 9. The position of Bidkhan rocks on A/CNK-A/NK Plot (Shand 1943) diagram.

4) Lithology of Bidkhan can be divided into 5 parts: Central part, middle volcaninc ring or belt, svsfacies, "unit C_g" external volcanogenic part, dykes and intrusive rocks and quartz deposits. $C_g$unit forms the heights of Bidkhan region. Heights are practically insurmountable due to their outward steep slope. C_g unit is extended in all directions and also extends to far distances from Bidkhan region. Based on the existing evidence, nomenclature of Bidkhan under the name of "Bidkhan's Caldera" is a controversial subject. Simply existing the morphology similar to a caldera cannot be a good reason for this nomenclature. On the contrary, there are numerous evidences that make the caldera of Bidkhan ambiguous. C_g unit is not a

Lahar but rather a conglomerate and a cutting, although, determination of this very subject also requires research and academic projects to be carried out. In addition to the common mechanism for the creation of caldera, there is a possibility that formation of dome shaped structure for Bidkhan is caused by uplift and positioning of an intrusive body. With this presumption, the igneous outcrops existing in Bidkhan region can be considered the product of the same igneous source activities. According to the age of the available units in Bidkhan's caldera and their method of deformation, it can be concluded that late alpine orogeny phases cause the creation of the current structural condition of Bidkhan region (Figure 14).

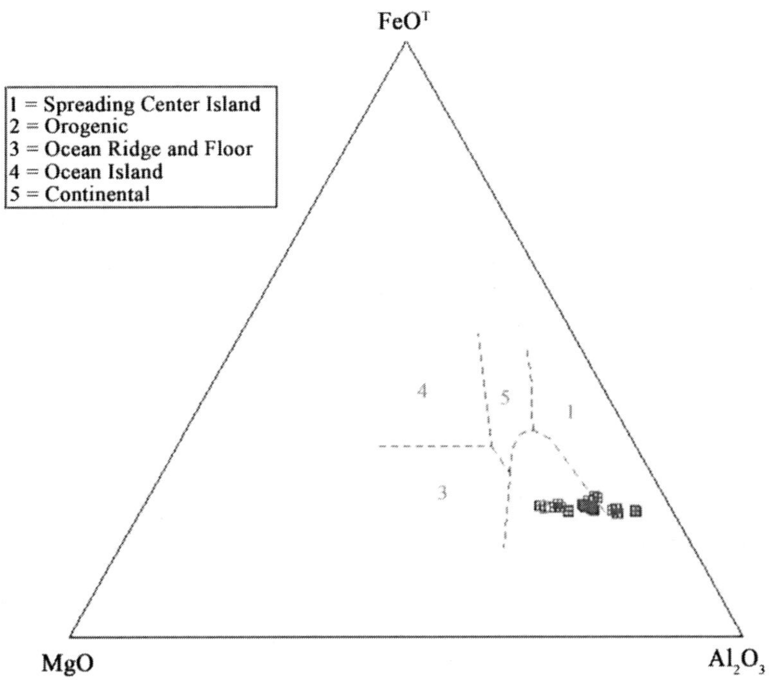

Figure 10. The position of Bidkhan rocks in relation to the source tectonic processes for magmatism.

Figure 11. The process of malahitization in the rocks of Bidkhan region and close view of azurite, malachite, and formation of geode in pores of rocks.

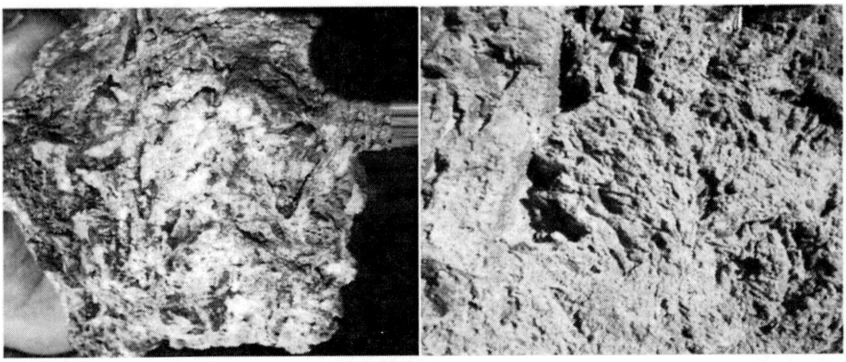

Figure 12. Pores caused by subsidence and filled with silica, oxides and hydroxides of iron and dissolution of all minerals and formation of quartz grains and silicic veinlets.

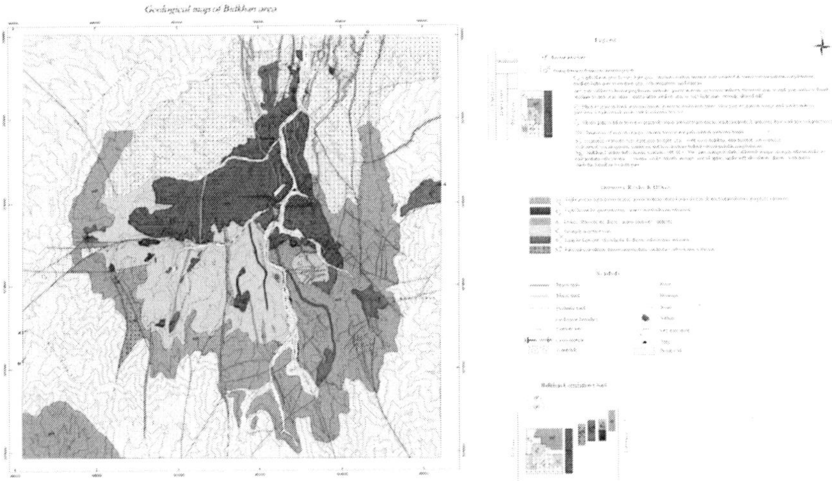

Figure 13. Geological map of bidkhan area (scale 1:25,000).

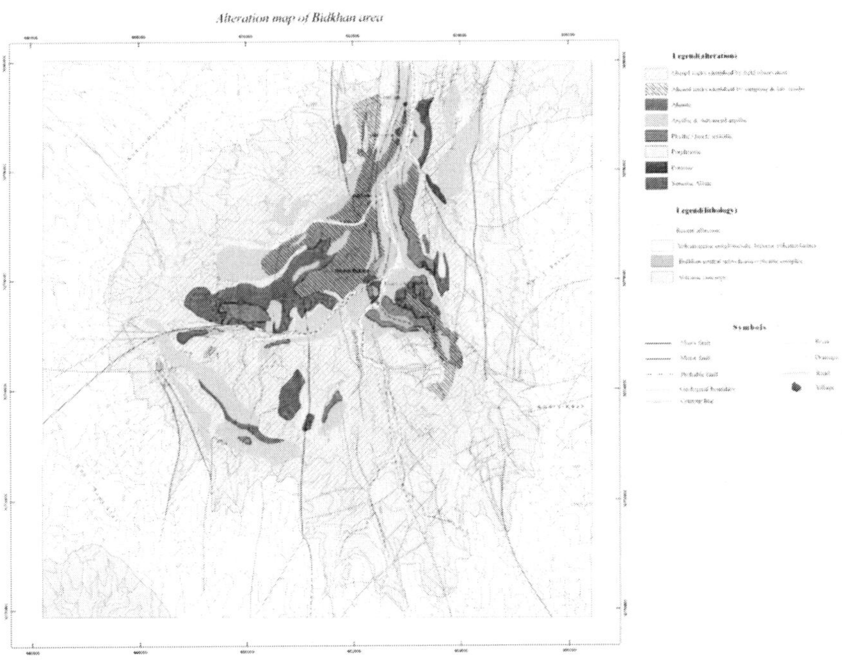

Figure 14. Alteration map of bidkhan area (scale 1:25,000).

ACKNOWLEDGEMENTS

The authors thank Islamic Azad University South Tehran branch for support of this research. This paper presents the research project entitled "Rapture and Land Deformation Statistical Analysis for Earthquake along the Selected Active Fault in Central Iran" is extracted.

REFERENCES

1. Berberian, M. and King, G.C.P. (1981) Towards a Paleogeography and Tectonic evolution of Iran. Canadian Journal of Earth Sciences, 18, 210-265.

2. Meyer, B. and Le Dortz, K. (2007) Strike-Slip Kinematics in Central and Eastern Iran: Estimating Fault Slip-Rates Averaged over the Holocene. Tectonics, 26, TC5009, 20 p.

3. Walker, R.T. (2006) A Remote Sensing Study of Active Folding and Faulting in Southern Kerman Province, S.E. Iran. Journal of Structural Geology, 28, 654-668.

4. Walker, R.T., Gans, P., Allen, M.B., Jackson, J., Khatib, M., Marsh, N. and Zarrinkoub, M. (2009) Late Cenozoic Volcanism and Rates of Active Faulting in Eastern Iran. Geophysical Journal International, 177, 783-805.

5. Walker, R. and Jackson, J. (2004) Active Tectonics and Late Cenozoic Strain Distribution in Central and Eastern Iran. Tectonics, 23, TC5010.

6. Ashofteh, A. and Adib, A. (2014) Primitive Studies for Copper Deposit in Bidkhan Copper Mine with Regards to Wide Survey of Altered Areas and Mineral Indexes. Journal of Middle East Applied Science and Technology, 4, 86-90.

7. Feizi, F. and Ashofteh, A. (2014) Area Exploration Takhte Soleiman Based on Geophysical and Satellite Imagery. Journal of Middle East Applied Science and Technology, 3, 414-417.

8. Keshavarzi, R., Esmaili, D., Rezaei Kahkhaei, M., Asghar Mokhtari, M. and Jabari, R. (2014) Petrology, Geochemistry and Tectonomagmatic

Setting of Neshveh Intrusion (NW Saveh). Open Journal of Geology, 4, 177-189.

9. VaeziNejad, S., Tofigh, M. and Marandi, S. (2011) Zonation and Prediction of Land Subsidence (Case Study-Kerman, Iran). International Journal of Geosciences, 2, 102-110

10. Jafari, H. and Yazdi, A. (2014) Radioactive Anomalies in 1:50,000 Dehbakri Sheet, South of Kerman Province, Iran. Open Journal of Geology, 4, 399-405.

11. SabokKhiz, H., MemarKocheBagh, A. and Hekmatian, M. (2015) Area Exploration Based on Airborne Radiometric Geophysical Data by Surfer Software at Rizab (the South East of Ardakan). Journal of Biodiversity and Environmental Sciences, 6, 761-767.

12. Adib, A., Afzal, P. and Heydarzadeh, K. (2014) Site Effect Classification Based on Microtremor Data Analysis Using Concentration-Area Fractal Model. Nonlinear Processes in Geophysics (NPG), 22, 53-63.

Chapter 3

Petrology, Geochemistry and Tectonomagmatic Setting of Neshveh Intrusion (NW Saveh)

Reza Keshavarzi[1], Dariush Esmaili[1], Mehdi Rezaei Kahkhaei[1], Mir Ali Asghar Mokhtari[2*], Reza Jabari[3]

[1]School of Geology, Faculty of Science, University of Tehran, Tehran, Iran

[2]Geology Group, Faculty of Science, University of Zanjan, Zanjan, Iran

[3]Geology group, Faculty of Science, Islamic Azad University of Science and Research of Tehran, Tehran, Iran

ABSTRACT

Neshveh intrusion is located in the NW of Saveh City and is a part of Orumieh-Dokhtar magmatic arc. This intrusion consists of quartz monzodiorite, granodiorite and granite that have intruded into the Eocene volcano-sedimentary rocks. This intrusion is high-K calk-alkaline and metaluminous and is classified as I-type granitoids. Field investigations along with petrographic and geochemical studies indicate that all phases of Neshveh intrusion are derived from a common magma source as a result of mineral differentiation. Different phases of this intrusion have low Mg#, Ni, Cr, Co and V which are indicative for higher evolution of magma during the magma ascent and before complete crystallization. All phases of Neshveh granitoid are characterized by LREE-rich patterns with high LREE/HREE ratio and negative Eu anomalies. Similarity of the mentioned patterns suggests a comagmatic source for these rocks and demonstrates the role of magmatic differentiation in their evolution. There are negative anomalies in the Nb and Ti along with positive anomalies of Rb, Ba, K and Pb on the spider diagrams. These anomalies are indicative for a subduction setting for magma source of these rocks. Geochemical studies indicate that the Neshveh granitoid is formed in a volcanic arc and active continental margin. In this base, it is assumed that this intrusion is formed as a result of Neo-Tethys oceanic lithosphere subduction beneath the Central Iran zone which is replaced in the Orumieh-Dokhtar magmatic arc.

KEYWORDS

Petrology, Geochemistry, Intrusion, Granitoid, Neshveh, Saveh

1. INTRODUCTION

Neshveh intrusion is located in the north west of Saveh and is a small part of the Orumieh-Dokhtar magmatic arc in the Alpine-Himalayan orogenic belt. Volcanic rocks are so common in the Orumieh-Dokhtar magmatic arc. These rocks have varieties in composition and tectonic setting which are varying from acidic to basic and continental to shallow marine environments. There are some intrusions also within the volcanic rocks. Acidic-intermediate volcanic and intrusive rocks are widespread in contrast to basic rocks that are probably indicative for the effect of continental crust in their formation.

Several studies had been carried out in the Saveh region, such as studies of [1] and [2] which were focused on the petrography and Petrochemistry of igneous rocks in the south west of Saveh, volcanic and plutonic rocks in the south of studied area. According to [3] , volcanic rocks around the Neshveh village consist of three parts: the oldest volcanic rocks include rhyolite, dacite and some andesite; pyroclastic rocks and lava flows along with intercalation of marl, sandstone and limestone with Lutetian age; and finally andesites and latites with porphyritic textures in the upper part with Priabonian age. The Eocene volcanic flows are widespread in the studied area, which are the result of hybrid crustal-mantle source. The hybrid source of these rocks is attributed to the ascending of basic magma with mantlic source and melting of the continental crust.

Neshveh intrusion located between 35°10' to 35°14' latitudes and 50°10' to 50°14' longitude at the 40 km north west of Saveh. This intrusion was intruded into the volcanic-sedimentary rocks of Eocene (Figure 1). This intrusion didn't investigate for detail petrology and geochemistry, yet. Also, no reliable analyses of trace elements have been published from this intrusion. This research is a detailed study on the petrology and geochemistry of major and trace elements and tectonomagmatic setting of the Neshveh intrusion.

2. GEOLOGY

Neshveh region which is located in the north west of Saveh is a small part of the Orumieh-Dokhtar magmatic arc within the Central Iran zone. All rocks crops out in this region have Cenozoic age and older rocks don't exposed. Based on Saveh geological map 1:100,000 [4] , the main rocks in this area consist of sedimentary, volcanic and intrusive rocks have Paleocene-Eocene age and later (Figure 2). These rocks composed of conglomerate equal to Fajan formation, volcanic-sedimentary rocks of Middle-Upper Eocene consist of intermediateacidic tuffs and lavas, agglomerate and intermediate-basic lavas along with the tuffaceous and sandy limestone and finally Upper Eocene ignimbrite, Lower Red Formation (conglomerate, sandstone, siltstone and red-brown marl), Qom Formation (limestone, marly limestone and marl), Upper Red Formation (conglomerate with red sandstone intercalation), Mio-Pliocene marl, conglomerate equal to Hezar Dareh Formation and quaternary alluvial deposits.

Several granitoid intrusions were intruded into the Eocene volcanic-sedimentary rocks. Dating of some of these intrusions by [1] by K-Ar method on the amphibole, biotite, K-feldspar and whole rock indicate 28 - 42 Ma (Table 1). The Silijard granite, Kalkhab tonalite-diorite, Neshveh granitoid (this study) and Niousht micro-granite are the most important intrusions in the northwest of Saveh.

In addition to mentioned intrusives, some sub-volcanic domes also exist in different parts of the northwest Saveh that have acidic-intermediate composition. These domes intruded into the Upper Eocene-Oligocene intrusives. In this base, these sub-volcanic intrusions attributed to the Post Oligocene.

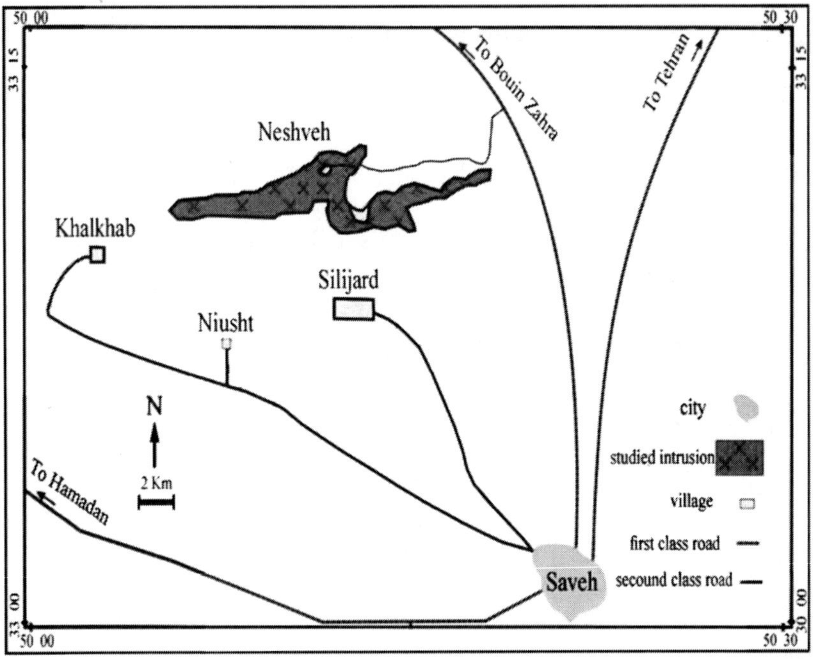

Figure 1. Geographic location of the Neshveh intrusion in the northwest Saveh.

There are some acidic and basic dykes in this area. Basic dykes with andesite-andesitic basalt composition are widespread in comparison to acidic dike with dacite-andesite composition. These dikes cut the Eocene volcanic rocks and Oligocene intrusives. In this base, their intrusions have been attributed to the post Oligocene.

3. RESEARCH METHODS

Our researches in this study consist of two parts: field and laboratory studies. Field studies include identifying the different phases of intrusion, relationship between them and host rock and finally sampling of different phases for laboratory studies. Laboratory investigations include preparing of 90 thin section and petrographic studies, analyzing of 9 samples by ICP-AES and ICP-MS for major, trace and rare earth elements at the ALS

Chemex (Table 2). Finally, these analyses were processed by using Excel, Minpet, GCDKit and Igpet programs.

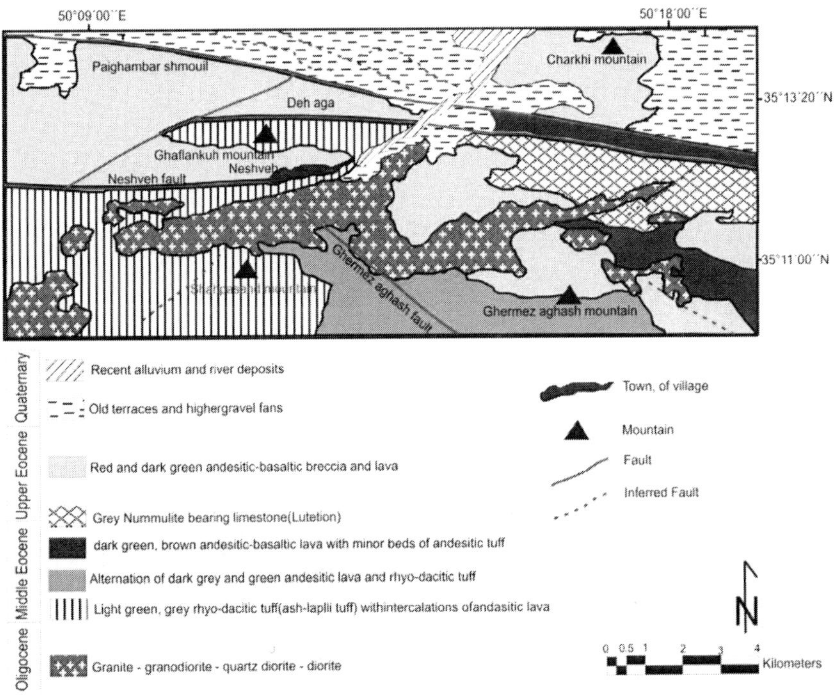

Figure 2. The geological map of studied area based on 1:100,000 geological map of Saveh [4] .

4. PETROGRAPHY

Neshveh granitic intrusion with an area about 22 km^2, consist of intrusive rocks in the mountains of Shahpasand and Zarou, north of Jafarabad Village, north of Aftabroo Qeshlaq village and south of Neshveh village. This intrusion crops out in the east-west trend and intruded into the Upper Eocene volcanic-sedimentary rocks (Figure 3). There are chloritic and argillic alterations present at the contact of intrusion with host rocks. Major faults include Kooshk Nosrat, Neshveh and QermezAghaj with NW-SE trend cut the Neshveh granitic intrusion and caused to crushing and fracturing the intrusion (Figure 3). Moreover, this intrusion cut by a series of andesiticbasaltic dikes. Some xenoliths of the Eocene volcanic rocks present at the marginal parts of the intrusion, which their dimension and frequency decrease to the interior part. In addition, there are some mafic micro-granular enclaves in different parts of the Neshveh intrusion which are characterized by rounded shapes. These enclaves mineralogically composed of the monzodiorite and diorite.

Figure 3. (a) View from the Neshveh intrusion and its contact with Eocene volcanic-sedimentary rocks (view to the south); (b) Close view from the fracturing in the Neshveh intrusion as a result of tectonic.

Table 1. The result of dating analysis on the some intrusions around the Saveh [1] .

Biotites	37 - 41 Ma
Amphiboles	31 - 38 Ma
Alkali Feldspars	35 - 42 Ma
Whole rocks	28 - 39 Ma

Based on Petrographic studies of 90 samples from different parts of the Neshveh intrusion, it was subdivided into two compositional zones: quartz monzogabbro in the West and quartz monzodiorite, granodiorite and granite in the East (Figure 4). The contacts are gradational but the rock types are distinct and easily recognized both in outcrop and in aerial photos and satellite images. Compositionally, dikes include andesite-basaltic andesite. Samples Neshveh intrusion has medium to coarse grained granular texture which is locally accompanied by porphyroidic texture with plagioclase phenocrysts, micro-granular and graphic texture.

The quartz monzogabbro occupies about 50% of the pluton and is medium to coarse-grained with various textures; some samples show inter-granular and poiklitic textures, while others show hetero-granular texture. The inter-granular texture is constituted by grains of clinopyroxene which occupy the angular interstices between plagioclase crystals. The quartz monzogabbro consists dominantly of plagioclase (51.5 - 55.6 modal %), clinopyroxene (20.2% - 25.7%), K-feldspar (9% - 14.1%), quartz (7.9% - 11.7%), and subordinate opaque minerals (2.3% - 3.5%) and apatite (0.6%).

The quartz monzodiorite surrounds the granodiorite and granite. To the west, quartz monzodiorite is transitional to more mafic rocks, the quartz monzogabbro while to the east it forms the margin of the pluton. The quartz monzodiorite is generally medium-grained and characterized by equi-granular texture. It consists dominantly of plagioclase (41.1% - 52.4 modal %), K-feldspar (9% - 19%), quartz (10.9% - 18.4%), hornblende (6.9% - 16.3%), subordinates clinopyroxene (0% - 12.1%), biotite (2.5% - 5%) and opaque minerals (2.3% - 3%) (Table 1). Accessory minerals such as titanite and apatite are rare.

About 20% of the Neshveh intrusion is constituted by granodiorite, emplaced within the quartz monzodiorite hosted by volcanic rocks. These rocks are mainly included plagioclase (40% - 45%), quartz (approximately 20%), alkali feldspar (15% - 20%), hornblende (~10%) and biotite (~5%). Granites are the next order and generally emplaced at the eastern part of the intrusion. These rocks are predominantly of medium-grained granular texture, but porphyroidic texture with plagioclase phenocrysts present also. Granites composed of plagioclase (35% - 40%), quartz (~25%), alkali feldspar (~30%), hornblende (~5%) and biotite (~1%).

In the all type of rocks at the studied area, plagioclase present as euhedral and subhedral crystals and are of andesine and oligoclase type. Some of them show zoning (Figure 5). These minerals underwent low grade alteration to sericite and sometimes epidote and calcite (Figure 5). Alkali feldspars (orthoclase) indicate argillic alteration in some samples. At the granitic samples, perthite texture was formed as a result of sub-solidus processes. Quartz presents as fine to medium-grained xenomorphic crystals between the other minerals. Graphic texture was formed as a result of quartz and orthoclase in the granites and occasionally in granodiorites (Figure 5). In some samples, quartz shows wavy extinction indicating tectonic stresses on the intrusion. Mafic minerals in the Neshveh intrusion are mainly hornblende and biotite is less. Hornblende present as euhedral and subhedral crystals at the all phases of Neshveh intrusion (Figure 5). Some crystals of hornblende, altered to actinolite which sometimes accompanied by biotite and iron hydroxides. Biotite present at the all phases of this intrusion which its content increase at the granites. Clinopyroxene is observed with low frequency in the quartz-monzodiorite which mainly has been replaced by actinolite. The rests of the clinopyroxene are found as islands form within the actinolite (Figure 5). Sphene, opaque, zircon and apatite are the minor minerals in the Neshveh intrusion and secondary minerals include sericite, epidote, actinolite, chlorite and clay minerals.

As mentioned before, there are some medium-basic dykes within the Neshveh intrusion that compositionally are andesite and basaltic andesite. They have porphyry, microlitic porphyry, inter-granular and intersertal textures phenocrysts of plagioclase, clinopyroxene and hornblende in a fine-grained to microlitic matrix (Figure 5). Some plagioclase crystals show zoning and classified as andesine-labradorite. Some plagioclase crystals altered to sericite and some clinopyroxene crystals show alteration to chlorite. Fine-grained matrix underwent alteration to chlorite.

Table 2. The result of samples analysis from the studied area by ICP-AES and ICP-MS methods (major elements in % and trace elements in ppm).

Rock Type	Andesitic Basalt	Granodiorite					Granite	Qz Monzodiorite	
Sample No.	SN05	SN15	SN17	SN31	SN48	SN44	SN52	SN10	SN11
SiO_2	49	62.7	62.5	62.3	64	65.1	69.4	56.5	59.6
TiO_2	1.12	0.57	0.51	0.59	0.53	0.49	0.32	0.73	0.71
Al_2O_3	14.65	16.15	15.45	15.75	15.65	14.8	13.8	17.1	16.7
Fe_2O_3	13.4	6.03	5.07	5.98	5.17	3.89	3.05	8.66	7.66
MnO	0.22	0.15	0.13	0.1	0.09	0.05	0.07	0.14	0.14
MgO	5.09	1.78	2.05	2.58	1.98	1.52	1.1	3.26	2.48
CaO	7.89	4.81	3.38	4.15	3.8	4.84	2.27	7.16	5.89
Na_2O	4.28	3.49	3.64	3.36	3.41	3.47	3.06	3.23	3.59
K_2O	1.14	2.73	3.46	2.95	3.26	2.9	4.19	1.85	2.31
P_2O_5	0.15	0.19	0.15	0.19	0.16	0.15	0.09	0.23	0.2
Total	96.94	98.6	96.34	97.95	98.05	97.21	97.35	98.86	99.28
V	506	128	89	118	87	76	47	241	194
Cr	15.9	10	10	10	10	10	10	20	10
Ni	19	8	5	6	5	5	5	14	8
Co	15	12.8	8	12	10.2	5	4	23.8	17.4
Cu	19	47	22	59	17	5	8	53	48
Zn	98	76	80	40	51	23	44	51	75
Ga	17.6	16.2	15	15.3	15.3	14.4	13.2	17.5	17.7
Sn	8	1	1	1	1	2	1	2	1
W	1	2	2	2	2	1	1	1	2
Ba	354	675	701	618	665	759	808	465	548
Sr	271	449	347	381	352	353	248	503	424
Rb	29.5	71.4	74.2	63.5	78.2	38.1	90.8	48.7	60.9
Nb	1.5	6.6	7.9	6.5	8.2	7.4	8.7	4.8	6.6
Y	22	20.4	21.6	17.5	19.7	18.4	16.2	20.9	23
Zr	81	121	134	113	145	145	124	83	119
Cs	0.71	1.22	1.49	1.19	1.74	0.4	1.47	1.95	2.34
Hf	2.3	3.6	3.8	3.1	4.1	4.1	3.8	2.4	3.5
Ta	0.1	0.5	0.5	0.5	0.5	0.5	0.7	0.3	0.5
Th	0.35	5.55	5.89	4.98	6.18	5.91	8.42	3.17	5.11
U	1.87	1.43	1.3	1.05	0.97	1.56	1.78	0.71	1.23
La	23.2	24.2	16.8	13.1	13.1	16	19.7	14	15.3
Ce	39.5	45.7	32.6	25.6	26.6	31.6	34.9	28.4	30.8
Pr	3.93	5.24	4.07	3.3	3.53	3.85	3.9	3.64	3.85

Nd	13.2	19.3	16.2	13.3	14.2	14.9	14.2	14.9	15.4
Sm	2.9	3.76	3.7	3.1	3.43	3.21	2.84	3.38	3.53
Eu	1.53	0.95	0.91	0.98	0.9	0.88	0.6	0.99	0.94
Gd	3.77	3.91	4.15	3.4	3.69	3.65	2.99	3.67	3.87
Tb	0.62	0.56	0.64	0.54	0.59	0.55	0.46	0.59	0.62
Dy	4.14	3.48	3.88	3.22	3.56	3.26	2.8	3.73	4
Ho	0.83	0.71	0.83	0.66	0.74	0.69	0.6	0.77	0.8
Er	2.53	2.12	2.41	1.99	2.28	2.12	1.89	2.18	2.45
Tm	0.36	0.32	0.37	0.29	0.34	0.32	0.29	0.32	0.34
Yb	2.37	2.14	2.44	1.93	2.29	2.16	2.06	2.24	2.39
Lu	0.39	0.36	0.37	0.29	0.36	0.33	0.34	0.33	0.4
Mo	2	2	2	2	2	2	2	2	2
Pb	0.5	13	21	10	15	6	17	7	11
Tl	0.5	0.5	0.5	0.5	0.5	0.5	0.5	0.5	0.5

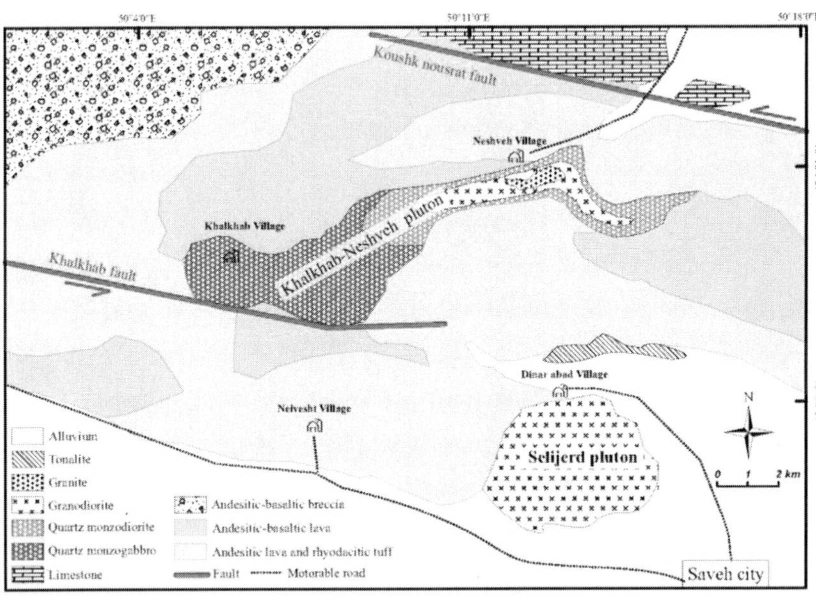

Figure 4. Simplified geological map of NW Saveh intrusions (Based on satellite data and the geological map of Saveh in 1:100,000 scale; Galamgash and Fonudi, 1998).

Figure 5. (A) Graphic texture in the granites and granodiorites (XPL light); (B) Plagioclase crystal that underwent epidote, sericite and chlorite alteration from the center (XPL light); (C) A plagioclase with zoning (XPL light); (D) Sericitized plagioclase crystals (XPL light); (E) Clinopyroxene bearing microgranular texture in the quartz-monzodiorite (XPL light); (F) Porphyritic texture includes plagioclase phenocrysts within the plagioclase, amphibole and pyroxene bearing fain-grained matrix (XPL light). (Gr: graphic, Plg: plagioclase, Qz: quartz, Hb: hornblende, Cpx: clinopyroxene).

5. GEOCHEMISTRY

For studying of geochemical characteristics of Neshveh intrusion, 8 samples from plutonic rock and one sample from basaltic andesite dikes were analyzed at the ALS Chemex laboratories for major, trace and rare earth elements by ICP-MS and ICP-AES analysis method. The results of this analysis are presented in Table 1.

Based on K_2O + Na_2O versus SiO_2 diagram [5] and [6] and the AFM triangular diagram [5] , all rocks units of the Neshveh intrusion are calc-alkaline character (Figure 6). At the SiO_2 against K_2O diagram [7] , they are classified as high potassium calc-alkaline rocks (Figure 7). Based on calcic-alkali index [8] , all samples Neshveh intrusion have Calc-alkali nature (Figure 7). On the ASI versus A/NK diagram [9] , all samples are classified as metaluminous rocks (Figure 7). K_2O/Na_2O ratio is more than one in the studied rocks.

Granitic rocks have been classified as M, I, S, A and H-type by some researchers [e.g., [10] -[15] . Several evidences suggest that the Neshveh intrusion is I-type and possibly Andean type: 1-Frequency of hornblende, magnetite, biotite and titanite in theses rocks and absence of muscovite, garnet, cordierite, andalusite and sillimanite. 2-SiO_2 content varies between 56.5% - 69.4% in these rocks. 3-Presence of mafic micro-granular enclaves in different parts of the intrusion and the absence of micaceous xenoliths [16] . 4-Absence of normative corundum in the samples and metaluminous nature of them. 5-Absence of pegmatites. 6-Decreasing P_2O_5 versus SiO_2 [17] (Figure 8). 7-Based on SiO_2 vs. K_2O [15] and ACNK versus SiO_2 diagrams [18] , the Neshveh intrusion classified as I-type granitoids (Figure 8). [19] classified granitoids into 7 main groups. Based on this classification, Neshveh intrusion is classified as amphibole-bearing calc-alkaline granitoids (ACG).

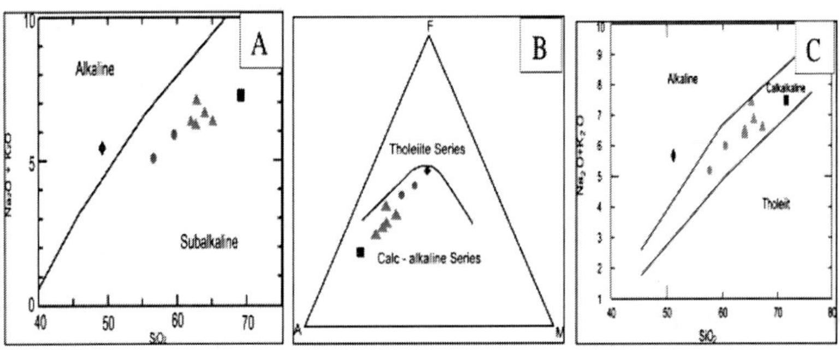

Figure 6. Location of the samples from studied area on the: (A) SiO_2versus Na_2O+K_2O diagram [5] , (B) AFM diagram [5] , and (C) SiO_2versus Na_2O+K_2O diagram [6] .

On the variation diagrams, with increasing SiO_2, contents of Al_2O_3, P_2O_5, CaO, TiO_2, MgO, MnO and Fe_2O_3 are decreased and K_2O and Na_2O demonstrate increasing trend. These trends may reflect the crystal fractionation process in the evolution of the Neshveh intrusion. The Nd/Sr vs. La/Nb diagram [20] , is useful indicator for fractionation of plagioclase in the magma. Based onFigure 8, crystallization and fractionation of plagioclase led to increasing of the Nd/Sr ratio in the magma while the La/Nb ratio will be constant.

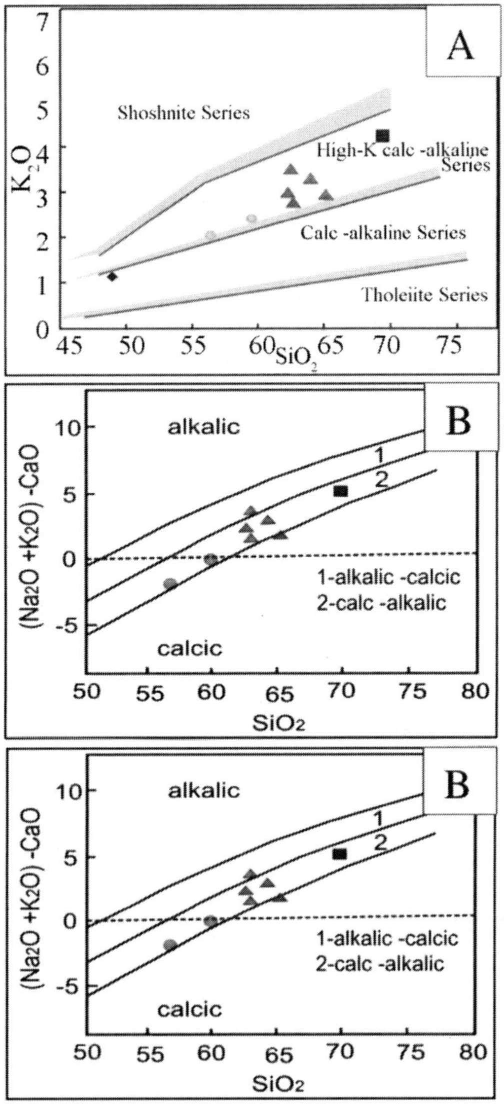

Figure 7. Location of the samples from studied area on the: (A) SiO_2 versus K_2O diagram [7] ; (B) SiO_2 versus (Na_2O+K_2O)-CaO diagram [8] , and (C) ASI versus A/NK diagram [9] .

21] exhibited a method to recognition of the three main processes of partial melting, fractional crystallization and magma mixing in the magmas. In this method, an incompatible element is used against a consistent element. According to this method, the consistent element put in the horizontal axis and the incompatible element is in the vertical axis. If the trend of samples is a smooth and straight and almost horizontal line, fractional crystallization is the probable process in the magma evolution. If the trend is a smooth and straight and almost vertical line, partial melting is the effective process in the evolution of magma. In the cases that the trends of samples show a concave curved, this trend indicating that the magma mixing is the most probable process in the evolution of magma. In the V vs. Rb and Ba diagrams (Figure 9), despite the very small scattering, samples of the Neshveh intrusion has relatively horizontal trend indicating that the main mechanism for evolution of the Neshveh intrusion was the fractional crystallization.

Based on geochemical studies, Neshveh granitoid can be classified as sodic granite ($Na_2O > K_2O$) and has low values of Mg#, Ni, Cr, Co and V. Lower contents of these elements indicative for evolution of magma during ascending and before complete crystallization [22] .

The primitive mantle normalized spider diagrams for different units of Neshveh intrusion demonstrate negative anomalies of HFSEs including Nb, Ti and P, along with positive anomalies in LILEs including Rb, Ba, Th, K and Pb (Figure 10). The negative anomalies of Ti and Nb are controlled by titanium bearing minerals such as titanite, ilmenite, rutile, garnet and some amphiboles. With increasing pressure, the solubility of titanium bearing minerals reduces in aqueous fluids [24] . Therefore, theses minerals that are rich in HFSEs will be stable during partial melting process at depths greater than 30 km that is led to negative anomalies in the melt [24] . The negative anomaly in P indicative for low content of apatite in the studied rocks. As mentioned before, in the I-type granitoids, phosphorus act as a consistent element and fractionation this element in the early stages of magma- crystallization, led to negative anomalies in the crystalized rocks. The sample from dikes has negative Th anomalies (Figure 9). Based on mineralogical and geochemical composition of dykes; it seems that their parental magma source is the mantle. Since the Th content in the mantle is higher than the crust, this negative anomaly indicating that this magma rise up rapidly and underwent very low crustal contamination.

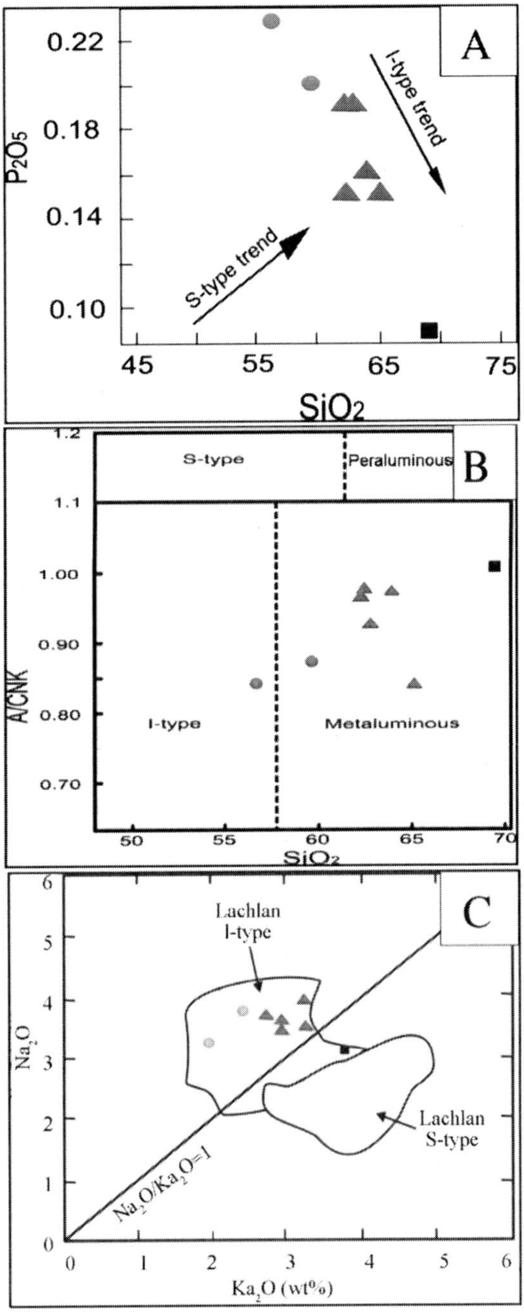

Figure 8. (A) Decreasing trend of P_2O_5 versus SiO_2 [17] . Location of the samples from studied area on the: (B) SiO_2 versus A/CNK [18] , and (C)SiO_2 versus Na_2O [15] .

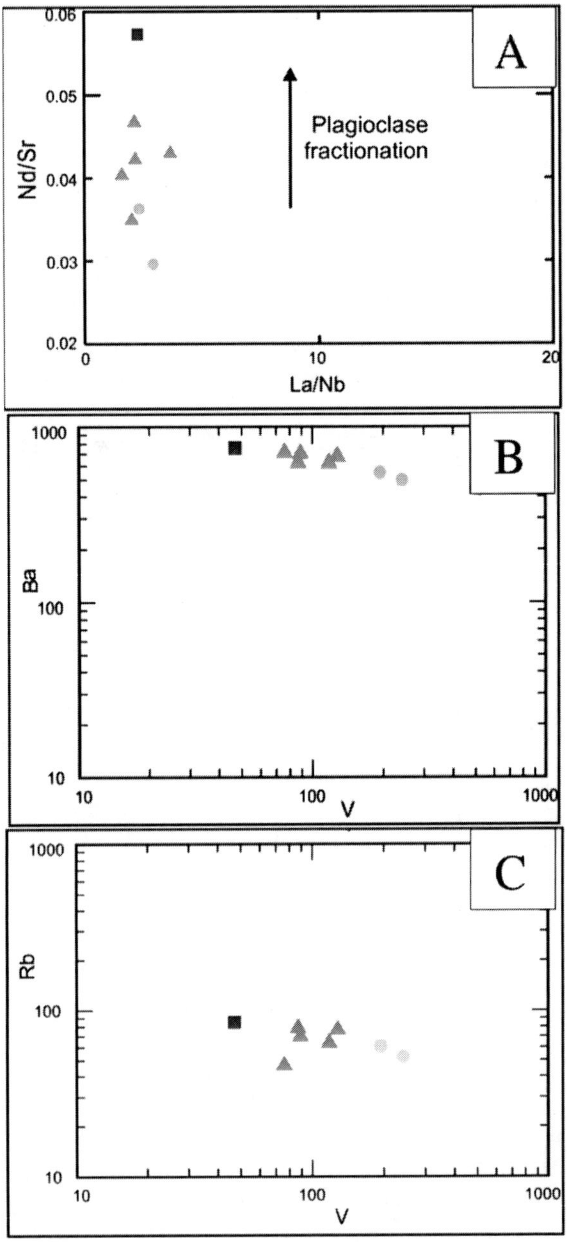

Figure 9. (A) Increasing trend in the Nd/Sr versus La/Nb which is indicative for plagioclase fractionation; (B) and (C) Decreasing trend of Rb and Ba versus V which is indicative for fractional crystallization on the evolution of Neshveh intrusion.

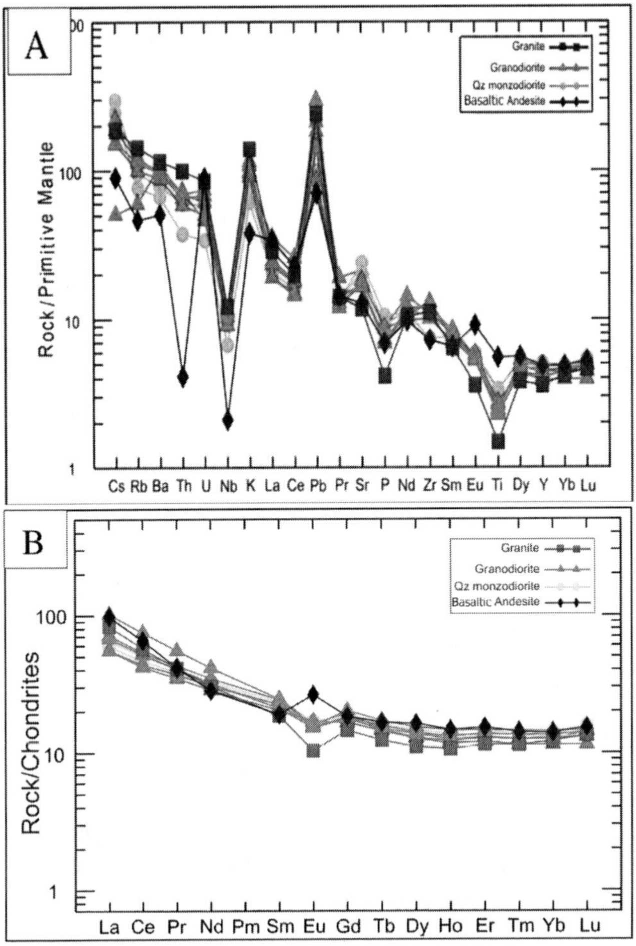

Figure 10. (A) Primitive mantle [23] normalized spider diagram for the studied rocks; (B) Chondrite normalized [23] REE patterns for the studied rocks.

Fluids and melts derived from the subducted oceanic slab, led to the metasomatism of upper mantle wedge and thereupon, negative anomalies of Nb and Ta [25] -[27] . Positive anomalies in Pb, K, Rb and Ba are attributed to metasomatism of mantle wedge by fluids derived from the subducted slab or/and contamination with continental crust [28] [29] .

The Chondrite-normalized REE patterns in the different phases of Neshveh intrusion show a LREE rich pattern with a high LREE/HREE ration (Figure 10).

All phases of Neshveh intrusion indicate negative anomalies Eu while sample from the dikes show a positive anomaly in Eu (Figure 10). Negative anomalies in Eu are related to the feldspar fractionation during magma crystallization or remaining of feldspar at the source. If these negative anomalies are associated with negative anomalies in Sr, fractional crystallization of plagioclase is responsible for these anomalies. While, the negative anomalies in Eu are associated with negative anomalies in Ba, K-feldspar fractionation has the main role in this respect [25] . In the studied rocks, negative anomalies of Eu occur along with Sr, which refers to the plagioclase fractionation. Positive anomaly in Eu from the dike sample is indicative for the absence of plagioclase fractionation during the magma crystallization and high concentration of this mineral in these rocks.

6. TECTONOMAGMATIC SETTING

Besides the geochemical characteristics of rare elements such as spider diagrams, tectonomagmatic setting discrimination diagrams also demonstrate a subduction setting relation for the Neshveh granitoid.

Based on the [30] diagram that is offered with considering, two cationic parameters include $R2 = 6Ca + 2Mg + Al$ and $R1 = 4Si - 11(Na + K) - 2(Fe + Ti)$, samples of Neshveh intrusion put into the pre-plate collision setting (Figure 11). On the [31] triangular diagrams that are drowned based on immobile elements include Ta and Hf and mobile element of Rb, samples of studied intrusion put into the volcanic arc (VA) setting (Figure 11)

Based on the Rb versus Y + Nb and Rb versus Ta + Yb diagrams [32] samples from the Neshveh intrusion put in the volcanic arc granite (VAG) setting (Figure 12). On the Th/Hf versus Ta/Hf, Th/Ta versus Yb and Ta/Yb versus Th/Ta diagrams [33] , all samples from the Neshveh intrusion put into the active continental margins are (Figure 13). The sample from the dikes is situated in the within plate setting (Figure 13).

Based on the AFM diagram [34] , the studied rocks located in the continental arc granites (CAG) setting (Figure 14), indicating that these granitoids originated from the subduction of an oceanic crust under a continental crust at the active continental margins. The parental magma was medium to high K calc-alkaline.

The Th/Yb versus Ta/Yb diagram [35] was used to recognition the source magma for Neshveh intrusion. According to this diagram, the studied samples are located within the calc-alkaline Andean granitoids (Figure 15). Moreover, the variation trend of samples on this diagram indicates the evolution of magma by fractional crystallization.

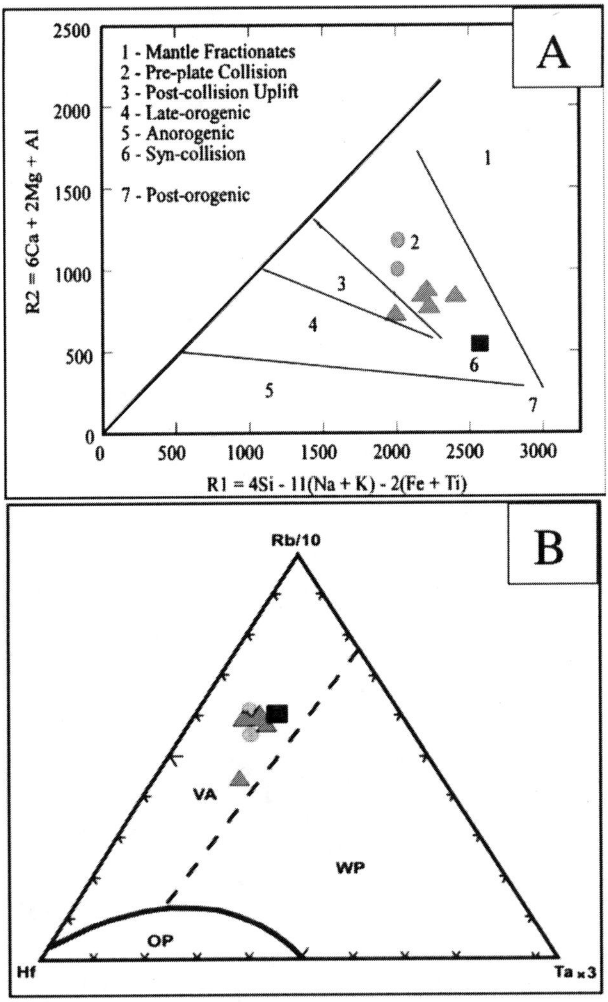

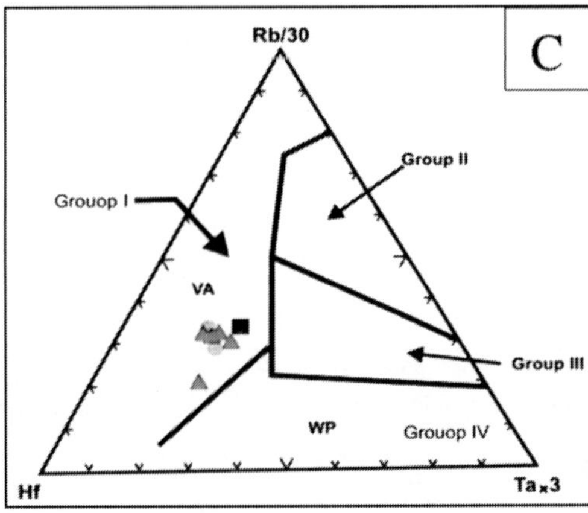

Figure 11. (A)Location of the Neshveh intrusion samples on the two cationic parameters of R1 and R2 [29] . (B) and (C) Location of the Neshveh intrusion samples on the Hf-Ta-Rb triangular diagrams [30] .

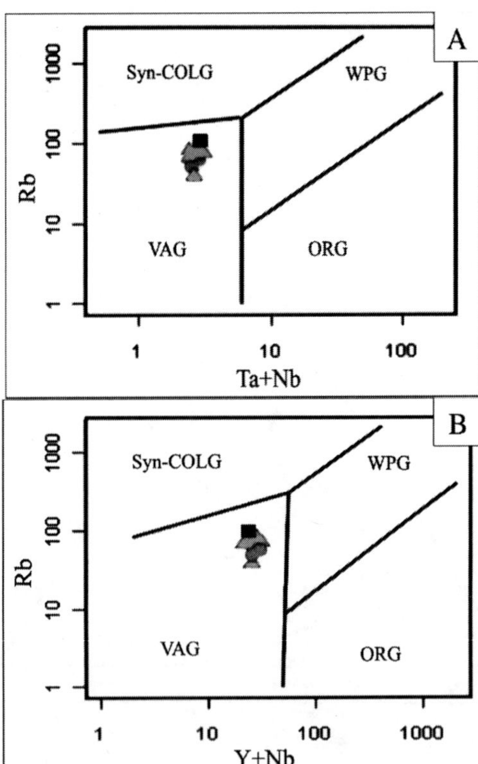

Figure 12. Location of the Studied samples on the: (A) Rb versus Ta+Yb diagram [31] ; and (B) Rb versus Y+Nb diagram [31] .

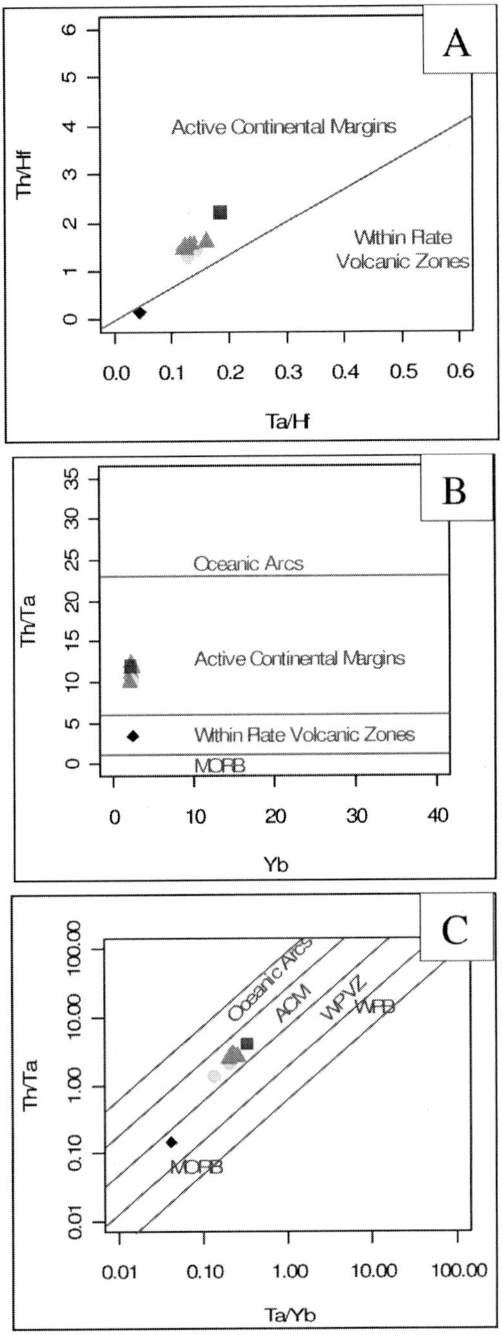

Figure 13. Location of the Studied samples on the: (A) Th/Hf versus Ta/Hf; (B) Yb versus Th/Ta; and (C) Th/Ta versus Ta/Yb diagrams [32]

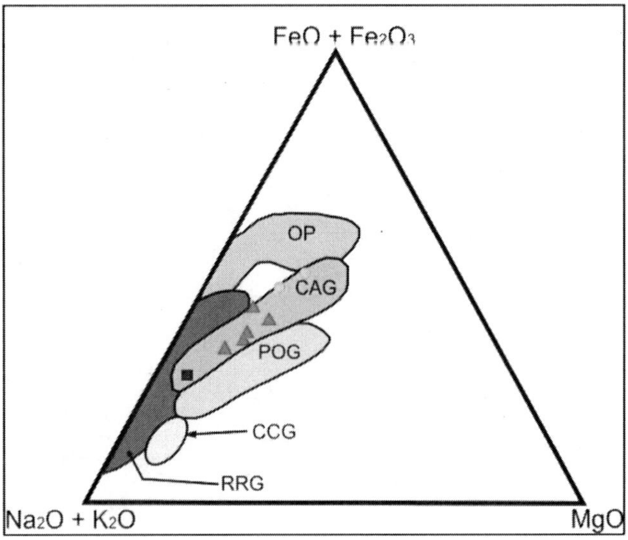

Figure 14. Location of the studied samples on the AFM diagram [33] .

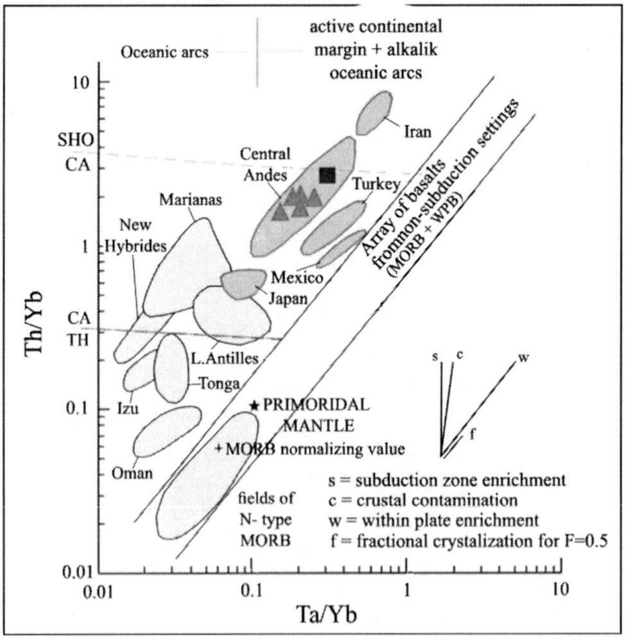

Figure 15. Location of the Neshveh intrusion sample on the Th/ Yb versus Ta/Yb diagram [34] .

As mentioned in the previous parts, based on [19] classification, Neshveh intrusion can be classified as ACG granitoids. According to him, ACG granitoids are hybrid type (mantle + crust) and are related with subduction settings. In this base, it is supposed a subduction setting to Neshveh intrusion.

7. CONCLUSION

Field studies, petrographic and geochemical investigations indicate that the Neshveh granitoid consists of the quartz-monzodiorite, granodiorite and granite. Field relations and geochemical studies demonstrate that all of these rocks are related to a single parent magma which underwent fractional crystallization. The Neshveh granitic intrusion has high-K calc-alkaline nature. This intrusion is metaluminous and classified as I-type granitoids. Negative anomalies of Nb and Ti along with positive anomalies of Rb, Ba, K, and Pb in the spider diagrams are indicative for the formation of this intrusion in relation with subduction setting. In terms of geodynamic setting, this intrusion is classified as volcanic arc granites (VAG) and active continental margin granites (CAG) which have been formed as a result of Neotethys oceanic crust subduction beneath the Central Iran continental crust.

ACKNOWLEDGEMENTS

This study was financially support by Research Deputy of Tehran University for the proposal entitled Petrography and Petrology of Neshveh intrusion (northwest Saveh). Our gratitude is expressed to their helps.

REFERENCES

1. Caillat, C., Dehlavi, P. and Martel, J.B. (1978) Geologie de la region de Saveh (Iran). Contribution a l'etude du volcanism et du plutonism tertiares de la zone de I Iran central. PhD Thesis, Grenoble, 325 Pages.

2. Helmi, F. (1991) Petrology and geochemistry of igneous rocks in Niousht area (NE Saveh). M.Sc. Thesis, University of Tehran, Iran (in Persian).

3. Moinvaziri, H. (1999) Introduction to the Magmatism of Iran. Tarbiat Moallem University, Tehran (in Persian).

4. Galamgash, J. and Fonudi, M. (1998) Explanatory Text of Saveh. Geological Map 1:100,000. Geological Survey of Iran, Tehran.

5. Irvine, T.N. and Baragar, W.R.A. (1971) A Guide to Chemical Classification of the Common Volcanic Rock. Canadian Journal of Earth Sciences, 8, 523-548.

6. Kuno, H. (1968) Differentiation of Basalt Magmas. In: Hess, H.H. and Poldervaart, A., Eds., Basalts: The Poldervaart Treatise on Rocks of Basaltic Composition, Interscience Publishers, New York, 623-688.

7. Peccerillo, A. and Taylor, S.R. (1976) Geochemistry of Eocene Calc Alkaline Volcanic Rocks from Kastamonu Area, Northern Turkey. Contribution to Mineralogy and Petrology, 58, 63-81.

8. Frost, B.F., Barnes, G.G., Collins, W.J., Arculus, R.J., Ellis, D.J. and Frost, C.D. (2001) A Geological Classification for Granitic Rocks. Journal of Petrology, 42, 2033-2048.

9. Shand, S.J. (1943) Eruptive Rocks. 2nd Edition, John Wiley, New York.

10. Chappell, B.W. and White, A.J.R. (1974) Two Contrasting Granite Types. Pacific Geology, 8, 173-174.

11. Collins, W.J., Beams, S.D., White, A.J.R. and Chappell, B.W. (1980) Nature and Origin of A-Type Granites with Particular Reference to South Eastern Australia. Contribution to Mineralogy and Petrology, 80, 189-200.

12. Whalen, J.B., Currie, K.L. and Chappell, B.W. (1987) A-Type Granites: Geochemical Characteristics, Discrimination and Petrogenesis. Contribution to Mineralogy and Petrology, 95, 407-419.

13. Castro, A., Monero-Venatas, I. and De La Rosa, J.D. (1991) H-Type (Hybrid) Granitoids: A Proposed Revision of the Granite Type Classification and Nomenclature. Earth Science Reviews, 31, 237-253.

14. Pitcher, W.S. (1993) The Nature and Origin of Granite. Blackie Academic and Professional Edition. Shopman and Hall, London.

15. Chappell, B.W. and White, A.J.R. (2001) Two Contrasting Granite Types. 25 Years Later. Australian Journal of Earth Sciences, 48, 489-499.

16. Didier, J., Duthou, J.L. and Lameyre, J. (1982) Mantle and Crustal Granites: Genetic Classification of Orogenic Granites and the Nature of Their Enclaves. Journal of Volcanology and Geothermal Research, 14, 125-132.

17. Chappell, B.W. and White, A.J.R. (1992) I- and S-Type Granites in the Lachlan Fold Belt. Transactions of the Royal Society of Edinburgh: Earth Sciences, 83, 1-26.

18. Chappell, B.W. and White, A.J.R. (1983) Granitoid Types and Their Distribution in the Lachlan Fold Belt, Southeastern Australia. Geological Society of America Memoirs, 159, 21-34.

19. Barbarin, B. (1999) A Review of the Relationships between Granitoid Types, Their Origins and Their Geodynamic Environments. Lithos, 46, 605-626.

20. Bourdon, E., Eissen, J.P., Monzier, M., Rovin, C., Martin, H., Cotten, J. and Hall, M.L. (2002) Adakite-Like Lavas from Antisana Volcano (Ecuador): Evidence for Slab Melt Metasomatism beneath the Andean Northern Volcanic Zone. Journal of Petrology, 43, 199-217.

21. Cocherie, A. (1986) Systematic Use of Trace Element Distribution on Patterns in Log-Log Diagrams for Plutonic Suites. Geochimica et Cosmoshimica Acta, 50, 2517-2522.

22. Woodhead, J.D. and Johnson, R.W. (1993) Isotopic and Trace Element Profile across the New Britain Island Arc, Papua New Guines. Contributions to Mineralogy and Petrology, 113, 479-491.

23. Sun, S.S. and McDonough, W.F. (1989) Chemical and Isotopic Systematics of Oceanic Basalts: Implications for Mantle Composition and Processes. In: Saunders, A.D. and Norry, M.J., Eds., Magmatism in the Ocean Basins, Vol. 42, Geological Society Special Publication, London, 313-345.

24. Glenn, A.G. (2004) The Influence of Melt Structure on Trace Element Partitioning near the Peridotite Solidus. Contributions to Mineralogy and Petrology, 147, 511-527.

25. Wilson, M. (2007) Igneous Petrogenesis. Chapman and Hall, London.

26. Sajona, F.G., Maury, R.C., Bellon, H., Cotton, J. and Defant, M. (1996) High Field Strength Elements Enrichment of Pliocene-Pleistocene Island Arc Basalts Zamboanga Peninsula, Western Mindanao (Philippines). Journal of Petrology, 37, 693-726.

27. Chappell, B.W. (1999) Aluminium Saturation in Iand S-Type Granites and the Characterization of Fractionated Haplogranites. Lithos, 46, 535-551.

28. Kamber, B.S., Ewart, A., Collerson, K.D., Bruce, M.C. and McDonald, G.D. (2002) Fluid-Mobile Trace Element Constraints on the Role of Slab Melting and Implications for Archean Crustal Growth Models. Contributions to Mineralogy and Petrology, 144, 38-56.

29. Atherton, M.P. and Ghani, A.A. (2002) Slab Breakoff: A Model for Caledonian, Late Granite Syncollisional Magmatism in the Orthotectonic (Metamorphic) Zone of Scotland and Donegal, Ireland. Lithos, 62, 65-85.

30. Batchelor, R.A. and Bowden, P. (1985) Petrogenetic Interpretation of Granitoid Rock Series Using Multicationic Parameters. Chemical Geology, 48, 43-55.

31. Harris, N.B.W., Pearce, J.A. and Tindle, A.G. (1986) Geochemical Characteristics of Collision Zone Magmatism. Collision Tectonic, Geological Society of American Bulletin, Special Publication, 19, 67-81.

32. Pearce, J.A. (1996) Sources and Settings of Granitic Rocks. Episodes, 19, 120-125.

33. Gorton, M.P. and Schandl, E.S. (2002) From Continents to Island Arc: A Geochemical Index of Tectonic Setting for Arc-Related and within Plate Felsic to Intermediate Volcanic Rocks. Canadian Mineralogist, 38, 1065-1073.

34. Bowden, P., Batchelor, R.A., Chappelle, B.W., Didier, J. and Lameyer, J. (1984) Petrological, Geochemical and Source Criteria for the Classification of Granitic Rocks: A Discussion. Physics of the Earth and Planetary Interiors, 35, 1-11.

Chapter 4

Storage and Evolution of Mafic and Intermediate Alkaline Magmas beneath Ross Island, Antarctica

Kayla Iacovino[1,*], Clive Oppenheimer[1], Bruno Scaillet[2] and Philip Kyle[3]

[1]Department of Geography, University of Cambridge, Cambridge CB2 3EN, UK;

[2]Institut des Sciences de la Terre d'Orléans, Université d'Orléans–CNRS–BRGM, 45071 Orléans Cedex 2, France and

[3]Department of Earth and Environmental Science, New Mexico Institute of Mining and Technology, Socorro, NM 87801, USA

ABSTRACT

We present the results of phase equilibrium experiments carried out on basanite and phonotephrite lavas from Ross Island, Antarctica. Experiments were designed to reproduce the P–T–X–fO2 conditions of deep and intermediate magma storage and to place constraints on the differentiation of each of the two predominant lava suites on the island, which are thought to be derived from a common parent melt. The Erebus Lineage (EL) consists of lava erupted from the Erebus summit and the Dry Valley Drilling Project (DVDP) lineage is represented by lavas sampled by drill core on Hut Point Peninsula. Experiments were performed in internally heated pressure vessels over a range of temperatures (1000–1150°C) and pressures (200–

400 MPa), under oxidized conditions (NNO to NNO + 3, where NNO is the nickel–nickel oxide buffer), with XH_2O of the H_2O–CO_2 mixture added to the experimental capsule varying between zero and unity. The overall mineralogy and mineral compositions of the natural lavas were reproduced, suggesting oxidizing conditions for the deep magma plumbing system, in marked contrast to the reducing conditions (QFM to QFM – 1, where QFM is the quartz–fayalite–magnetite buffer) in the Erebus lava lake. In basanite, crystallization of spinel is followed by olivine and clinopyroxene; olivine is replaced by kaersutitic amphibole below ~1050°C at intermediate water contents. In phonotephrite, the liquidus phase is kaersutite except in runs with low water content (XH_2Ofluid < 0·2) where it is replaced by clinopyroxene. Experimental kaersutite compositions suggest that the amphibole-bearing DVDP lavas differentiated below 1050°C at 200–400 MPa and NNO + 1·5 to NNO + 2. Olivine- and clinopyroxene-bearing EL lavas are consistent with experiments performed above 1050°C and pressures around 200 MPa. The plagioclase liquidus at <1–2 wt % H_2O suggests extremely dry conditions for both lineages (XH_2Ofluid approaching zero for EL, ~0·25 for DVDP), probably facilitated by dehydration induced by a CO_2-rich fluid phase. Our results agree with previous studies that suggest a single plumbing system beneath Ross Island in which DVDP lavas (and probably other peripheral volcanic products) were erupted through radial fractures associated with the ascent of parental magma into the lower crust. The longer travel time of the DVDP lavas through the crust owing to lateral movement along fractures and the lack of a direct, sustained connection to the continuous CO_2-rich gas flux that characterizes the main central Erebus conduit is probably responsible for the lower temperatures and slightly wetter conditions and hence the change in mineralogy observed.

KEYWORDS

alkaline volcanism, basanite, CO_2, phase equilibria, phonotephrite, Ross Island

1. INTRODUCTION

Ross Island, Antarctica (77°32'S, 167°10'E) is part of the Erebus Volcanic Province (EVP) and comprises five predominant volcanic centers. The currently active and volumetrically dominant Erebus volcano is surrounded by the now extinct subsidiary volcanic centers of Mts Bird, Terror, Terra Nova, and Hut Point Peninsula (Fig. 1). Erebus is known for its persistently active phonolitic lava lake, which has been at the focus of recent active monitoring and petrological studies as it represents a window into the shallow regions of the alkaline intraplate volcanic plumbing system. Recent investigations employing experimental petrological work (Moussallam *et al.*, 2013) and seismic tomography (Zandomeneghi *et al.*, 2013) have placed tight constraints on the physical and petrological properties of the shallow plumbing system and phonolitic lava lake including its geometry, temperature, pressure, and oxygen fugacity. These studies have the potential to refine the results of numerical models developed for the shallow Erebus system, such as those of Burgisser *et al.*(2012) and Molina *et al.* (2012). The intermediate and deep plumbing system beneath Erebus (>8 km depth) has been investigated via study of a suite of melt inclusions (Oppenheimer *et al.*, 2011) and mixed volatile solubility experiments (Iacovino *et al.*, 2013), which have placed constraints on the depths of degassing and the equilibrium fluid contents of primitive and intermediate composition magmas. Little is known, however, about the specific $P–T–X–fO_2$ conditions of magma storage and differentiation processes at these depths. Understanding the deep system at Erebus is crucial because the constant supply of deeply derived CO_2-rich gases combined with a sustained energy and mass input into the lava lake suggests a direct link between the phonolitic lava lake and the mantle source of the parental magmas via a deep mafic plumbing system (Kyle*et al.*, 1992; Oppenheimer & Kyle, 2008; Oppenheimer *et al.*, 2011).

Gas emissions from the Erebus lava lake are rich in CO_2 (~30 mol %) and comparatively low in H_2O (~65 mol %; Oppenheimer *et al.*, 2011; Burgisser*et al.*, 2012), consistent with the volatile compositions deduced from melt inclusions for the deeper system. The longevity of the lava lake itself is hypothesized to reflect, in part, the constant fluxing of heat and CO_2-rich fluid (Oppenheimer *et al.*, 2011) that originates from depths greater than those recorded by the most primitive melt inclusions (Iacovino *et al.*, 2013*a*). Indeed, many separate studies have concluded that various aspects of the Erebus system are controlled by the influx of heat and CO_2, including polybaric fractional crystallization throughout the magmatic system (Oppenheimer *et al.*, 2011), the bimodal eruptive behavior in the lava lake (Oppenheimer & Kyle, 2008), shallow magma chamber processes (Moussallam *et al.*, 2013), and fumarolic degassing and the formation of ice caves and ice towers (Curtis & Kyle, 2011). CO_2 flushing has been invoked to explain the characteristics of degassing and melt inclusions at other volcanoes including Stromboli (Allard, 2010), Vesuvius (Marianelli*et al.*,

2005), Etna (Spilliaert *et al.*, 2006; Shinohara *et al.*, 2008; Taglianiet *al.*, 2012) and volcanoes in southeastern Guatemala (Walker *et al.*, 2003) that sit atop large sedimentary carbonate sequences, the devolatilization of which could provide the necessary excess carbon (Iacono-Marziano *et al.*, 2009). As Erebus has no such shallow carbon reservoir and probably derives all of its volatiles from the same mantle source responsible for magma production, we propose Erebus as the archetype for mantle-sourced carbon-controlled volcanism on Earth.

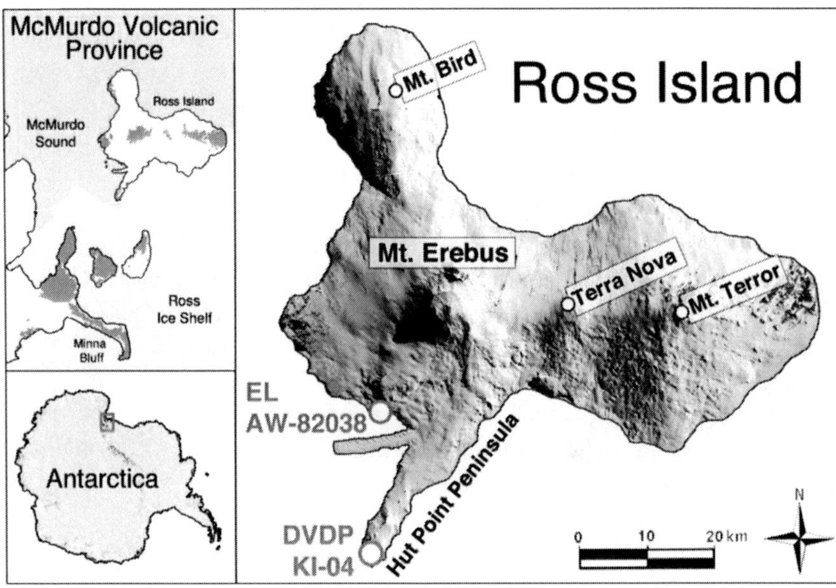

Figure 1. Map showing the location of Ross Island within the McMurdo Volcanic Province, Antarctica. Erebus volcano is surrounded by the peripheral volcanic cones Bird, Terra Nova, Terror, and Hut Point Peninsula. The locations of the DVDP and EL samples collected for this experimental study are indicated.

Subsidiary extinct volcanic centers and cones on Ross Island are distributed radially around Erebus at *c.* 120° from each other. This symmetrical distribution has been hypothesized to be the result of radial fractures created by the pressure of updoming mantle material on the crust directly beneath Mount Erebus (Kyle & Cole, 1974; Kyle *et al.*, 1992). The rare earth element patterns in primary Ross Island basanitoids indicate the generation of a basanitic parent magma through low-degree partial melting of the mantle source (Sun & Hanson, 1975, 1976). Major and trace element modeling suggests that lavas from Erebus [the Erebus Lineage (EL)] and the peripheral Hut Point Peninsula [the Dry Valley Drilling Project (DVDP) lineage] both evolved from a common basanite parent melt, which underwent two distinct fractional crystallization trends (Kyle, 1981;Kyle *et al.*, 1992). So far experimental work has focused on the evolved magmas (Moussallam *et al.*, 2013) and no experimental work is available to constrain storage and differentiation conditions of mafic lavas at Erebus (or other subsidiary volcanoes).

In this experimental study, one lava from each of the two Ross Island lava lineages (EL phonotephrite AW-82038 and DVDP basanite KI-04) were chosen as starting materials for equilibrium crystallization experiments to establish the P–T–X–fO_2 conditions necessary to produce the observed natural rock series. Phase equilibrium experiments are commonly used to constrain the pre-eruptive storage conditions of magmas, as well as their liquid line of descent (Sack *et al.*, 1987; Moore & Carmichael, 1998; Freise*et al.*, 2003; Martí *et al.*, 2008; Andújar *et al.*, 2010; Di Carlo *et al.*, 2010). Although non-equilibrium processes such as magma mixing and repeated replenishment of the system are important to consider, high pressure–temperature experimentation nevertheless illuminates the fundamental processes that control magma evolution from mantle to surface, allowing us to evaluate to what extent crystal–liquid equilibrium has been attained or maintained at depth.

Experiments were performed over a range of temperatures (1000–1150°C) and at pressures between 200 and 400 MPa to reproduce the intermediate region of the magma plumbing system where much early crystallization (differentiation of a basanitic parent to form intermediate lavas) is thought to occur. A temperature of 1000°C was chosen as a thermal lower bound, as the temperature of the surficial lava lake, which should be the coolest region of the plumbing system, has been constrained to ~1000°C by several studies (Kyle, 1977; Caldwell & Kyle, 1994; Dunbar *et al.*, 1994). The resulting experimental charges thus span the entire range of compositions seen in the natural lavas, allowing elucidation of the evolutionary processes occurring throughout the Ross Island magmatic plumbing system. With this approach, our aim has been two-fold: (1) to experimentally establish the conditions necessary to derive the distinct Erebus and DVDP lineages from a hypothesized single basanite parent; (2) to place constraints on the intermediate Erebus and DVDP plumbing systems with respect to pressure, temperature, oxygen fugacity, and dominant volatiles (H_2O and CO_2).

Experiments with H_2O–CO_2 fluid compositions in the range of $XH_2O = 0$–1 not only help elucidate the role of H_2O in controlling phase equilibria, but also constrain the effect of CO_2 on crystallization processes via melt dehydration. Given that carbon dioxide is known to play an important role in the genesis of alkaline magmas, in both the mantle (Spera, 1984) and crust (Kaszuba & Wendlandt, 2000), it is important to understand its role at Erebus throughout the magmatic system (from mantle to surface).

2. MINERALOGY AND PETROLOGY OF ROSS ISLAND LAVAS

The majority of lavas on Ross Island are represented by two strongly undersaturated sodic differentiation lineages: the Erebus Lineage (EL), made up of lavas erupted from Erebus volcano, and the Dry Valley Drilling Project (DVDP) lineage, made up of lavas on the Hut Point Peninsula and sampled via drill coring. The rarer Enriched-Iron Series (EFS) lavas are less silica-undersaturated and constitute only a small volume of Ross Island rocks. The EFS will not be discussed in this study. The DVDP and Erebus lineages are similar in terms of whole-rock chemistry as demonstrated by the similar variation in major element evolution of the DVDP and EL lavas with the exception of P_2O_5 (Fig. 2), the variation of which is probably being controlled by apatite crystallization. Major element mass balance and trace element modeling of DVDP and EL rocks (Kyle, 1981; Kyle *et al.*, 1992) suggests fractional crystallization from a common basanitic parent melt, with slight differences in differentiation conditions resulting in the two similar yet distinct lineages.

The mineralogy of the EL and DVDP lineage rocks from Ross Island has been extensively documented by Kyle *et al.* (1992) and Kyle (1981), respectively, whose observations are summarized below. These lineages share the same assemblage with two notable exceptions. First, kaersutitic amphibole is present in almost all DVDP lavas (except in basanites where it is rare) but is absent in EL lavas. Second, olivine is present in all EL lavas and in DVDP basanites, but occurs only as xenocrysts in more evolved DVDP lavas. In the EL rocks, olivine ranges in composition from Fo_{88} in basanites to Fo_{51} in phonolites. Olivine xenocrysts are rare and occur only in the most primitive lavas. In the DVDP lavas, olivine occurs only in basanites as euhedral to subhedral phenocrysts and in the groundmass.

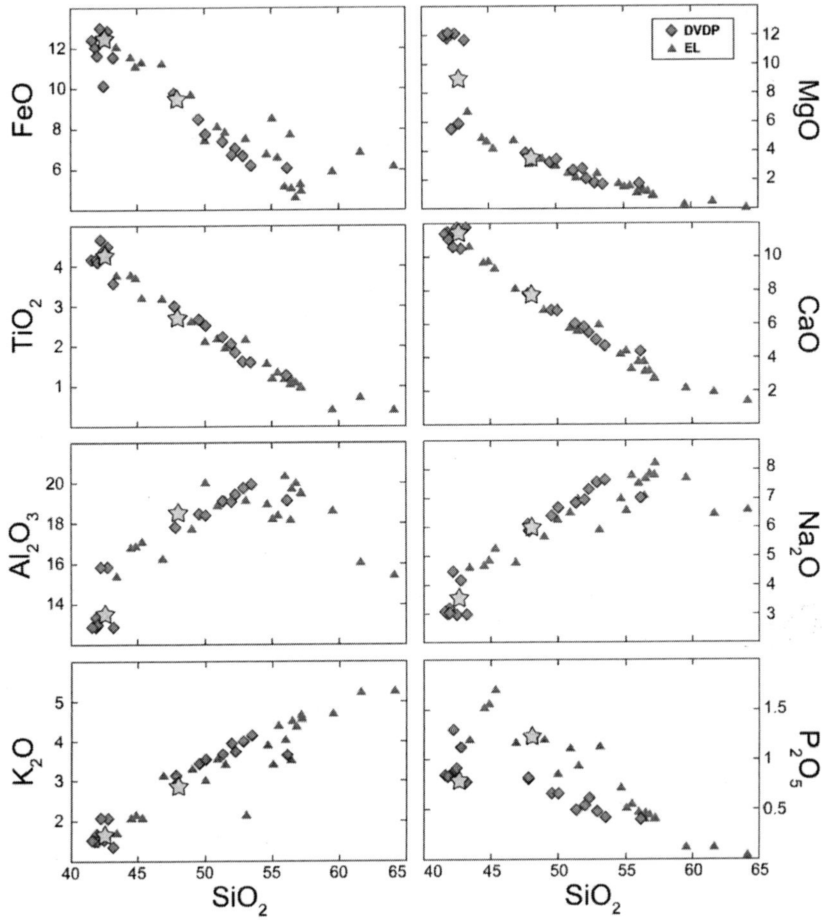

Figure 2. Variation diagrams illustrating the evolution of the DVDP lineage (red diamonds; Kyle, 1981) and Erebus Lineage (EL; blue triangles; Kyle *et al.*, 1992). For whole-rock compositions the major element totals are normalized to 100%. One lava sample from each of these two lineages (red star: DVDP KI-04; blue star: EL AW-82038) was used as the starting material for the experimental work.

Clinopyroxene is ubiquitous in all EL and DVDP lavas and occurs as euhedral to subhedral phenocrysts. Pyroxenes with green cores, which show irregular rims, possibly owing to resorption, are occasionally found in DVDP lavas. In the EL lavas, clinopyroxenes show oscillatory zoning in basanites and intermediate lavas, but no zoning in phonolites.

Kaersutite is absent in EL lavas but is an important phase in all DVDP lavas except the basanites where it is rare. In some phonotephrites, kaersutite ranges from euhedral grains to those showing strong resorption and oxidation. It was suggested by Kyle (1981), based on previous experimental work, that temperature and PH_2O probably exert a strong control on the appearance or absence of kaersutite in DVDP and Erebus lineages, respectively.

Feldspar is the modally dominant phenocryst phase in EL lavas (up to 40% of the mode in anorthoclase-bearing phonolites) with compositions ranging from An_{72} to An_{59} in basanites, becoming increasingly albitic in the more evolved lavas. The feldspar in the EL phonolite is primarily anorthoclase. In the DVDP lava, feldspar occurs mostly as microphenocrysts or in the groundmass and is mainly labradorite (An_{50-60}) in basanites, and andesine to oligoclase in all other lava types.

Feldspathoids occur in the groundmass of the EL lavas, as nepheline, which contains up to 24% normative Ne. Microphenocrysts of apatite are common in most rocks of both lineages and are often found as inclusions in kaersutite and magnetite in the DVDP rocks. Pyrrhotite is present in all rocks from both lineages and occurs as small round blebs, suggesting the presence of an immiscible sulfide liquid in the Ross Island magmas. Opaque Fe–Ti oxides are common in both lava suites and are typically titanomagnetite as phenocrysts and in the groundmass, plus rare ilmenite phenocrysts.

3. EXPERIMENTAL METHODS

3.1 Starting materials

To investigate the evolution of the Ross Island magmas and the two dominant lava lineages, two samples were chosen, one from each lineage (Table 1): AW-82038 is an Erebus Lineage phonotephrite lava (SiO_2 = 48·01 wt %; MgO = 3·18 wt %) from Turk's Head (Kyle *et al.*, 1992) representative of intermediate EL lavas; KI-04 is a basanite (SiO_2 = 43·06 wt %; MgO = 8·92 wt %) collected from Hut Point Peninsula for this study and represents the less evolved magmas of the DVDP lineage. Figure 3illustrates our starting material compositions relative to those of a suite of melt inclusions in Erebus Lineage rocks, which provides a representative cross-section through the evolutionary chemical trend of lavas on Ross Island (Oppenheimer *et al.*, 2011).

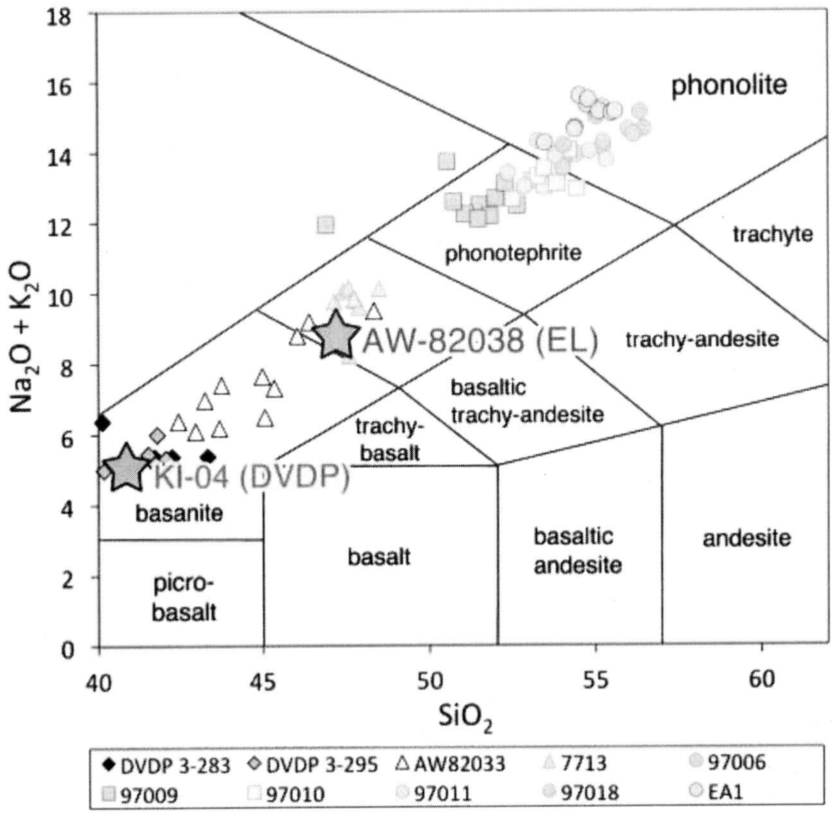

Figure 3. Total alkalis vs silica (TAS) diagram showing the compositions of the starting materials (stars) relative to the fractional crystallization sequence seen in Erebus Lineage rocks represented by a suite of melt inclusions [data from Oppenheimer*et al.* (2011)].

The AW-82038 phonotephrite [also referred to as Ne hawaiite by Kyle*et al.* (1992)] is weakly porphyritic (<8% phenocrysts) with a phenocryst assemblage of plagioclase, olivine, clinopyroxene, Fe–Ti oxides, apatite and rare accessory nepheline. AW-82038 is consistent with the fractionation of 12% olivine, 35% clinopyroxene, 7% Fe–Ti oxides, and 1% apatite from a basanitic parent, yielding a 43% residual melt (basanite DVDP 2-105.53; Kyle *et al.*, 1992).

Table 1: Composition of AW-82038 and KI-04 starting glass, whole-rock and representative analyses of phenocrysts and microphenocrysts in the natural rock

AW-82038	Whole-rock[1]	St. Gls[2]	Ol-6	Ol-10	Plag-5	Plag-12	Ap-1	Sp-1	gnd Ol-5	gnd Plag-5
SiO_2	48·01	48·03	38·72	38·47	54·10	53·69	0·26		36·27	55·46
TiO_2	2·77	2·84			0·14	0·15		26·18		0·16
Al_2O_3	18·56	18·12			28·54	29·24		4·34		28·56
FeO*	9·49	9·59	20·47	22·54	0·35	0·31	0·33	63·15	29·08	0·39
MnO	0·22	0·23	0·28	0·45	0·00	0·04	0·10	0·80	0·64	0·03
MgO	3·18	3·08	41·79	39·84	0·03	0·03	0·26	5·29	35·09	0·05
CaO	7·65	7·57	0·30	0·30	11·11	11·53	55·11	0·04	0·44	10·62
Na_2O	5·99	6·04			4·90	4·66	0·07	0·04		5·20
K_2O	2·88	3·07			0·62	0·50				0·70
P_2O_5	1·24	1·41					42·02			
Total	98·29	99·54	101·72	101·80	99·82	100·21	98·16	100·04	101·74	101·19
Mg#	37	36	78	76					68	
Or					2·3	1·8				2·6
An					69·8	71·9				67·5
Ab					27·9	26·3				29·9

KI-04	Whole-rock[1]	St. Gls[2]	Ol-6	Ol-9	Aug-5	Aug-24	Plag-1	Plag-4
SiO_2	43·06	43·15	38·04	37·62	47·09	42·35	49·91	50·10
TiO_2	4·26	4·43			2·57	4·99	0·23	0·18
Al_2O_3	13·80	13·94			4·92	9·16	29·95	30·91
FeO	9·73	12·35*	17·61	19·83	7·98*	6·99*	0·66	0·53
Fe_2O_3	2·87							
MnO	0·19	0·17	0·22	0·31	0·25	0·10	0·00	0·00
MgO	8·92	8·46	43·56	40·66	13·34	12·10	0·10	0·09
CaO	11·07	11·16	0·29	0·39	22·49	22·70	14·11	14·62
Na_2O	3·64	3·81			0·46	0·45	3·41	3·31
K_2O	1·63	1·68					0·35	0·29
P_2O_5	0·83	0·85						
Total	98·61	97·68	100·00	98·99	99·10	99·06	98·72	100·04
Mg#	55	55	82	79	75	76		
Or							1·2	1·0
An							81·1	82·2
Ab							17·7	16·8

Glass analyses are given in wt % and normalized to 100% with original analysis totals given. Mineral analyses are not normalized; original analysis totals are given. St. Gls, starting glass after vitrification; Ol, olivine; Plag, plagioclase; Ap, apatite; Sp, spinel; gnd Ol, groundmass olivine; gnd Plag, groundmass plagioclase.

Total iron given as FeO; Mg# = 100[Mg/(Mg + Fe)] in mol %.

[1]Whole-rock composition. AW-82038 analyzed by Kyle *et al.* (1992); KI-04 analyzed in this study, using X-ray fluorescence (XRF).

[2]Electron microprobe (EMP) analyses of glass composition after vitrification in 1 atm furnace, but before run in pressure vessel. AW-82038 vitrified at an oxygen fugacity of NNO + 1 [analysis from Iacovino *et al.* (2013)]; KI-04 vitrified in air (analysis from this study).

In the KI-04 basanite, phenocrysts constitute around 12 vol. % and include olivine (6 vol. %), plagioclase feldspar (2 vol. %), augite (2 vol. %), apatite (1 vol. %), and titanomagnetite (<1 vol. %). Although kaersutitic amphibole is common in almost all DVDP lavas, it is rare in DVDP basanites and is absent in KI-04. Crystal distribution is relatively homogeneous throughout the sample. This sample is consistent with 17% fractionation from basanitic parent DVDP 2-105.53 yielding an 83% residual melt.

The lavas were crushed to a powder and melted in a Pt crucible in a 1 atm furnace at 1300°C. Sample AW-82038 was held in the furnace for ~30 min under a stream of nearly pure CO_2 (corresponding to an oxidation state near the Ni–NiO buffer, NNO) and then drop-quenched into cold water [sample prepared by Iacovino *et al.* (2013*a*)]. Sample KI-04 was held in furnace for ~1 h in air (i.e. extremely oxidizing conditions) and then removed from the furnace and immediately placed into cold water to quench. The resulting glassy samples were extracted from their crucibles, re-crushed, and re-melted under identical conditions to ensure a homogeneous and crystal-free glass. A portion of each starting material was polished and analyzed by electron microprobe to verify that no compositional changes had occurred during melting (see Table 1).

3.2 Experimental techniques

Between 30 and 60 mg of glassy starting material powder was added to $Au_{80}Pd_{20}$ capsules with lengths of ~20 mm, an outer diameter of 2·9 mm, and an inner diameter of 2·5 mm. Liquid H_2O was added via a microsyringe (~0–6 mg), and CO_2 was added as silver oxalate powder ($Ag_2C_2O_4$, ~0–18 mg). The X_{H2O} in the H_2O–CO_2 mixture of pre-run charges was varied at steps between zero and unity (i.e. X_{H2O} = 0, 0·25, 0·5, 0·75, and 1), and the fluid proportion in the charge was always ~10% to ensure fluid saturation (see Supplementary Data Appendix 1;supplementary data are available for downloading athttp://www.petrology.oxfordjournals.org). Once filled, capsules were weighed, crimped at their open end, arc welded shut with a graphite electrode, and then subsequently re-weighed to check for any mass loss or gain during welding. After welding, capsules were placed in an oven at 110°C for several minutes and then re-weighed to check for any leaks in the capsule. As a third method to check for leaks, some capsules were held in a beaker of oil at 300°C for a few seconds. If bubbles were observed escaping from the capsule, it was considered to have leaks and was thus discarded. After completion of the experiment, capsules were weighed to check that no fluid loss occurred during the run. Any capsules that showed a change in mass >0·2 mg either after welding or after heating were considered to have leaked at least part of their fluid phase and were discarded.

Experiments were run in vertically working internally heated pressure vessels (IHPVs; see Holloway, 1971) at the Institut des Sciences de la Terre Orléans (ISTO) at isothermal and isobaric conditions between 1000 and

1150°C and between 200 and 400 MPa with the oxygen fugacity of the vessel at about one log unit above the Ni–NiO buffer (NNO + 1; Table 2). Each run contained 5–10 sample capsules loaded into a vertically working IHPV equipped with a double-wound Kanthal furnace. An Ar–H_2 gas mixture loaded sequentially at room temperature was used as the pressure medium with the Ar/H_2 ratio fixed so as to achieve an experimental fO_2 in the vessel of NNO + 1 (Scaillet *et al.*, 1992), although the fO_2 within each experimental charge varied slightly depending on its specific H_2O/CO_2fluid mixture (see below). It is noteworthy that, because the experimentalfO_2 is lower than that imposed during melting of the sample KI-04 starting material, some additional H_2O was created in the capsule during KI runs via the reduction of Fe^{3+}. Thus, even nominally 'dry' runs will contain some H_2O (typically < 1–2 wt %), and this has been taken into account for the calculation of experimental oxygen fugacity. Pressure was measured with a transducer calibrated against a Heise–Bourdon tube gauge with an accuracy of±20 bars. Temperature was measured with two chromel–alumel (Type K) thermocouples, which allow for precise control over the length of the hotspot (~4 cm) to within ±5°C. Experiments were run for 17–24 h and then quenched by dropping the sample into the lower, cold part of the pressure vessel (<40°C), which gives a quench rate of >100°C s^{-1} (Roux & Lefevre, 1992). The very low viscosities of our mafic alkalic glasses facilitated the attainment of equilibrium even given these relatively short run times. In similar studies on mafic melts, the attainment of equilibrium was observed in runs of similar duration (Di Carlo *et al.*, 2006; Feig *et al.*, 2010; Pichavant *et al.*, 2014).

4. ANALYTICAL TECHNIQUES

4.1 Fluid phase composition

For capsules that did not leak, the conventional weight-loss method was employed as a first-order measure of the composition of the equilibrium H_2O–CO_2 fluid phase (see Behrens *et al.*, 2009). After the first post-run weighing, the capsule was submerged in a liquid nitrogen bath, thus freezing any liquid water or water vapor but allowing the CO_2 to remain as a gas. The capsule was then punctured to allow for the escape of the CO_2gas, and a hissing noise and bubbles at the puncture site were typically observed. After warming at room temperature for up to 1 min, the capsule was weighed to determine the mass of CO_2 lost after puncture. This was then recorded as the mass of CO_2 in the fluid phase. The capsule was then placed in a drying oven at 110°C for several minutes to evaporate the remaining H_2O and then weighed to determine the mass of the H_2O fluid. During this step, capsule weight was checked periodically until its mass remained constant. It should be noted that this method is an indirect measure of the composition of the fluid phase in our runs and can be prone to significant error owing to uncertainties in the weighing of the capsules before and after opening. Behrens *et al.* (2009) estimated an error in XH_2O determination of around ±0·03 using this method. The composition of the fluid phase was also calculated by mass balance, where the mass of dissolved H_2O or CO_2 in the glass [as measured by Fourier transform IR (FTIR) spectroscopy] is subtracted from the amount of fluid (liquid H_2O or CO_2 as silver oxalate) added to the capsule (seeSupplementary Data Appendix 1). A comparison of fluid compositions calculated both by weight loss and by mass balance shows reasonable agreement, with a correlation coefficient between the two methods of $R = 0{\cdot}79$ and an average error in XH_2O determination between the two methods of ±0·11. Here, calculation of the fluid phase composition by weight loss is preferred, with the mass-balance method used for sample KI-07, for which weight-loss data were not obtained. After fluid mass measurement, the AuPd capsule was peeled away, and the sample was mounted in epoxy and polished for analysis. Selected samples were doubly polished for analysis by FTIR spectroscopy.

4.2 SEM and electron microprobe

Charges were initially studied using a JEOL WINSET JSM 6400 scanning electron microscope (SEM) at ISTO for preliminary phase identification, modal abundance determination and textural analysis. Most phases could be identified by visual inspection, and several crystals from each phase were analysed using energy-dispersive spectrometery (EDS). The EDS analyses are not reported here as they were used only for mineral identification.

Representative SEM images from each sample were digitally processed to measure total crystallinity and the modal abundance of all phases. Images were converted to grayscale and a threshold was applied and adjusted to isolate each phase. The total area of each phase was then calculated as a percentage of the total image. Any areas of the image that contained no sample (holes, cracks, etc.) were not included in the total sample area.

Table 2: Experimental conditions and run products

Run	Experiment	*P* (MPa)	*T* (°C)	*X*H2O[in 1]	XH_2O[fluid 2]	H_2O glass (wt %)[3]	CO_2 glass (ppm)[3]	fO_2 (ΔNNO)[4]	Phases (vol. %)[5]
AW 82038 phonotephrite experiments									
PV-4B	AW-19B	200	1000	0·18	0·14	0·49	—	0·9	Gl 5, Cpx 10, Plag 81, Tit 3, Ap 1
PV-4B	AW-20B	200	1000	0·49	0·35	1·79	1077	1·7	Gl 81, Kaers 16, Tit 3
PV-4B	AW-21	200	1000	0·72	0·56	3·22 (23)	931	2·1	Gl 87, Kaers 9, Tit 4
PV-4B	AW-22	200	1000	0·88	0·74	4·04 (37)	817	2·4	Gl 87, Kaers 8, Tit 5
PV-4B	AW-23	200	1000	1·00	1·00	5·55 (10)	0	2·6	Gl 96, Tit 4
PV-3	AW-06	200	1050	—	—	1·42 (7)	1298 (137)	—	Gl 92, Plag 7, Tit < 1
PV-3	AW-17	200	1050	0·75	0·43	3·14 (11)	443 (17)	1·2	Gl 99, Tit 1
PV-3	AW-18	200	1050	0·87	0·71	4·00[6]	—	1·6	Gl 99, Tit 1
PV-3	AW-10	200	1050	1·00	1·00	5·57 (11)	0	1·9	Gl 99, Tit 1, Ap < 1
PV-2	AW-14	200	1100	0·78	0·20	1·86	831	1·2	Gl 99, Tit 1
PV-2	AW-13	200	1100	0·91	0·68	4·27 (20)	937	2·3	Gl 99, Tit 1
PV-8	AW-24B	250	1100	0·21	0·13	1·14 (1)	1700 (74)	1·0	Gl 100
PV-11	AW-42	300	1000	0·21	0·22	0·73 (3)	1145 (107)	1·6	Gl 78, Kaers 13, Plag 2, Tit 7
PV-11	AW-41	300	1000	0·49	0·49	1·78	1180	2·3	Gl 80, Kaers 15, Tit 5
PV-11	AW-40	300	1000	0·72	0·64	2·75 (34)	1975 (329)	2·6	Gl 94, Kaers 2, Tit 4
PV-11	AW-39	300	1000	0·86	0·86	4·30 (16)	1227 (366)	2·8	Gl 91, Tit 9, Ap < 1
PV-11	AW-38	300	1000	1·00	1·00	7·26 (22)	0	3·0	Gl 95, Tit 5
PV-6	AW-29	300	1050	0·18	—	1·01 (3)	1653 (150)	—	Gl 93, Plag 7
PV-6	AW-33	300	1050	0·73	0·71	2·94 (6)	1702 (358)	1·9	Gl 99, Tit 1
PV-6	AW-34	300	1050	1·00	1·00	5·90 (78)	0	2·2	Gl 97, Tit 3
PV-10	AW-37	300	1100	0·21	0·14	0·17[6]	—	1·2	Gl 99, Tit 1
PV-5	AW-25	300	1100	0·53	0·25	2·25 (9)	1186 (97)	1·7	Gl 99, Tit < 1
PV-5	AW-27	300	1100	0·88	0·67	3·03 (1·13)	1622 (311)	2·6	Gl 99, Tit < 1
PV-5	AW-26	300	1100	0·75	0·43	3·59 (3)	982 (59)	2·2	Gl 99, Tit < 1
PV-5	AW-28	300	1100	1·00	1·00	6·65 (85)	0	3·0	Gl 99, Tit < 1
PV-15	AW-43	400	1000	0·21	0·18	0·02[6]	—	1·7	Gl 36, Cpx 22, Plag 35, Tit 7
PV-15	AW-44	400	1000	0·49	0·51	2·21	—	2·6	Gl 75, Kaers 19, Tit 6
PV-15	AW-45	400	1000	0·70	0·71	4·04 (13)	3375 (824)	2·9	Gl 80, Kaers 13, Tit 7
PV-15	AW-46	400	1000	0·86	0·74	4·76 (76)	3439 (134)	2·9	Gl 87, Kaers 9, Tit 4
PV-15	AW-47	400	1000	1·00	1·00	6·84 (49)	0	3·2	Gl 74, Tit 26

KI-04 basanite experiments								
PV-7	KI-09	200	1050	0·19	0·17	0·73 (17)	1154 (271)	0·4	Gl 70, Cpx 23, Tit 7
PV-7	KI-10	200	1050	0·75	0·55	1·52 (11)	1233 (137)	1·4	Gl 53, Cpx 44, Tit 3 + rare Ol
PV-7	KI-11	200	1050	0·70	0·60	2·18[6]	—	1·4	Gl 60, Kaers 37, Tit 4
PV-7	KI-12	200	1050	0·86	0·78	3·55[6]	—	1·7	Gl 65, Ol 6, Cpx 25, Tit 5
PV-2	KI-05	200	1100	1·00	—	—	—	—	All quench
PV-2	KI-06	200	1100	0·89	—	—	—	—	Quench + rare Ol
PV-2	KI-08	200	1100	0·58	—	—	—	—	Gl 86, Ol 11, Tit 2
PV-2	KI-07	200	1100	0·76	0·58	2·90 (8)	1131 (100)	2·2	Gl 93, Ol 4, Tit 3
PV-8	KI-14	250	1100	0·17	0·20	0·45[6]	—	1·4	Gl 52, Ol 26, Cpx 23
PV-8	KI-15	250	1100	0·49	0·43	1·91 (2)	3074 (627)	2·1	Gl 91, Ol 9, Tit < 1
PV-8	KI-16	250	1100	0·69	0·54	3·33	4338	2·3	Gl 72, Ol 27, Tit 1
PV-8	KI-17	250	1100	0·87	0·77	3·84[6]	—	2·6	Gl 58, Ol 21
PV-8	KI-18	250	1100	1·00	1·00	—	—	2·8	All quench
PV-10	KI-23	300	1100	0·21	0·14	1·11 (4)	2628 (257)	1·2	Gl 80, Ol 15, Q 5
PV-10	KI-22	300	1100	0·49	0·26	2·13[6]	—	1·8	Gl 92, Ol 8, Tit < 1
PV-10	KI-21	300	1100	0·70	0·48	2·79 (58)	2798 (573)	2·3	Gl 91, Ol 9, Tit < 1
PV-10	KI-20	300	1100	0·87	0·61	2·21 (31)	1256 (364)	2·5	Gl, Q + rare Ol
PV-10	KI-19	300	1100	1·00	1·00	—	—	3·0	All quench
PV-12	KI-28	300	1150	0·21	0·21	0·59[6]	—	0·9	Gl 98, Ol 2, Tit < 1
PV-12	KI-27	300	1150	0·49	0·32	1·80[6]	—	1·3	Gl 99, Ol 1, Tit < 1
PV-12	KI-26	300	1150	0·70	0·43	—	—	1·6	All quench
PV-12	KI-25	300	1150	0·86	0·83	—	—	2·2	All quench
PV-12	KI-24	300	1150	1·00	1·00	—	—	2·3	All quench
PV-13	KI-29	400	1000	0·21	0·13	—	—	1·4	Subsolidus
PV-13	KI-30	400	1000	0·48	0·31	—	—	2·1	Subsolidus
PV-13	KI-31	400	1000	0·71	0·59	2·72[6]	—	2·7	Gl 38, Kaers 60, Tit 2
PV-13	KI-32	400	1000	0·85	0·80	3·05 (5)	1601 (139)	3·0	Gl 57, Kaers 41, Tit 2

PV-13	K1-33	400	1000	1·00	1·00	4·30[6]	0	3·2	Gl 38, Kaers 59, Tit 3
PV-14	KI-36	400	1100	0·72	0·53	3·38[6]	—	2·6	Gl 95, Ol 5, Tit < 1
PV-14	KI-37	400	1100	0·85	0·80	3·56[6]	—	3·0	Gl 29, Q 51, Ol 20
PV-14	KI-38	400	1100	1·00	1·00	—	—	3·2	All quench

[1]Mole fraction of H_2O in the H_2O–CO_2 fluid added to the experimental capsule prior to running at *P* and *T*.
[2]Mole fraction of H_2O in the equilibrium H_2O–CO_2 fluid phase during the run, measured by weight-loss method. Sample KI-07 calculated by mass balance.
[3]Weight per cent H_2O or ppm CO_2 dissolved in the glass, measured via FTIR spectroscopy. Standard deviation shown in parentheses.
[4]Calculated oxygen fugacity of the experiment in terms of ΔNNO (log units from the Ni–NiO buffer). (See text for details.)
[5]Phase proportions calculated as volume per cent; Gl, glass; Ol, olivine; Cpx, clinopyroxene; Kaer, kaersutite; Fsp, feldspar; Ap, apatite; Tit, titanomagnetite; Q, quench crystals.
[6]Volatile concentration calculated by mass balance using values in Table 3. Corrected values using the calibration lines in Supplementary Data Fig. A2.2 are reported here.

Electron microprobe (EMP) analyses of the major element compositions of the glass and crystal phases in the natural samples, starting glasses and experimental run products were made primarily using a Cameca SX-100 electron microprobe at the University of Cambridge. Some samples were analyzed on a Cameca SX-100 at the University of Arizona or on a Cameca SX-50 at the Università degli Studi di Padova. Glass was measured using an accelerating voltage of 15 keV, a beam current of 3 nA, a defocused beam size of 10 µm, and a peak counting time of 10 s, analyzing Na first to minimize any measurement error owing to Na migration. Crystal phases were analyzed with an accelerating voltage of 15 keV, a beam current of 10 nA, a focused beam with a diameter of 1 µm, and a peak counting time of 20 s. In each experimental charge, typically 20–30 analyses were performed on glass and 8–12 analyses were performed on each crystal phase.

4.3 FTIR spectroscopy

Whenever possible, the H_2O and CO_2 concentrations in the glass phase of each experimental charge were measured using a Thermo Scientific Nicolet iN10 MX mapping FTIR system at the US Geological Survey in Menlo Park, California. Transmission IR spectra of doubly polished glass chips ranging in thickness from ~20 to 150 µm were obtained in the 6000–1000 cm^{-1} wavenumber range using a KBr beamsplitter with a variable aperture size between 25 and 500 μm^2. The sample stage was continuously purged with dry air, and a background measurement was made before the collection of each spectrum to minimize and correct for the contribution of atmospheric H_2O and CO_2. Moderately to highly crystal-rich charges were characterized with FTIR mapping to identify crystal-free regions for the analysis of dissolved volatiles.

The concentrations of OH, H_2O, CO_2, and CO_3 species were calculated with the Beer–Lambert Law (Ihinger *et al.*, 1994), using near-IR (4500 and 5200 cm^{-1}), total water (3500 cm^{-1}), molecular H_2O (1640 cm^{-1}), CO_2(2345 cm^{-1}), and carbonate (1430 and 1525 cm^{-1}) bands. The total water 3500 cm^{-1} band was always preferred for retrieving H_2O concentrations; in five samples the peak at this band was unusable and so near-IR bands were used. The H_2O concentrations in samples measured with both the total water and near-IR bands agree well, with an average standard deviation of 0·57 wt % and an R^2 value of 0·83.

Sample thicknesses were measured to an accuracy of ±2 μm using a digital tabletop micrometer equipped with a fine tip that allowed for measurement at multiple spots on the sample surface. Additional sample thicknesses for some samples were also measured using an ADE Phaseshift Microxam surface-mapping microscope (with white light interferometer) to an accuracy of ±0·5–2 μm. In very thin samples, the presence of interference fringes in reflectance IR spectra allowed for the measurement of the sample thickness in the precise spot that was analyzed with transmitted light, following the method of Nichols & Wysoczanski (2007, and references therein; see Supplementary Data Appendix 2 for details).

Glass densities were calculated for each sample using the model of Lange & Carmichael (1987), assuming an H_2O molar volume of 22·9 cm^3 mol^{-1}(Ochs & Lange, 1999). Molar absorption coefficients strongly depend on glass composition (Pandya *et al.*, 1992). Because of variation in the crystallinity of our samples, and therefore significant variation in glass chemistry, an attempt was made to use absorption coefficients appropriate for each sample. All absorption coefficients used in this study are listed in Table 3. Where possible, the model of Dixon & Pan (1995) was used to calculate ε_{1525} and ε_{1430} (carbonate absorption) for single samples. This calculation can only be applied to glasses with an Na/(Na + Ca) ratio between 0·25 and 0·5 (in our case, most glasses classified as phonotephrite and tephriphonolite fall in this range). For glasses with Na/(Na + Ca) < 0·25 or > 0·5, ε_{1525} and ε_{1430} values were selected from the literature (see Table 3). (See Supplementary Data Appendix 2 for details of FTIR spectral background fitting, absorbance measurements, sample thicknesses, and glass density calculations.)

Dissolved volatile concentrations could not be measured in nine crystalline samples (29–75 vol. % glass) where glass pools were too small to analyze by FTIR spectroscopy. A further six glassy samples (95–99 vol. % glass) could not be successfully prepared for FTIR analysis. For these 15 samples, H_2O and CO_2 contents were calculated using the mass-balance method (see Supplementary Data Appendices 3 and 4 for details), and the 35 samples for which FTIR measurements were obtained were used as internal standards. The standard error for these calculations is large at ±1·4 wt % and ±1000 ppm for calculated H_2O and CO_2 values, respectively.

Table 3: Absorption coefficients used to determine concentrations of dissolved volatiles in experimental charges via FTIR spectroscopy

Composition and wavenumber (cm^{-1})	Species	Coefficient [l (mol cm)$^{-1}$]	Reference
Phonolite			
1640	H_2O_m	25 ± 3	Dixon *et al.*, 1994
2345	CO_2	890	Morizet *et al.*, 2002
3500	H_2O^{tot}	64 ± 8	Cioni, 2000
Tephriphonolite			
1430	CO_3	Calculated	Dixon & Pan, 1995
1525	CO_3	Calculated	Dixon & Pan, 1995
1640	H_2O_m	25 ± 3	Dixon *et al.*, 1994
3500	H_2O^{tot}	64 ± 8	Cioni, 2000
4500	OH	0·62 ± 0·02	Behrens *et al.*, 2009
5200	H_2O_m	1·02 ± 0·06	Behrens *et al.*, 2009
Phonotephrite			
1430	CO_3	Calculated	Dixon & Pan, 1995
1525	CO_3	Calculated	Dixon & Pan, 1995
1640	H_2O_m	25 ± 3	Dixon *et al.*, 1994
3500	H_2O^{tot}	63·9 ± 5·4	Behrens *et al.*, 2009
4500	OH	0·62 ± 0·02	Behrens *et al.*, 2009
5200	H_2O_m	1·02 ± 0·06	Behrens *et al.*, 2009
Basanite			
1430	CO_3	375 ± 20	Fine & Stolper, 1986
1525	CO_3	375 ± 20	Fine & Stolper, 1986
1640	H_2O_m	25 ± 3	Dixon *et al.*, 1994
3500	H_2O^{tot}	63·9 ± 5·4	Behrens *et al.*, 2009
4500	OH	0·58 ± 0·02	Cocheo, 1993
5200	H_2O_m	0·56 ± 0·04	Cocheo, 1993

4.4 Oxygen fugacity

The fO_2 of all IHPV runs was controlled by loading the pressure vessel with sufficient H_2 to achieve an apparatus fO_2 of ~NNO + 1. The fO_2 inside each experimental capsule will vary depending on the H_2O/CO_2 ratio of the fluid phase contained therein. Assuming that the fH_2 of the pressure vessel will be the same as the fH_2 inside an experimental capsule owing to the very fast diffusion rates of hydrogen, we can calculate the fO_2 inside each capsule following the method of Iacovino *et al.* (2013*a*). First, because the partial pressure of H_2 is fixed for each experiment, we can calculate the fH_2 of the pressure vessel for each run using the equation

$$fH_2 = \phi P_{H2} \tag{1}$$

where φ is the fugacity coefficient of hydrogen for the *P* and *T* of a given experiment (Shaw & Wones, 1964), and P_{H2} is the partial pressure of hydrogen gas added to the IHPV at the start of the run. If the XH_2O^{fluid} is known, we can also then calculate the fH_2O inside the capsule as

$$fH_2O^{inside} = fH_2O^{pure} XH_2O^{fluid} \tag{2}$$

where fH_2O^{pure} is the fugacity of pure H_2O vapor at the experimental *P*and *T* conditions, assuming ideal mixing between H_2O and CO_2 in the fluid [calculated with the modified Redlich–Kwong equation of state ofHolloway (1977) as modified by Flowers (1979)]. Now, it is possible to calculate the oxygen fugacity for each experiment as follows:

$$fO_2^{inside} = \left(\frac{fH_2O^{inside}}{fH_2 K_F}\right)^2 \tag{3}$$

where K_F is the equilibrium constant of formation for H_2O (Robie *et al.*, 1979). Oxygen fugacities in our experiments are reported in Table 2 and range from a ΔNNO of +0·3 to +3·2.

5. RESULTS

5.1 Textural and mineralogical observations

Optical and electron microscope images revealed homogeneously distributed euhedral crystals in most experimental charges, with crystals ranging in size from <5 μm to ~80 μm in length. Phases identified include olivine, clinopyroxene, plagioclase, amphibole, Fe–Ti oxides, apatite, glass and fluid inclusions. In the most highly crystalline charges, glass tends to collect in pools and small subhedral crystals ('quench' texture) are ubiquitous. Quench texture is especially common in KI samples with high H_2O contents, probably owing to the extremely low-viscosity melt allowing for the rapid crystallization upon quench. In samples where crystals were too small to analyze by electron microprobe (<25 μm²), phases were analyzed quantitatively for identification purposes only. Those analyses are not reported here.

5.2 Phase relationships

The equilibrium phase relations for the AW and KI experiments are shown in Figs 4 and 5. In both compositions, titanomagnetite is ubiquitous in nearly all experimental charges independent of $P/T/H_2O$ and is probably stabilized owing to the relatively oxidizing conditions of our experiments. Experiments are plotted in terms of the mole fraction of H_2O in the H_2O–CO_2 fluid added to the capsule, XH_2O^{in}.

In AW, kaersutite is the liquidus phase at median to high water contents, crystallizing below about 1025°C. Amphibole is common in the AW experiments and is stable to melt H_2O contents as low as ~1 wt %. A reaction relationship between kaersutite and clinopyroxene is inferred to exist around $XH_2O^{fluid} = 0{\cdot}1$–$0{\cdot}2$, below which clinopyroxene is the dominant calcic ferromagnesian phase (experiments AW-19B and AW-43). As with kaersutite, clinopyroxene is restricted to below about 1025°C. Plagioclase feldspar is the dry liquidus phase and is restricted to low water contents, with its stability field increasing with decreasing temperature. Phase equilibria at 300 MPa are similar to those at 200 MPa [the stability field for clinopyroxene at 300 MPa (Fig. 4b) is inferred based on that at 200 MPa (Fig. 4a)]. Although apatite was not always detected, it is inferred to crystallize below 1100°C at both pressures. Its stability is mostly constrained by water-saturated experiments, but apatite is probably stable but undetectable in drier runs (see Harrison & Watson, 1984). This hypothesis is supported by the steady decrease in the P_2O_5 content of residual liquids in phonotephrite experiments. The polybaric section at 1000°C (Fig. 4c) shows that in the pressure range 200–400 MPa amphibole is always the liquidus phase, followed by Pl and Cpx, in that order.

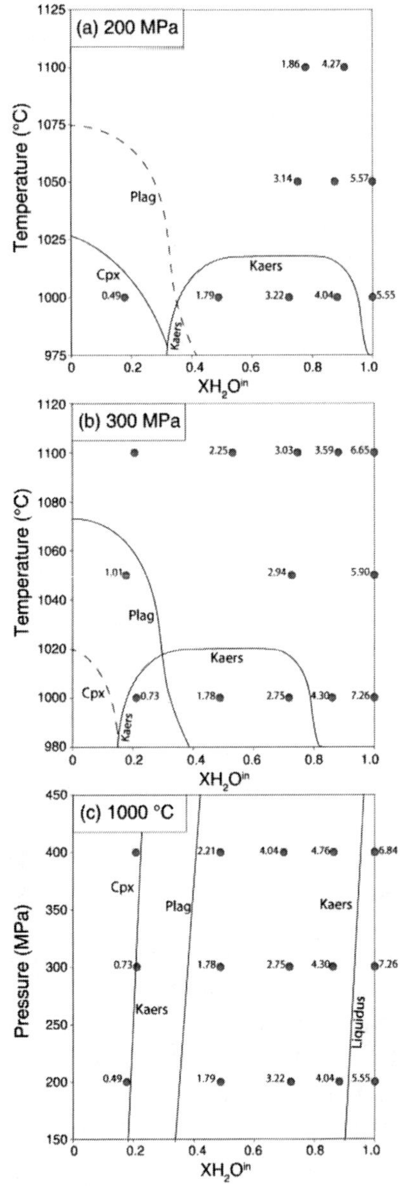

Figure 4. Isothermal (a, b) and isobaric (c) phase diagrams illustrating the stability fields of different phases in AW experiments. Numbers next to points indicate wt % H_2O dissolved in the glass of the experimental charge as measured by FTIR spectroscopy. Although not shown, titanomagnetite is ubiquitous in all charges. Apatite was not always detected but is inferred to be stable in runs below ~1100°C. A dashed line indicates an inferred phase boundary.

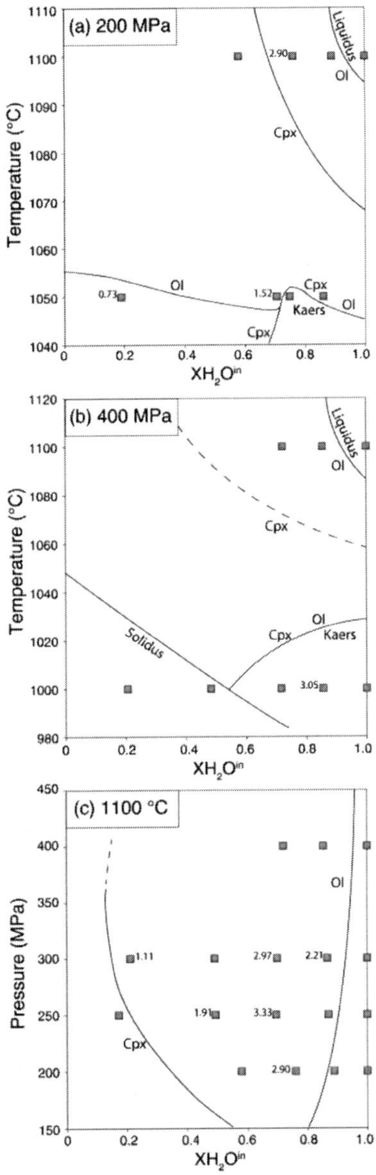

Figure 5. Isothermal (a, b) and isobaric (c) phase diagrams illustrating the stability fields of different phases in KI experiments. Numbers next to points indicate wt % H_2O dissolved in the glass of the experimental charge as measured by FTIR spectroscopy. Although not shown, titanomagnetite is ubiquitous in all charges. Apatite was not always detected but is inferred to be stable in runs below ~1100°C. A dashed line indicates an inferred phase boundary.

In KI experiments at 200 MPa (Fig. 5a), olivine is the liquidus phase followed by clinopyroxene and kaersutite. Although loosely constrained, the liquidus is located above 1100°C except at very high H_2O contents (XH_2O^{in} approaching unity), where it drops to ~1075°C. A peritectic-like relationship exists at intermediate melt water contents and 1050°C at 200 MPa where olivine is replaced by kaersutite at low temperatures. At these*P–T* conditions, clinopyroxene replaces kaersutite as the dominant ferromagnesian phase at low and high melt water contents where it coexists with olivine, except at very low melt water contents (XH_2O^{in}approaching zero) where clinopyroxene is the only stable silicate phase. The phase relationships observed at 400 MPa (Fig. 5b) are broadly similar to those at 200 MPa except that the kaersutite stability field is enlarged towards H_2O-rich conditions, this mineral being in reaction relationships with both olivine and clinopyroxene as temperature decreases. The polybaric section at 1100°C (Fig. 5c) illustrates that olivine is the liquidus mineral in the pressure range 200–400 MPa, for a large range of fluid compositions (melt water content), followed by clinopyroxene at $XH_2O_{in} < 0{\cdot}25$.

5.3 Experimental phase compositions

Compositions of the experimental mineral phases and residual glasses are given in Table 4. Chemical analyses for all natural and experimental phases are given in Supplementary Data Appendix 5. Titanomagnetite, apatite, and some clinopyroxene crystals were too small to be analyzed by electron microprobe (<25 μm^2). In six charges, clinopyroxenes were large enough to be analyzed, and their compositions are reported in Table 4. In most samples, the $K_D^{Fe–Mg}$ values of experimental clinopyroxenes, which range from 0·31 to 0·46 and average 0·36 (calculated using Fe^{2+} as*X*FeO), are similar to the $K_D^{Fe–Mg}$ values observed in natural KI-04 basanite clinopyroxenes, which average 0·40, suggesting equilibrium crystal growth. Overall, clinopyroxene is characterized by elevated contents of Al_2O_3 and TiO_2, which do not show any obvious correlation with experimental parameters in the explored range.

Table 4: Compositions of experimental phases

	Phase	SiO_2	TiO_2	Al_2O_3	FeO^T	MnO	MgO	CaO	Na_2O	K_2O	P_2O_5	Total	Mol %
AW 82038 experiments													
AW-6	PT (9)	48·03 (0·74)	2·84 (0·24)	18·12 (0·62)	9·60 (0·89)	0·23 (0·05)	3·08 (0·28)	7·57 (0·38)	6·04 (0·12)	3·08 (0·11)	1·41 (0·22)	99·20 (0·59)	Mg no. 36·4
	plag (4)	50·38 (0·9)	1·75 (0·30)	22·74 (1·26)	5·29 (0·85)	0·13 (0·04)	1·62 (0·34)	8·83 (0·31)	5·00 (0·20)	1·89 (0·28)	n.d.	97·64 (0·54)	$An_{61}Al_{31}Or_8$
AW-10	PT (7)	49·45 (0·3)	2·64 (0·07)	18·74 (0·26)	8·26 (0·36)	0·25 (0·04)	2·73 (0·07)	7·62 (0·07)	5·98 (0·09)	3·08 (0·11)	1·26 (0·03)	95·00 (0·69)	Mg no. 37·1
AW-13	PT (9)	48·84 (0·44)	2·74 (0·07)	18·64 (0·24)	8·89 (0·27)	0·23 (0·04)	2·84 (0·10)	7·64 (0·11)	5·89 (0·11)	3·00 (0·06)	1·30 (0·05)	96·08 (0·80)	Mg no. 36·3
AW-14	PT (7)	48·75 (0·49)	2·72 (0·08)	18·59 (0·15)	8·94 (0·34)	0·18 (0·04)	2·89 (0·08)	7·63 (0·13)	5·99 (0·12)	2·99 (0·07)	1·31 (0·06)	97·08 (0·72)	Mg no. 36·6
AW-17	PT (7)	49·29 (0·26)	2·63 (0·08)	18·75 (0·17)	8·43 (0·36)	0·21 (0·02)	2·85 (0·07)	7·59 (0·19)	5·94 (0·17)	3·03 (0·07)	1·29 (0·06)	97·52 (0·43)	Mg no. 37·7
AW-18	PT (7)	49·60 (0·41)	2·56 (0·14)	18·76 (0·28)	8·21 (0·16)	0·21 (0·04)	2·89 (0·11)	7·60 (0·08)	5·96 (0·15)	2·97 (0·09)	1·24 (0·05)	96·41 (0·51)	Mg no. 38·5
AW-19B	P (4)	53·74 (0·87)	1·53 (0·04)	19·46 (0·10)	4·05 (0·29)	0·15 (0·07)	1·03 (0·11)	4·21 (0·23)	8·29 (0·23)	5·87 (0·09)	1·68 (0·21)	96·63 (0·70)	Mg no. 31·1
	cpx (4)	45·09 (1·3)	3·45 (0·51)	7·75 (0·45)	8·01 (0·47)	0·34 (0·02)	11·19 (0·30)	21·17 (0·71)	1·16 (0·13)	0·26 (0·17)	n.d.	98·43 (0·65)	$En_{36}Wo_{49}Fe_{15}$
	plag (1)	55·16	1·06	22·61	3·8	0·1	1·64	9·72	5·07	1·26	n.d.	100·43	$An_{65}Al_{30}Or_5$
AW-20B	T (10)	55·30 (0·64)	1·72 (0·04)	19·86 (0·49)	4·25 (0·25)	0·20 (0·07)	1·30 (0·09)	4·15 (0·33)	8·05 (0·13)	4·74 (0·23)	0·44 (0·19)	94·76 (1·30)	Mg no. 35·3
	kaers (7)	37·38 (0·91)	6·33 (0·35)	12·68 (0·41)	10·32 (0·55)	0·23 (0·04)	12·12 (0·36)	11·33 (0·64)	2·76 (0·05)	1·18 (0·08)	0·12 (0·10)	94·48 (2·32)	Mg no. 67·7
AW-21	T (10)	53·93 (0·44)	2·01 (0·07)	20·07 (0·36)	4·79 (0·28)	0·20 (0·08)	1·56 (0·09)	5·80 (0·19)	7·36 (0·15)	3·80 (0·08)	0·48 (0·19)	94·24 (0·84)	Mg no. 36·7
	kaers (6)	36·75 (0·56)	5·96 (0·28)	13·58 (0·44)	10·92 (0·33)	0·23 (0·06)	11·97 (0·23)	11·92 (0·09)	2·64 (0·08)	1·09 (0·08)	0·25 (0·33)	95·35 (0·77)	Mg no. 66·1
AW-22	T (12)	52·51 (0·58)	2·37 (0·07)	19·69 (0·17)	5·23 (0·21)	0·22 (0·05)	2·24 (0·10)	7·12 (0·14)	6·57 (0·13)	3·47 (0·11)	0·59 (0·17)	92·11 (0·71)	Mg no. 43·3
	kaers (10)	37·56 (0·69)	5·27 (0·23)	13·00 (0·37)	10·15 (0·17)	0·16 (0·05)	13·22 (0·22)	12·16 (0·30)	2·57 (0·08)	1·09 (0·06)	0·22 (0·30)	95·42 (0·71)	Mg no. 69·9
AW-23	T (11)	51·91 (0·53)	2·57 (0·04)	19·62 (0·25)	5·41 (0·29)	0·20 (0·07)	2·52 (0·16)	7·46 (0·23)	6·37 (0·15)	3·32 (0·11)	0·62 (0·10)	90·83 (0·60)	Mg no. 45·4
AW-24B	PT (11)	49·77 (0·63)	2·85 (0·05)	18·97 (0·23)	7·36 (0·25)	0·19 (0·06)	3·00 (0·15)	7·47 (0·14)	6·25 (0·12)	2·95 (0·07)	1·18 (0·10)	93·35 (0·89)	Mg no. 42·1
AW-25	PT (5)	48·80 (0·38)	2·85 (0·04)	18·80 (0·44)	9·23 (0·06)	0·23 (0·04)	2·95 (0·03)	7·59 (0·08)	5·87 (0·17)	2·80 (0·12)	0·88 (0·04)	95·45 (0·45)	Mg no. 36·3
AW-26	PT (6)	48·54 (0·38)	2·80 (0·06)	18·49 (0·17)	9·34 (0·21)	0·23 (0·06)	2·81 (0·07)	7·58 (0·10)	5·87 (0·09)	3·03 (0·05)	1·31 (0·03)	96·49 (0·52)	Mg no 34·9
AW-27	PT (27)	48·54 (0·36)	2·87 (0·03)	18·95 (0·44)	9·52 (0·24)	0·25 (0·03)	2·91 (0·04)	7·62 (0·05)	5·72 (0·35)	2·83 (0·12)	0·79 (0·04)	93·53 (0·66)	Mg no. 35·2
AW-29	PT (12)	48·44 (0·49)	3·10 (0·10)	18·03 (0·56)	8·26 (0·53)	0·25 (0·09)	3·10 (0·17)	7·72 (0·31)	6·49 (0·14)	3·35 (0·18)	1·25 (0·12)	94·84 (0·76)	Mg no. 40·1
	plag (3)	51·84 (1·4)	1·37 (0·76)	24·80 (2·60)	4·14 (2·30)	0·10 (0·08)	1·07 (0·67)	9·72 (1·01)	4·83 (0·21)	1·40 (0·58)	n.d.	99·27 (0·61)	$An_{65}Al_{29}Or_6$
AW-33	PT (6)	49·07 (0·4)	2·61 (0·05)	18·41 (0·18)	8·63 (0·33)	0·24 (0·08)	2·74 (0·03)	7·70 (0·29)	6·18 (0·11)	3·19 (0·11)	1·20 (0·27)	94·66 (0·85)	Mg no. 36·2
AW-34	PT (6)	48·78 (0·79)	2·85 (0·05)	18·78 (0·15)	8·37 (0·29)	0·23 (0·05)	2·56 (0·08)	7·88 (0·09)	6·07 (0·12)	3·06 (0·05)	1·42 (0·08)	90·41 (1·05)	Mg no. 35·3
AW-37	PT (10)	48·39 (0·77)	3·02 (0·07)	18·64 (0·31)	8·76 (0·26)	0·25 (0·06)	2·95 (0·08)	7·69 (0·13)	5·94 (0·11)	3·05 (0·05)	1·30 (0·06)	96·86 (1·47)	Mg no. 37·5
AW-38	PT (6)	49·97 (0·51)	2·68 (0·07)	19·64 (0·21)	7·92 (0·13)	0·19 (0·07)	2·82 (0·03)	7·58 (0·12)	5·11 (0·26)	2·96 (0·06)	1·14 (0·06)	92·42 (0·98)	Mg no. 38·8
AW-39	PT (11)	51·30 (0·55)	2·40 (0·07)	19·95 (0·30)	6·49 (0·20)	0·19 (0·06)	2·70 (0·16)	7·47 (0·20)	5·65 (0·19)	3·12 (0·11)	0·71 (0·04)	94·25 (0·92)	Mg no. 42·6
AW-40	PT (11)	50·51 (0·62)	2·46 (0·07)	19·55 (0·28)	7·20 (0·11)	0·23 (0·05)	2·76 (0·10)	7·42 (0·17)	5·89 (0·11)	3·15 (0·06)	0·83 (0·06)	95·74 (1·23)	Mg no. 40·6
AW-41	T (12)	53·73 (0·82)	1·72 (0·05)	20·18 (0·38)	5·97 (0·14)	0·22 (0·06)	1·57 (0·07)	5·25 (0·14)	6·86 (0·14)	4·03 (0·08)	0·46 (0·03)	95·79 (1·40)	Mg no. 32·0
	kaers (5)	38·19 (0·3)	6·08 (0·21)	13·61 (0·17)	11·44 (0·31)	0·24 (0·02)	11·45 (0·26)	11·79 (0·33)	2·73 (0·05)	1·14 (0·04)	n.d. (0·00)	96·66 (0·25)	Mg no. 64·1
AW-42	T (9)	52·77 (0·46)	1·63 (0·05)	20·30 (0·30)	6·16 (0·24)	0·23 (0·07)	1·94 (0·03)	4·25 (0·11)	7·37 (0·21)	4·68 (0·07)	0·67 (0·06)	97·88 (0·95)	Mg no. 35·9
	kaers (2)	38·78 (0·8)	7·03 (0·36)	13·15 (0·12)	11·38 (0·35)	0·28 (0·05)	11·27 (0·09)	11·86 (0·40)	2·57 (0·08)	1·32 (0·04)	n.d.	97·63 (0·45)	Mg no. 63·8
	plag (4)	55·57 (2)	0·69 (0·34)	25·75 (1·28)	1·42 (0·61)	0·05 (0·03)	0·28 (0·13)	9·15 (1·42)	5·96 (0·25)	1·14 (0·27)	n.d.	100·03 (1·58)	$An_{60}Al_{36}Or_4$
AW-43	PT (1)	52·00	2	21·12	5·96	0·22	1·21	4·95	7·62	4·05	0·86	98·55	Mg no. 27·0
	cpx (4)	45·37 (0·66)	2·46 (0·50)	6·68 (1·09)	8·72 (0·40)	0·38 (0·02)	11·44 (0·46)	21·63 (0·71)	1·01 (0·12)	0·08 (0·01)	n.d.	97·77 (0·68)	$En_{36}Wo_{49}Fe_{15}$
	plag (3)	56·35 (1)	0·91 (0·58)	24·12 (1·13)	2·58 (1·43)	0·08 (0·05)	0·84 (0·68)	7·97 (0·50)	5·36 (0·32)	1·87 (0·13)	n.d.	100·09 (0·65)	$An_{57}Al_{35}Or_8$
AW-44	T (11)	54·57 (0·8)	1·63 (0·05)	20·59 (0·32)	5·29 (0·13)	0·17 (0·06)	1·28 (0·06)	4·71 (0·12)	7·07 (0·16)	4·30 (0·08)	0·38 (0·04)	96·74 (1·46)	Mg no. 30·2
	kaers (6)	38·20 (0·41)	5·92 (0·19)	13·58 (0·16)	12·34 (0·19)	0·28 (0·03)	11·92 (0·16)	11·44 (0·20)	2·75 (0·07)	1·27 (0·04)	0·21 (0·19)	97·93 (0·65)	Mg no. 63·3
AW-45	T (4)	53·03 (0·89)	1·91 (0·05)	20·69 (0·26)	5·55 (0·23)	0·21 (0·01)	1·68 (0·02)	6·34 (0·06)	6·95 (0·15)	3·37 (0·05)	0·26 (0·04)	92·94 (1·32)	Mg no. 35·1
	kaers (5)	37·03 (0·83)	5·06 (0·26)	14·44 (0·20)	12·11 (0·71)	0·24 (0·01)	11·42 (0·43)	11·91 (0·55)	2·73 (0·15)	1·21 (0·11)	n.d.	96·18 (0·67)	Mg no. 62·7
AW-46	B-TA (9)	52·98 (0·87)	2·18 (0·09)	20·49 (0·67)	5·54 (0·39)	0·20 (0·08)	2·00 (0·21)	7·10 (0·40)	5·68 (0·29)	3·16 (0·19)	0·66 (0·21)	93·74 (1·86)	Mg no. 39·2
	kaers (4)	38·54 (0·58)	4·58 (0·10)	14·07 (0·36)	10·64 (0·27)	0·21 (0·05)	13·28 (0·16)	12·23 (0·18)	2·66 (0·03)	1·15 (0·06)	n.d. (0·00)	97·41 (0·38)	Mg no. 69·0
AW-47	B-TA (10)	51·80 (0·2)	2·53 (0·04)	20·16 (0·26)	6·56 (0·12)	0·22 (0·07)	2·79 (0·07)	7·31 (0·07)	4·66 (0·23)	3·17 (0·07)	0·79 (0·05)	92·26 (0·43)	Mg no. 43·2

	Phase	SiO_2	TiO_2	Al_2O_3	FeO^T	MnO	MgO	CaO	Na_2O	K_2O	P_2O_5	Total	Mol %
KI-04 basanite experiments													
KI-06	B (3)	45·92 (0·21)	4·37 (0·02)	14·42 (0·11)	10·22 (0·05)	0·25 (0·04)	7·81 (0·18)	13·03 (0·31)	3·35 (0·32)	0·64 (0·01)	n.d.	90·87 (0·03)	Mg no. 57·7
	ol (5)	40·67 (0·08)	0·09 (0·01)	0·03 (0·01)	9·41 (0·25)	0·25 (0·02)	49·09 (0·23)	0·35 (0·01)	0·01 (0·01)	0·01 (0·00)	n.d.	99·93 (0·13)	Fo_{90}
KI-07	B (15)	44·61 (0·38)	4·37 (0·09)	14·41 (0·20)	10·61 (0·40)	0·17 (0·04)	7·69 (0·19)	11·55 (0·25)	3·93 (0·11)	1·74 (0·13)	0·92 (0·04)	96·88 (0·63)	Mg no. 56·4
	ol (3)	38·33 (1·35)	0·65 (0·45)	0·50 (0·38)	14·11 (2·45)	0·22 (0·01)	45·11 (2·17)	0·49 (0·10)	0·10 (0·02)	0·03 (0·01)	n.d.	99·75 (1·09)	Fo_{85}
KI-08	B (16)	44·04 (0·49)	4·54 (0·13)	14·46 (0·44)	11·51 (0·61)	0·16 (0·05)	7·11 (1·31)	11·59 (0·46)	3·90 (0·14)	1·79 (0·08)	0·91 (0·05)	98·24 (0·79)	Mg no. 52·4
	ol (5)	39·56 (0·52)	0·21 (0·13)	0·12 (0·11)	12·93 (0·55)	0·27 (0·02)	46·09 (0·59)	0·40 (0·03)	0·02 (0·01)	0·01 (0·00)	n.d.	99·63 (0·22)	Fo_{86}
KI-09	PT (9)	49·66 (0·37)	2·87 (0·06)	18·78 (0·37)	7·01 (0·38)	0·14 (0·05)	3·13 (0·09)	6·42 (0·14)	7·01 (0·14)	3·76 (0·08)	1·22 (0·11)	94·67 (0·72)	Mg no. 44·3
	cpx (4)	44·34 (0·82)	3·44 (0·37)	8·29 (1·03)	7·82 (0·57)	0·12 (0·04)	12·35 (0·76)	20·82 (0·73)	0·65 (0·08)	0·18 (0·12)	n.d.	98·06 (0·18)	$En_{39}Wo_{47}Fe_{14}$
KI-10	B (8)	47·27 (0·79)	3·65 (0·26)	16·32 (1·20)	7·96 (0·38)	0·20 (0·08)	5·89 (1·61)	9·28 (1·34)	5·44 (0·57)	2·60 (0·32)	1·39 (0·23)	92·05 (1·04)	Mg no. 56·9
	ol (3)	38·82 (0·18)	0·19 (0·07)	0·17 (0·14)	17·98 (0·40)	0·33 (0·01)	41·25 (0·49)	0·71 (0·28)	0·02 (0·00)	0·02 (0·00)	n.d.	99·51 (0·22)	Fo_{80}
	cpx (5)	43·08 (0·46)	4·38 (0·22)	9·98 (0·31)	8·46 (0·43)	0·12 (0·03)	11·14 (0·21)	20·96 (0·33)	0·82 (0·09)	0·18 (0·05)	n.d.	99·23 (0·45)	$En_{36}Wo_{49}Fe_{15}$
KI-11	B (5)	47·87 (0·26)	3·78 (0·06)	16·71 (0·24)	8·25 (0·22)	0·20 (0·07)	4·86 (0·20)	9·18 (0·09)	5·35 (0·22)	2·34 (0·13)	1·47 (0·09)	92·99 (0·25)	Mg no. 51·2
	kaers (5)	37·68 (1·47)	5·75 (0·38)	12·08 (0·15)	11·79 (2·45)	0·11 (0·03)	15·25 (0·31)	11·66 (0·70)	2·51 (0·07)	1·07 (0·06)	n.d.	97·97 (1·22)	Mg no. 70·0
KI-12	B (11)	45·59 (0·48)	4·54 (0·08)	16·01 (0·15)	8·54 (0·30)	0·16 (0·11)	5·89 (0·34)	11·32 (0·38)	4·88 (0·13)	1·90 (0·20)	1·17 (0·11)	91·56 (0·98)	Mg no. 55·2
	ol (7)	38·65 (0·32)	0·14 (0·04)	0·09 (0·06)	11·64 (0·16)	0·25 (0·03)	47·16 (0·50)	0·43 (0·23)	0·05 (0·02)	0·01 (0·01)	n.d.	98·48 (0·36)	Fo_{88}
	cpx (4)	44·23 (0·98)	3·05 (0·28)	5·92 (0·47)	6·95 (0·49)	0·12 (0·01)	14·45 (0·48)	22·75 (0·14)	0·65 (0·12)	0·07 (0·01)	n.d.	98·26 (0·52)	$En_{42}Wo_{47}Fe_{11}$
KI-14	B (7)	44·62 (0·39)	5·11 (0·06)	15·75 (0·28)	10·25 (0·44)	0·21 (0·13)	5·58 (0·17)	10·56 (0·14)	4·68 (0·13)	2·20 (0·12)	1·03 (0·09)	96·57 (1·00)	Mg no. 49·3
	ol (4)	38·05 (0·16)	0·16 (0·02)	0·09 (0·06)	20·19 (0·31)	0·32 (0·05)	40·21 (0·47)	0·53 (0·13)	0·05 (0·04)	0·03 (0·01)	n.d.	99·71 (0·26)	Fo_{78}
KI-15	B (15)	43·55 (0·37)	4·94 (0·08)	14·37 (0·24)	11·28 (0·32)	0·19 (0·06)	6·73 (0·22)	12·11 (0·28)	4·06 (0·12)	1·81 (0·14)	0·97 (0·09)	93·70 (0·69)	Mg no. 51·5
	ol (8)	38·72 (0·13)	0·11 (0·02)	0·04 (0·02)	16·69 (0·14)	0·26 (0·06)	42·84 (0·27)	0·41 (0·02)	0·01 (0·01)	0·01 (0·01)	n.d.	99·17 (0·29)	Fo_{82}
KI-16	B (15)	43·81 (0·30)	4·83 (0·09)	14·35 (0·23)	11·52 (0·23)	0·19 (0·12)	6·73 (0·30)	12·20 (0·48)	4·03 (0·12)	1·40 (0·20)	0·95 (0·08)	94·20 (0·59)	Mg no. 51·0
	ol (8)	39·25 (0·27)	0·13 (0·03)	0·07 (0·03)	15·27 (0·08)	0·26 (0·03)	43·90 (0·23)	0·40 (0·03)	0·07 (0·02)	0·02 (0·01)	n.d.	99·42 (0·21)	Fo_{84}
KI-17	B (1)	44·80	4·43	14·88	11·17	0·2	6·68	11·66	3·84	1·73	0·57	95·16	Mg no. 51·6
	ol (4)	39·42 (0·52)	0·11 (0·02)	0·05 (0·02)	13·51 (0·29)	0·24 (0·03)	45·76 (0·27)	0·33 (0·01)	0·02 (0·02)	0·00 (0·01)	n.d.	99·58 (0·92)	Fo_{86}
KI-20	B (5)	45·05 (0·41)	4·44 (0·10)	14·40 (0·11)	11·23 (0·17)	0·21 (0·02)	7·74 (0·60)	12·31 (0·88)	3·11 (0·34)	0·91 (0·27)	0·60 (0·04)	92·55 (0·61)	Mg no. 55·1
KI-21	B (8)	43·60 (0·36)	4·69 (0·07)	14·17 (0·09)	11·77 (0·23)	0·15 (0·04)	8·03 (0·09)	11·38 (0·19)	3·62 (0·05)	1·66 (0·11)	0·93 (0·07)	95·28 (0·69)	Mg no. 54·9
	ol (4)	39·00 (0·44)	0·12 (0·03)	0·05 (0·03)	14·38 (0·33)	0·23 (0·02)	46·75 (0·97)	0·34 (0·03)	0·02 (0·02)	0·00 (0·01)	n.d.	101·01 (0·68)	Fo_{85}
KI-22	B (10)	43·42 (0·43)	4·74 (0·10)	14·46 (0·21)	11·71 (0·34)	0·20 (0·05)	7·32 (0·25)	11·60 (0·13)	3·81 (0·09)	1·82 (0·06)	0·92 (0·06)	96·02 (1·19)	Mg no. 52·7
	ol (8)	39·07 (0·45)	0·12 (0·09)	0·13 (0·23)	15·90 (0·83)	0·26 (0·04)	45·24 (0·97)	0·52 (0·46)	0·03 (0·03)	0·01 (0·01)	n.d.	101·36 (0·72)	Fo_{83}
KI-23	B (10)	43·63 (0·61)	4·95 (0·08)	15·04 (0·30)	11·26 (0·31)	0·19 (0·05)	6·49 (0·12)	11·49 (0·12)	4·11 (0·07)	1·88 (0·04)	0·96 (0·03)	97·21 (1·30)	Mg no. 50·7
	ol (5)	38·71 (0·37)	0·13 (0·01)	0·05 (0·01)	18·74 (0·19)	0·26 (0·04)	43·25 (0·23)	0·51 (0·01)	0·01 (0·01)	0·01 (0·01)	n.d.	101·74 (0·69)	Fo_{80}
	cpx (6)	43·57 (1·60)	4·44 (0·54)	11·33 (1·99)	8·29 (1·21)	0·17 (0·01)	9·82 (1·89)	17·67 (2·74)	1·60 (0·78)	0·79 (0·48)	n.d.	97·96 (1·25)	$En_{36}Wo_{47}Fe_{17}$
KI-26	B (6)	43·76 (0·58)	4·48 (0·11)	13·75 (0·19)	11·45 (0·49)	0·16 (0·07)	9·33 (1·38)	11·04 (1·59)	3·49 (0·17)	1·64 (0·68)	0·90 (0·04)	96·65 (0·35)	Mg no. 59·2
KI-27	B (6)	43·48 (0·63)	4·48 (0·04)	14·63 (0·21)	10·79 (0·17)	0·15 (0·05)	9·06 (0·15)	11·74 (0·11)	4·09 (0·09)	1·57 (0·15)	n.d.	92·24 (0·82)	Mg no. 59·9
KI-28	B (6)	44·61 (0·41)	4·55 (0·05)	14·25 (0·16)	9·89 (0·11)	0·21 (0·04)	8·98 (0·07)	11·40 (0·08)	3·76 (0·12)	1·69 (0·11)	0·66 (0·02)	95·99 (0·33)	Mg no. 61·8
KI-31	PT (3)	51·87 (0·66)	2·68 (0·16)	18·59 (0·33)	8·22 (0·31)	0·23 (0·05)	2·98 (0·08)	7·70 (0·05)	5·19 (0·13)	2·55 (0·01)	n.d.	92·06 (0·66)	Mg no. 39·2
	kaers (8)	39·82 (0·73)	4·88 (0·17)	11·94 (0·49)	11·05 (0·45)	0·15 (0·05)	14·04 (0·40)	11·58 (0·14)	2·64 (0·06)	1·21 (0·05)	0·09 (0·10)	97·49 (0·44)	Mg no. 69·4
KI-32	PT (7)	49·92 (0·75)	3·22 (0·30)	17·06 (0·57)	8·94 (0·53)	0·23 (0·01)	4·43 (1·30)	9·14 (0·41)	4·19 (0·36)	2·18 (0·20)	0·69 (0·09)	91·46 (1·71)	Mg no. 46·9
	kaers (6)	39·67 (0·78)	4·04 (0·19)	11·91 (0·42)	10·45 (1·33)	0·18 (0·02)	14·94 (1·24)	11·84 (0·12)	2·58 (0·07)	1·24 (0·13)	0·15 (0·17)	97·09 (0·50)	Mg no. 71·7
KI-33	TB (7)	47·39 (0·70)	3·69 (0·06)	16·09 (0·20)	9·81 (0·32)	0·18 (0·05)	5·57 (0·27)	9·99 (0·15)	3·83 (0·14)	2·18 (0·09)	1·26 (0·06)	89·81 (1·41)	Mg no. 50·3
	kaers (4)	39·82 (0·82)	4·38 (0·07)	11·80 (0·56)	10·59 (0·35)	0·14 (0·04)	14·79 (0·31)	11·77 (0·27)	2·60 (0·05)	1·25 (0·05)	0·12 (0·19)	97·37 (0·20)	Mg no. 71·3
KI-36	B (10)	43·75 (0·61)	4·70 (0·08)	14·24 (0·28)	11·33 (0·27)	0·18 (0·08)	8·21 (0·31)	11·46 (0·35)	3·56 (0·10)	1·66 (0·11)	0·91 (0·07)	94·04 (1·40)	Mg no. 56·4
	ol (13)	39·22 (0·21)	0·09 (0·02)	0·04 (0·01)	15·30 (0·35)	0·24 (0·03)	45·80 (0·28)	0·33 (0·02)	0·02 (0·01)	0·01 (0·01)	n.d.	101·17 (0·57)	Fo_{84}
KI-37	ol (10)	39·72 (0·35)	0·09 (0·02)	0·04 (0·01)	14·17 (0·43)	0·22 (0·04)	46·99 (0·41)	0·28 (0·02)	0·01 (0·01)	0·01 (0·01)	n.d.	101·70 (0·66)	Fo_{86}

Analyses given in wt % and normalized to 100% with original analysis total given. Numbers in parentheses next to analyses are standard deviations; those next to phase names are number of analyses. n.d., not determined. Reported values for glasses are average compositions for each charge. FeO^T, total iron given as FeO. Glass name classifications: B, basanite; PT, phonotephrite; T, tephriphonolite; P, phonolite; B-TA, basaltic trachyandesite; TB, trachybasalt. Mg no. calculated as molar $MgO/(FeO^T + MgO)$.

Experimental olivine compositions in KI range from Fo_{77} to Fo_{90} (Table 4). No olivine crystals were observed in AW experimental samples. Experimental olivine Fo contents become more forsteritic with increasing H_2O^{melt} (Fig. 6), reflecting the increase in the melt fraction of the charge and a change in oxygen fugacity caused by an increase in the water activity in the melt (e.g. Feig *et al.*, 2010). Crystal–melt Fe–Mg partition coefficients (K_D) for most experimental olivines range from 0·33 to 0·39, with an overall average value (including outliers) of 0·33. Our olivines are expected to have a higher K_D value than the canonical 0·3 owing to the alkalinity and redox state of the charges. Using the model of Carmichael & Ghiorso (1990), which accounts for the effect of melt composition on K_D, we calculate that our experimental K_D values should range between 0·34 and 0·36, suggesting that the majority of the olivine crystals in our experiments are approaching equilibrium with their host melts. Four outliers, which do not appear to have attained equilibrium, are samples KI-06 (K_D = 0·22), KI-08 (0·22), KI-10 (0·52), and KI-36 (0·43). In samples KI-06 and KI-10, olivines were rare and small, suggesting that they grew late during the experiment, possibly during quench. Some olivines may have been affected by Fe loss to the experimental capsule, which is difficult to avoid at high T even given the relatively short run times of 24 h.

Amphibole is classified as sodic-kaersutite on the basis of 23 oxygens and crystallized in both phonotephrite and basanite experiments (compositions reported in Table 4). Al contents in kaersutites show a positive correlation with pressure, as documented for other systems (e.g.Prouteau & Scaillet, 2003) but in contrast to the results of experiments with Erebus phonolite (Moussallam *et al.*, 2013). Kaersutites synthesized at the lowest fO_2 conditions have markedly lower Mg# and higher Ti contents. At all pressures, the modal proportion of amphibole in AW runs increases as the water content of the melt decreases.

Plagioclase feldspar was observed only in the driest AW experiments and ranges from An_{57} to An_{65}. It was not found in any of the KI experiments.

5.4 Glass compositions

The compositions of residual experimental glasses highlight the evolution of the melts from the starting composition to intermediate and evolved compositions. FeO, MgO, CaO, and TiO_2 all show continuously decreasing trends with melt evolution reflecting the crystallization of pyroxene or kaersutite, olivine, plagioclase, and titanomagnetite. In contrast, Al_2O_3, and K_2O show continuously increasing trends. In detail, Harker-type plots clearly show the effect of the fractionating assemblage on the liquid lines of descent of each starting material (Figs 7 and 8).

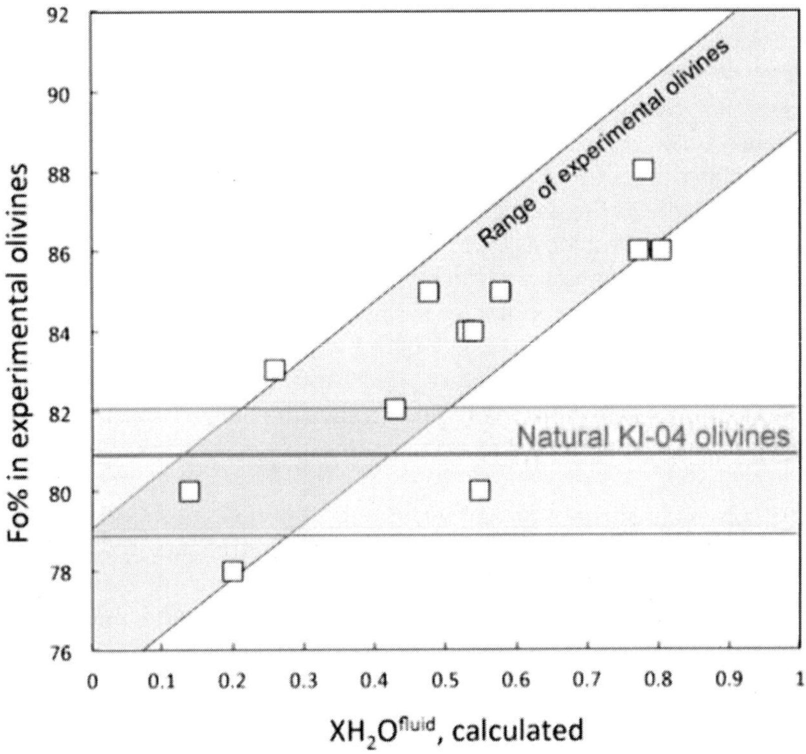

Figure 6. Chemical compositions of experimental olivines in terms of Fo content versus the calculated equilibriumXH_2O^{fluid}, illustrating a general positive correlation between the two variables. The Fo content of natural KI-04 olivines is indicated by the horizontal red line. The gray shaded region represents the average range of experimental Fo contents in olivines and respectiveXH_2O^{fluid} values. The intersection of this region with the red line thus indicates the range of expected XH_2O^{fluid} values for the natural KI-04 magma based solely on olivine composition at the investigated pressures (200–400 MPa).

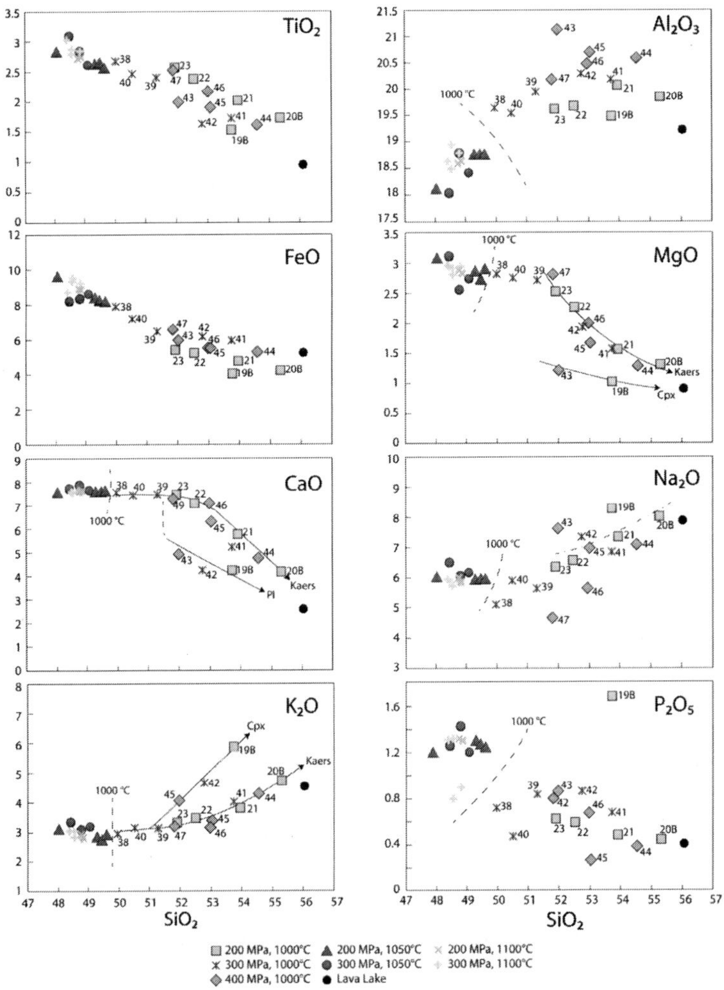

Figure 7. Harker variation diagrams for AW experimental glasses illustrating the production of a phonolite magma resembling that in the Erebus lava lake [the composition used byMoussallam *et al.* (2013); black dot] by crystallization of an AW-type magma. The liquids that best approximate the lava lake composition are the driest ones at 1000°C (<1·7 wt %) for all pressures. The closest experimental analogues are charges AW-19B and AW-20B at 200 MPa. Lower than natural FeO concentrations in these two runs could indicate that the experiments were too oxidized relative to the natural magmas. High experimental CaO contents may indicate that the production of lava lake phonolite requires lower temperatures, possibly around 975°C. The roles of kaersutite and Cpx ± Pl crystallization are clearly defined by two distinct evolutionary trends.

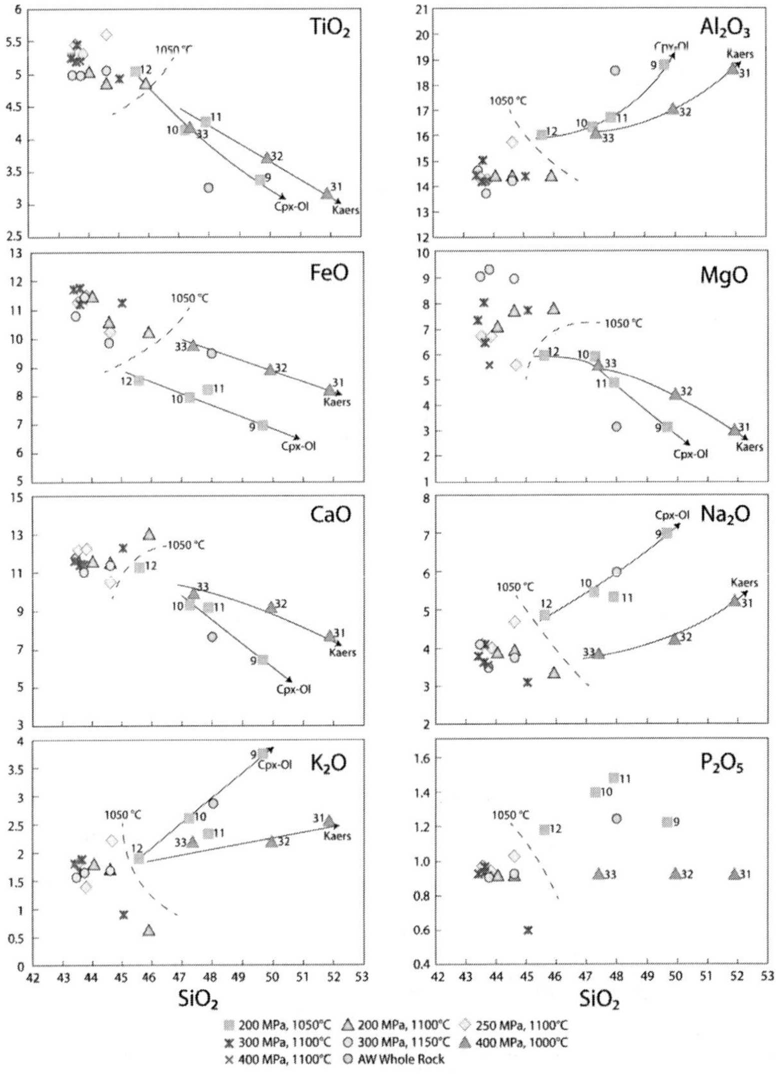

Figure 8. Harker variation diagrams for KI experimental glasses illustrating the production of intermediate AW-like magmas (represented here by the AW whole-rock; blue dot) from a KI-type parent. At low *T*, two evolutionary trends are apparent in which either Cpx ± Pl or kaersutite amphibole is the dominant fractionating phase. An AW-type daughter liquid is bracketed for several oxides by experiments KI-09 and KI-10, thus constraining the melt water content reached at the stage of AW magma production to between 0·7 and 1·5 wt %. FeO is too low in the experimental charges relative to natural AW whole-rocks, indicating that the experiments were more oxidized than the natural magmas.

KI-related melts in charges held above 1050°C show limited compositional evolution relative to the starting composition, except for FeO and MgO contents, which are affected by olivine and titanomagnetite crystallization to some extent. In contrast, charges held at or below 1050°C show the prominent roles of either kaersutite or Cpx–Ol-dominated assemblages. In both cases, derivative liquids produced by crystallization at these temperatures reach SiO_2 contents of 50–52 wt % (i.e. are more evolved than the AW starting composition), trending toward the phonotephritic to phonolitic fields. Charges in which kaersutite is stable are characterized by higher TiO_2, FeO, and CaO melt contents relative to those in which Cpx and Ol are stable (at the same SiO_2 content). In contrast, the crystallization of a Cpx ± Ol assemblage produces liquids with higher Al_2O_3, Na_2O, and K_2O contents relative to those in which kaersutite predominates. It is worth noting that the role of kaersutite increases at high pressure; charges crystallized at 400 MPa have up to 60 wt % amphibole, whereas at 200 MPa, amphibole does not exceed 37 wt % of the crystal cumulate (Table 2).

The relationship between the evolution of CaO and MgO in KI liquids is illustrated in Fig. 9. Such a plot is useful for illuminating the evolution of basaltic compositions *sensu lato* in which olivine and Ca-pyroxene (or amphibole) play a dominant role (e.g. Pichavant *et al.*, 2009, 2014). The KI trend can be divided into two segments: the first runs broadly parallel to the MgO axis, primarily reflecting depletion in MgO of the liquid owing to olivine crystallization. The second segment corresponds to liquids in which either kaersutite or Cpx ± Ol crystallize, defining either a kaersutite control line or the Cpx–Ol cotectic (at 1050°C), along which both CaO and MgO contents decrease. It is apparent that liquids lying on the Cpx–Ol cotectic are slightly less CaO-rich than those driven by kaersutite crystallization. Also shown in Fig. 9 are the Cpx–Ol cotectics experimentally determined for Stromboli and Vesuvius primitive basalts (Di Carlo *et al.*, 2006; Pichavant *et al.*, 2009, 2014), which, like the magma compositions investigated here, are mafic and alkali-rich. It is clear that basaltic liquids saturated in Cpx ± Ol and kaersutite define a common general and relatively narrow trend in such a diagram, despite the compositional differences existing between the three bulk-rock compositions used to establish the cotectics. Significant departure of a Cpx ± Ol-bearing mafic rock composition from this trend might indicate that accumulation or contamination processes of some sort have occurred.

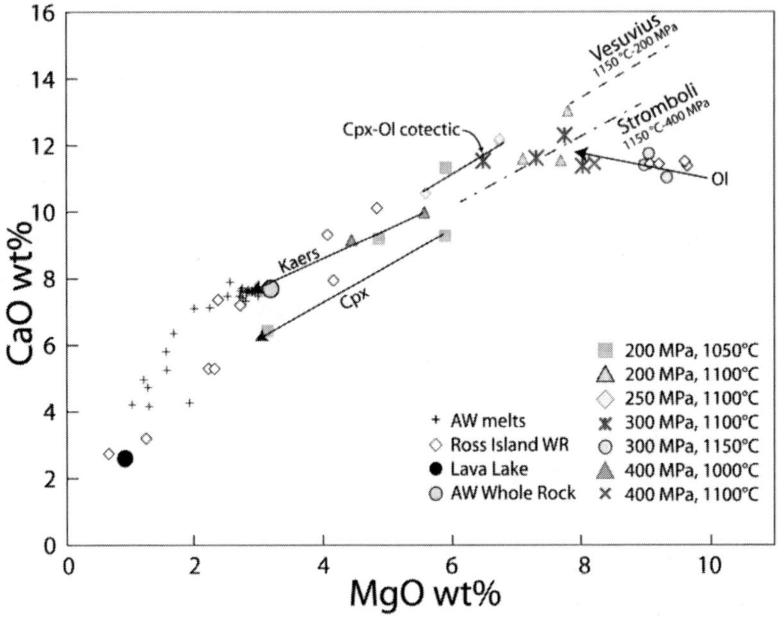

Figure 9. CaO vs MgO plot showing the compositions of cotectic and near-cotectic experimental glasses, which mimic natural whole-rock compositions from a suite of Ross Island lavas (open diamonds), useful in identifying trends in compositions dominated by the precipitation of olivine and Ca-pyroxene (or amphibole). KI runs are shown with different symbols in terms of *P–T* conditions. The Cpx–Ol cotectic at 1050°C is shown for KI melts. Also plotted are experimentally determined Cpx–Ol cotectics for melts from Stromboli (Pichavant*et al.*, 2009) and Vesuvius (Pichavant *et al.*, 2014). Despite compositional differences in the starting materials used in these three studies, mafic alkali-rich liquids saturated in Cpx ± Ol and kaersutite define a relatively narrow evolutionary trend in this diagram. KI melts are marked by initial Ol crystallization, followed by either Cpx ± Pl or Kaers.

AW-related charges produced liquids compositionally different from the starting material mostly at 1000°C. Liquids produced at higher temperatures cluster broadly around the bulk-rock AW composition. As observed for KI, two distinct trends appear depending on whether kaersutite or Cpx ± Pl is the dominant fractionating silicate. In both cases, the most evolved compositions reach the phonolitic field with SiO_2 in excess of 52–53 wt %. Kaersutite-saturated liquids in particular have higher

MgO and CaO contents, whereas liquids crystallizing Cpx ± Pl have higher K_2O (at the same SiO_2 content). For most major elements in AW experiments, there is no apparent effect of pressure, liquid compositions being governed mostly by variations in *T* and H_2O. The sole exception is Al_2O_3: liquids produced at 400 MPa are distinctly richer in Al_2O_3 relative to those at 200 MPa, the former having >20 wt % Al_2O_3 whereas at 200 MPa the majority of liquids have less than 20 wt %. This reflects in part the fact that, for the AW composition, kaersutite crystallization depletes the liquid less in Al_2O_3 relative to Cpx and Pl.

6. DISCUSSION

6.1 Pre-eruptive conditions of KI and AW magmas and the production of the phonolite lava lake

Overall, the experimental mineral phases reproduce those observed in the natural rocks, showing that our experiments correctly capture the natural system in terms of fO_2–T–H_2O conditions of evolution. In detail, experiments best reproducing the Ol composition of KI (Fo_{78-80}) are at around 1100°C and with less than 1·5 wt % dissolved H_2O (corresponding to $XH_2O^{fluid} < 0{\cdot}2$; charges KI-10, KI-14 and KI-23, Table 4), in the pressure range 200–300 MPa. Higher melt water contents would yield olivine that is too magnesian. The natural DVDP clinopyroxene composition was best reproduced at 200 MPa and 1050°C as shown in Fig. 10 (sample KI-12; $En_{42}Wo_{47}Fs_{11}$). Other experimental pyroxenes are more iron-rich (Fs_{14-17}) than those in the natural samples, although the most Fe-rich clinopyroxenes measured (KI-23) are possibly quench crystals, as indicated by high variability in Ca and Na, and high K. It should be noted that the experimental clinopyroxenes are in general rich in Al_2O_3 and Na_2O, consistent with those in the natural rocks (Kyle, 1981). This Al- and Na-rich character of Cpx does not therefore reflect crystallization at higher pressure (e.g. Kyle, 1981) but instead the Al- and Na-rich character of the melt from which the pyroxenes grew. Experiments between NNO + 1·5 and NNO + 2 produced kaersutite that most closely resembles the natural amphiboles in terms of Ti content (~5·5–6·5 wt % TiO_2) and average Mg# (65), although the natural amphiboles have a significant spread in Mg# (56–72) (Fig. 11). Most natural DVDP amphiboles contain 12·5–14 wt % Al_2O_3 (Kyle, 1981), resembling the compositional range of experimental AW amphiboles. Given the correlation between kaersutite Al contents and pressure (Fig. 12), this suggests that significant DVDP differentiation [in which the role of amphibole is important (Kyle, 1981)] occurred at 200–400 MPa. The An contents in experimental phonotephrite feldspars range from An_{57} to An_{65} and closely mimic the compositions of natural phonotephrite feldspars (Fig. 13). The fact that the Or contents of experimental plagioclases are higher than those of the natural rocks is probably due to contamination of the microprobe analyses by surrounding glass.

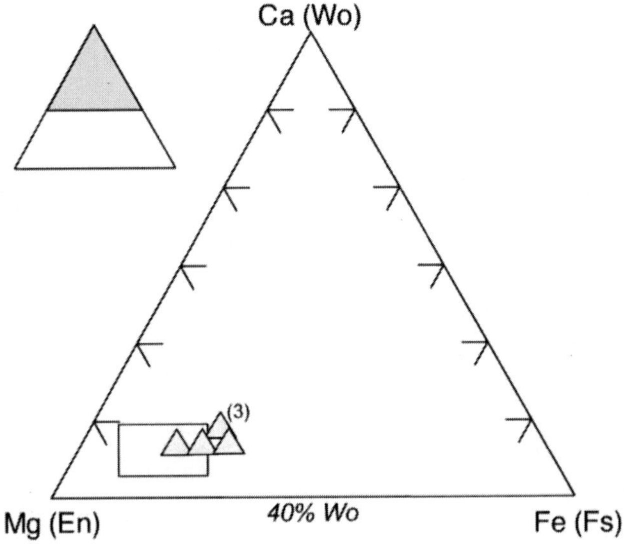

Figure 10. Enlarged ternary diagram (from 40% Wo) showing the compositions of experimental (symbols) and natural (yellow box) clinopyroxenes, plotted in terms of recalculated end-member compositions enstatite, ferrosilite and wollastonite. The number (3) indicates three samples with the same composition.

The experiments allow us to constrain possible petrogenetic paths for magma production at Erebus. We first consider the production of intermediate, AW-like, magmas from a KI-type parent. To this end, the AW bulk-rock has been plotted in Fig. 8 displaying KI-related liquids. As can be seen here, an AW-type liquid (48 wt % SiO_2) is bracketed for several oxides by charges run at 1050°C and 200 MPa (between KI-09 and KI-10), in particular for Na_2O, K_2O and CaO, and to a lesser extent Al_2O_3 and TiO_2. FeO, which is too low in the experiments (8 wt % vs 9·4 wt %), does not fit these trends, suggesting therefore that the experimental fO_2 is slightly too oxidized compared with nature. The corresponding melt water content reached at the stage of AW magma production is thus bracketed by that of those two charges, between 0·7 and 1·5 wt % (Table 2).

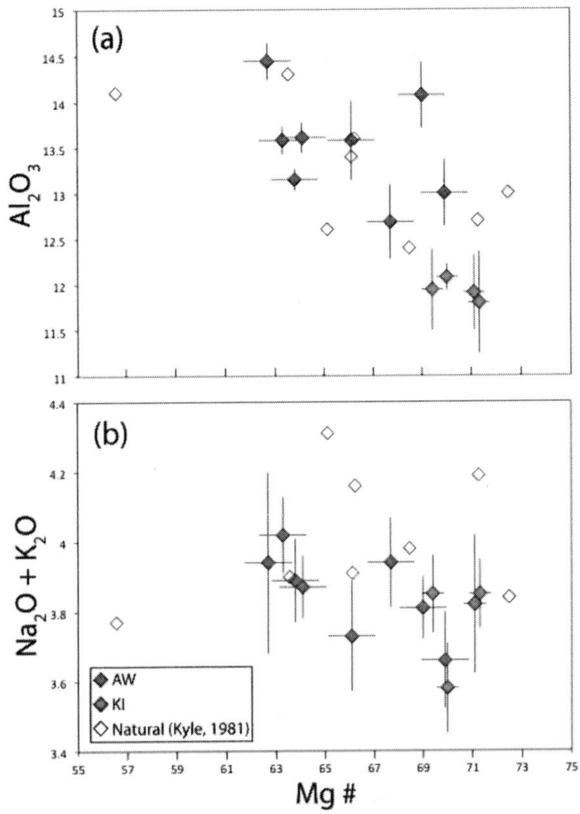

Figure 11. Chemical compositions of experimental (filled diamonds) and natural (open diamonds) kaersutite amphibole, plotted in terms of magnesium number versus Al_2O_3 (a) and Na + K (a). Natural kaersutites are from intermediate and evolved DVDP lavas fromKyle (1981). Error bars represent 1SD.

We next evaluate the conditions for the production of a phonolite magma similar to that of the lava lake by crystallization of an AW-type magma (Fig. 7). The experimental AW liquids define a trend toward the lava lake composition [taken as the composition used by Moussallam *et al.* (2013)], although that end-member is not bracketed by the experimental results. At all pressures, the liquids best approaching the lava lake are the driest ones at 1000°C (<1·7 wt %). In detail, the liquids obtained at 400 MPa are too Al_2O_3-rich relative to that of the lava lake, suggesting this is too high a pressure for phonolite production in the Erebus plumbing system. In contrast, the two driest liquids at 200 MPa (AW-19B and AW-20B) are close synthetic analogues for the lava lake, charge AW-19B having the observed mineralogy of the lava lake. Notably, the TiO_2 trend falls short, with experimental liquids being 0·5 wt % too rich in TiO_2 relative to the lava lake, suggesting a lower temperature for the natural system. FeO concentrations in those two driest 200 MPa charges are slightly below that of the lava lake, which could be taken to indicate that the experimentalfO_2 is slightly too high, as inferred above for the production of AW magmas. The MgO and CaO trends clearly distinguish the role of kaersutite versus that of Cpx ± Pl, suggesting again the latter assemblage as the dominant one during differentiation of the EL. It is in CaO content that the difference between the experimental liquids and the lava lake is greatest. Inspection of the experimental CaO trends suggests that the production of a magma with a CaO content similar to that of the lava lake requires temperatures below 1000°C, possibly around 975°C. This is in agreement with the work of Moussallam *et al.* (2013), who experimentally determined the liquidus of the phonolite lava lake under nearly dry conditions at 975–1000°C.

In summary, KI near-liquidus conditions are inferred to be around 1100°C and 1 ± 0·5 wt % H_2O at 200–300 MPa. Production of intermediate magmas at Erebus (AW-type) from a KI-type parent requires a temperature drop to around 1050°C at a pressure of 200 MPa, reaching slightly higher melt water contents but still below 1·5 wt %. Redox conditions appear to be moderately oxidized at this stage, possibly between those defined by the quartz–fayalite–magnetite (QFM) and NNO solid buffers, or just slightly above NNO. AW-type magma may in turn crystallize down to 975–1000°C, possibly in the pressure range 100–200 MPa, yielding a phonolitic liquid similar to that sustaining the lava lake, which is characterized by very low water contents and reduced conditions. As shown by Moussallam *et al.*(2014), the transition between moderately oxidized conditions in the crustal magma reservoir and the more reduced conditions typifying the lava lake may result from degassing processes involving melt–COHS fluid equilibrium (see below). Although the temperature of the lava lake has been inferred to lie somewhere between 950 and 1080°C (see Introduction), the present study, in combination with that of Moussallam*et al* (2013), consistently favors the lower bound of that temperature range.

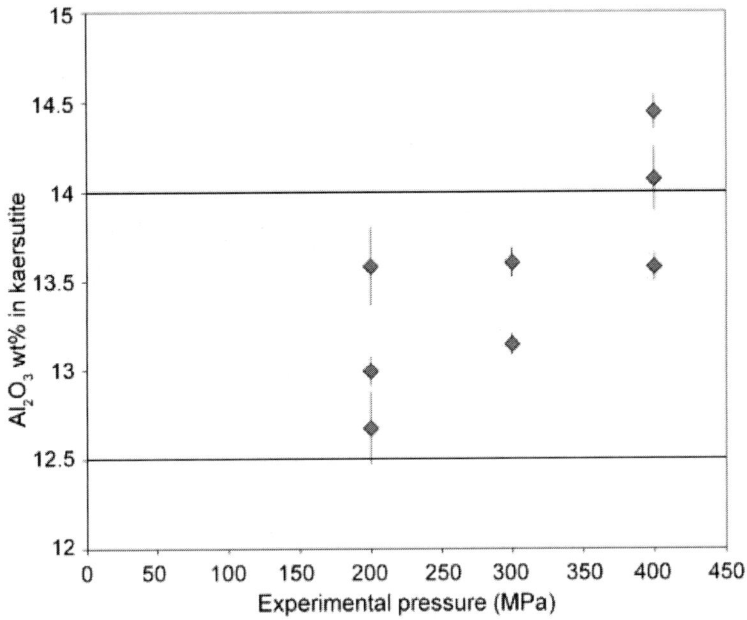

Figure 12. Compositions of experimental AW kaersutite illustrating the positive correlation between pressure and Al_2O_3 content. The majority of natural DVDP amphiboles have 12·5–14 wt % Al_2O_3 (grey box), suggesting that significant fractionation of DVDP magmas took place between 200 and 400 MPa. Vertical error bars represent 1SD. Horizontal error bars (±20 bar) are smaller than symbols.

6.2 Differences between EL and DVDP magmatic evolution

Although neither experimental series completely reproduced the phase assemblages of the natural Erebus starting lavas (i.e. occurrence of kaersutite in EL experiments whereas it is not present in the corresponding rocks), our results shed light on critical aspects of the Erebus system and elucidate key differences in the formation conditions of the two natural lava suites. Plagioclase is confined to very low H_2O–high CO_2 conditions, suggesting maximum melt H_2O concentrations of ~1–2 wt % for both EL and DVDP lava suites at the *P–T* conditions explored, consistent with melt inclusion volatile contents (Oppenheimer *et al.*, 2011). The presence of kaersutite in all except the most primitive natural DVDP lavas suggests higher water activities for the DVDP suite compared with the EL, with

XH_2O^{fluid} around 0·2, indicated by the co-precipitation of plagioclase and kaersutite in phonotephrite experiments. The peritectic-like relationship inferred between olivine (± clinopyroxene) and kaersutite allows us to place constraints on the relative temperatures of the EL and DVDP suites. All EL lavas plus DVDP basanites (the most primitive of which probably represents the parent melt for both suites) contain olivine and lack kaersutite. Conversely, all other DVDP lavas lack olivine and contain kaersutite. This is consistent with EL and primitive DVDP lavas being thermally buffered during the early and intermediate stages of differentiation to above 1050°C, with intermediate and evolved DVDP lavas undergoing a drop in equilibration temperature to below 1050°C.

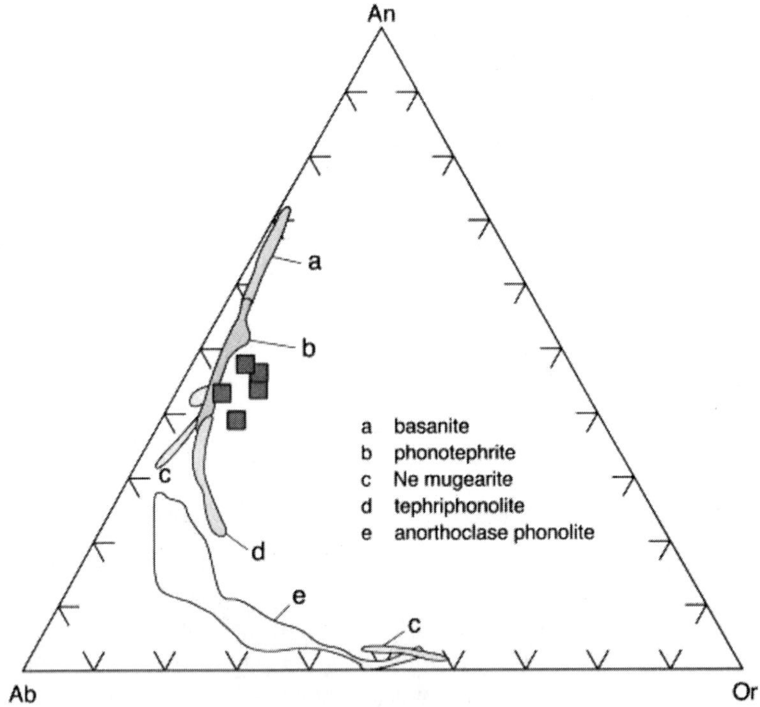

Figure 13. Ternary diagram showing the compositions of experimental plagioclase (purple squares) and natural plagioclase from EL lavas [shaded fields; data fromKyle *et al.* (1992)]. It should be noted that all experimental plagioclase crystals were very small and so their chemical analyses via EMP may be contaminated by the surrounding glass.

The chemical compositions of both experimental series (squares and diamonds) and natural EL melt inclusions (shaded fields) are shown in variation diagrams of their major oxides versus SiO_2 in Fig. 14. Because of the rarity of melt pools large enough to be measured in the DVDP lavas, no residual melt compositions exist in the literature.

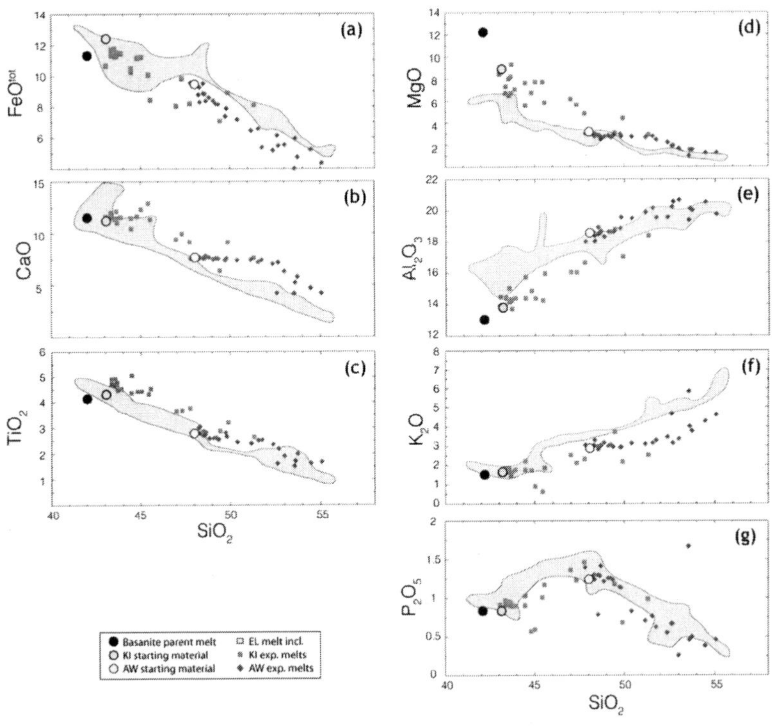

Figure 14. Harker variation diagrams showing the compositions of experimental residual glasses (KI-04, red squares; AW-82038, blue diamonds) compared with compositions of olivine-hosted melt inclusions in EL lavas [green fields; data fromOppenheimer *et al.* (2011)]. The parent basanite composition is shown as a black dot, and compositions of starting materials used in this study are shown as pale red (basanite KI-04) and pale blue (phonotephrite AW-82038) dots.

Many of the experimental glass compositions match the natural trends fairly well, but some are clearly offset, indicating a difference in natural versus experimental P–T–H_2O–fO_2 conditions for those charges.

Mismatch between natural and experimental glass compositions is notable in particular for CaO and MgO, and to a lesser extent for Al_2O_3. Higher than observed CaO values suggest insufficient fractionation of plagioclase and/or calcic clinopyroxene in the experiments compared with the natural lavas. The lack of plagioclase in the KI experiments and moderate plagioclase crystallization in the AW experiments may indicate that the liquid line of descent we have established experimentally is either hotter or, more probably, wetter than that of the natural system. Deep and intermediate magmas are not expected to be significantly cooler than the temperature of the surficial lava lake, which provides a thermal lower bound of ~975–1000°C (Kyle, 1977; Caldwell & Kyle, 1994; Dunbar *et al.*, 1994; Moussallam *et al.*, 2013). It is thus unlikely that significantly lower temperatures are responsible for enhanced calcic plagioclase crystallization in natural samples. For this reason, we suggest that the Ross Island magmas differentiated under lower water activities in our experiments, significantly constraining the natural XH_2O^{fluid} conditions of the DVDP and EL series to ≪≪ 0·25. The elevated MgO contents in the KI experiments relative to natural melt inclusions may reflect the lack of a cardinal phase consuming this element, in particular olivine, which in turn could be controlled by oxygen fugacity. Low experimental Al_2O_3 contents in KI melts may reflect insufficient plagioclase crystallization in addition to illustrating the dominant role of kaersutite fractionation in the DVDP magmas compared with EL melts, whose evolution is defined instead by Cpx fractionation.

Trends in the K_2O and P_2O_5 contents of natural EL glasses are distinct from those of EL and DVDP whole-rocks (Figs 2 and 14). Experimental K_2O melt contents mimic those of the whole-rock samples in that they are depleted relative to EL melt inclusions, probably reflecting the dominant role of kaersutite in our experiments and in natural DVDP rocks. The trends in P_2O_5 content illustrate well the onset of apatite saturation in experimental charges and natural EL melt inclusions. In EL lavas, apatite saturation occurs when the melts have differentiated such that the silica content is ≥48–49 wt %, after which P_2O_5 decreases with increasing SiO_2(Fig. 14g), in agreement with existing solubility models (Harrison & Watson, 1984). This suggests a hotter temperature of *c.* >1100°C for the primitive and intermediate EL magmas, and a lower temperature (indicated by the onset of apatite saturation) for the evolved lavas.

6.3 Redox constraints on Ross Island magmatism

The presence of kaersutite in our experimental samples is largely independent of H_2O content. The growth of amphibole in relatively dry experiments indicates that some other component is taking the place of OH^- in the crystal structure. The role of fluorine is usually invoked to explain the enhanced stability of amphibole, but this has not been detected in the kaersutite of the DVDP rocks (Kyle, 1981). Unusually low H contents have been measured in other natural kaersutites (notably those from Martian meteorites) suggesting that the amphibole does not require a hydrous environment to grow (Popp & Phillips, 1995) In addition to $Cl^-$and F^- occupying O3 site occupancies, oxy substitutions in kaersutite are favored by high Ti and Fe^{3+} contents (Popp & Taras Bryndzia, 1992) via the reactions

$$Fe^{2+} + OH^{-} = Fe^{3+} + O^{2-} \quad (4)$$

$$[6]Al^{3+} + OH^{-} = Ti^{4+} + O^{2-}. \quad (5)$$

This may explain the growth of kaersutite in our driest experiments, especially given the oxidizing experimental conditions, which will favor oxy substitutions via the reaction in equation (4). The presence of kaersutite in relatively dry natural samples thus may indicate the prevalence of oxidizing conditions for the DVDP lavas.

Given that the natural phase assemblages were reproduced under the relatively oxidizing experimental conditions employed, we infer a similarly oxidizing environment for the deep and intermediate Ross Island magmas. This is in direct contrast to the reducing conditions (QFM – 1·22) inferred for Erebus's phonolitic lava lake (based on oxide pairs; Kyle, 1977) but is consistent with a reducing upwards redox trend that can be inferred for the EL. Based on the experimental study of Moussallam *et al.* (2013), a phonolite reservoir at 200 MPa has an oxygen fugacity of QFM – 0·5 at 1000°C. Intermediate tephriphonolites at 300 MPa have an estimated oxygen fugacity of QFM – 0·44 at 1081°C [based on oxide pairs from sample 97009 from Kyle *et al.* (1992)]. Coupled chemical–physical modeling of conduit flow by Burgisser & Scaillet (2007) has shown that reducing upwards behavior is expected for some sulfur-bearing volcanic systems. This is confirmed by the systematic analyses of the Fe redox state of a suite of melt inclusions at Erebus, which have conclusively shown a strong reduction trend as pressure decreases (Moussallam *et al.*, 2014). In addition, the large CO_2 flux proposed as the mechanism for dehydrating Erebus magmas might also contribute to the reduction in oxidation state of the shallow Erebus system by reducing H_2O fugacity.

6.4 The case for CO_2-dominated volcanism at Erebus

The results of this study add to the compelling evidence that Erebus volcano is a remarkably CO_2-rich, H_2O-poor system. Melt inclusion volatile contents from Erebus maintain markedly high C/H ratios, even in intermediate and evolved melts. Conceptually, this can be explained by the presence of a CO_2-rich gas buffering the melt to higher dissolved CO_2concentrations, an interpretation consistent with previous modeling (Oppenheimer *et al.*, 2011) and experimental work (Iacovino *et al.*, 2013). As discussed above, the natural phase assemblages of the starting lavas in this study could be approximated only under extremely H_2O-poor conditions. Even for KI-type lavas, which are slightly wetter than the AW lavas, our experiments suggest that low-H_2O conditions dominate ($XH_2O^{fluid} \sim 0{\cdot}2$). More evolved kaersutite-bearing lavas of the DVDP lineage were best reproduced in terms of modal abundance of amphibole (<3–5%) only in experiments with higher H_2O contents (low-H_2O runs containing 15–19% amphibole), indicating a possible increase in water activity in evolved DVDP magmas. Given that the Erebus system is thought to be fluid-saturated, our results agree with the conclusions ofOppenheimer *et al.* (2011), who suggested that carbon-induced dehydration of Erebus magmas at various depths promotes crystal fractionation and magmatic differentiation.

Even though our driest experimental melts contain only 1–2 wt % H_2O, insufficient plagioclase crystallization and higher than observed CaO contents in our experimental glasses indicate that our experimentally established liquid line of descent is more water rich than the natural system. This suggests a maximum H_2O content of 1–2 wt % in Ross Island basanites and a maximum of ~1 wt % in EL phonotephrites. Moussallam*et al.* (2013) demonstrated that evolved Erebus melts (i.e. lava lake phonolite) must have extremely low water contents, with a maximum of ~0·5 wt % (but probably even below 0·2 wt %). The very tight constraints that these two studies place on the maximum water contents of Erebus magmas very closely match the trends in H_2O contents of Erebus melt inclusions (Oppenheimer *et al.*, 2011), which range from 1–2 wt % in basanite to ≤0·5 wt % in all other lavas, phonolites having ≤0·2 wt %.

Our results are consistent with the presence of a CO_2-rich fluid in the EL plumbing system but not in the DVDP. Given that both lineages are thought to be derived from the same parental melt (Kyle *et al.*, 1992), we suggest that a sustained connection of the EL plumbing system to its deep mantle source and a subsequent CO_2-rich fluid flux that characterizes the EL (and lack of such in the DVDP) is responsible for the differences in storage conditions and thus mineralogy of the two lava suites. The decoupled ascent of CO_2-rich vapor through the magmatic system of Erebus may explain the extreme dehydration of the EL melts, as CO_2 will degas early and deep and leach H_2O from the melt and into the vapor phase upon ascent (Métrich & Wallace, 2008). CO_2 fluxing through the EL system can also explain the sustained high-temperature regime of EL magmas inferred from phase equilibria (i.e. olivine–kaersutite stability) and would suggest a more reducing environment relative to the DVDP, consistent with enhanced kaersutite stability under oxidizing conditions.

6.5 One source, two lava lineages: pre-eruptive histories of EL and DVDP magmas

Based on our experimental results and the observations presented here, we propose two differentiation pathways from a single basanitic parent magma for the Ross Island lavas: a high-temperature pathway, followed by EL lavas, and a low-temperature pathway, followed by DVDP lavas. The parental magma beneath Erebus is a high-temperature (>1050°C) basanite that accumulated at high pressure, perhaps in the upper mantle (Kyle, 1981) or lower crust (Oppenheimer *et al.*, 2011). Both EL and DVDP basanites must have ascended rapidly (with little to no differentiation during ascent; Kyle, 1981), sourced directly from a high-temperature region near the crust–mantle boundary, as they both contain olivine phenocrysts. Rapid magma ascent is supported by the observation of mantle xenoliths within some basanites (see, e.g. Moore & Gullikson, 2013). The sustained high-temperature regime inferred for the EL (>1050°C) could be facilitated by the direct connection of the EL conduit to the hot parental magma storage region beneath Erebus. The sustained open-vent nature of Erebus volcano implies a continual resupply of heat and gas from a deep source region (Oppenheimer & Kyle, 2008;Oppenheimer *et al.*, 2009, 2011; Iacovino *et al.*, 2013*a*; Moussallam *et al.*, 2013). The difference in high- and low-temperature regimes, coupled with the fact that EL lavas have erupted continuously whereas DVDP lavas have stopped erupting, suggests that the DVDP plumbing system was a shorter-lived offshoot of the main Erebus conduit with little resupply of heat from the source region. This is consistent with the model proposed for the formation of the subsidiary Ross Island volcanic centers in which magma followed radial fractures created by the updoming of mantle material into the crust directly beneath Erebus (Kyle & Cole, 1974; Kyle*et al.*, 1992). More extensive dating of Hut Point Peninsula rocks would be necessary to precisely constrain the timing and longevity of the activity associated with DVDP lavas.

The DVDP magmas travelled along an offshoot of the main Erebus conduit, culminating at the surface on Hut Point Peninsula. During their rapid ascent, some olivine was removed from the DVDP basanites and injected into subsequently produced intermediate magmas as xenocrysts. Temperature decreased to below ~1050°C during differentiation of the intermediate and evolved magmas, evidenced by the lack of olivine phenocrysts and the presence of kaersutite in all subsequent lava types. This drop in temperature may be facilitated by the lack of continuous heat input from the source region coupled with long travel times through the shallow crust compared with the EL lavas.

7. CONCLUSIONS

Experiments were performed to constrain and compare the conditions of deep and intermediate magma storage and differentiation of EL and DVDP lavas. Our results indicate relatively dry conditions (<1–2 wt % H_2O) for both lineages, established by the plagioclase liquidus. The appearance of kaersutitic amphibole in natural intermediate and evolved DVDP lavas indicates an elevated water activity for the DVDP plumbing system compared with the EL, which does not contain kaersutite. The co-precipitation of kaersutite and plagioclase in natural DVDP lavas places very tight constraints on the H_2O content of the intermediate DVDP magmas, as phonotephrite experiments produce this assemblage only between XH_2O^{fluid} *c.* 0·15 and 0·35. Previous studies on kaersutite-bearing rocks have shown that kaersutite can be stabilized at such low water activities by the replacement of OH^- in the crystal structure by $F^-$and Cl^-, and via oxy-substitutions (see above). This, combined with the approximate reproduction of the natural phase assemblages in our experiments at NNO + 1·5 to NNO + 2·0, indicates that similarly oxidizing conditions may be present during deep and intermediate magma storage. We can also constrain temperature conditions for both lineages based on the presence or lack of olivine and kaersutite, for which our experiments indicate a peritectic-like relationship at intermediate melt water contents and ~1050°C. Olivine is present in all EL lavas and in DVDP basanites, but is absent in intermediate and evolved DVDP rocks. The opposite is true of kaersutite, which appears only in intermediate and evolved DVDP rocks. Because kaersutite is stable to very low melt water contents (with its stability field increasing with evolution of the melt), we infer that the lack of kaersutite in EL lavas and DVDP basanites is the result of sustained high temperatures during deep and intermediate magmatic differentiation (>1050°C) and that kaersutite is stabilized in DVDP magmas owing to a drop in equilibration temperature from above 1050°C in basanites to below 1050°C in the intermediate magmas. The differences in T, XH_2O^{fluid}, and possibly fO_2 between DVDP and EL magmas can be explained by the EL plumbing system being dominated by the flux of a CO_2-rich fluid (as hypothesized in previous studies) and the lack of such a fluid in the DVDP. This is consistent with the model of the formation of Erebus and its subsidiary volcanic centers.

FUNDING

Fieldwork in Antarctica was supported by the Office of Polar Programs (National Science Foundation) (ANT1142083). Experimental research was supported by the Environmental Research Council project 'DEMONS'; Labex Voltaire (ANR-10-LABX-100-10); and by the University of Cambridge Department of Geography Phillip Lake and William Vaughn Lewis grants.

ACKNOWLEDGEMENTS

Many thanks go to Nicole Métrich, Michael Carroll, Peter Ulmer, and an anonymous reviewer for helpful and extensive reviews of the original paper, and to Marie Edmonds for a detailed review on the portion of this paper that contributed to K.I.'s PhD thesis. We thank Bill McIntosh for assistance in sample collection on Ross Island; Marina Alletti, Juan Andújar, and Remí Champellier for assistance with experiments; Ida Di Carlo for assistance on the SEM; Nial Peters for his FTIR spectra background fitting code; Jacob Lowenstern for use of the mapping FTIR system and for advice on analyzing experiments; Iris Buisman, Raul Carampin, and Kenneth

Domanik for assistance with electron microprobe work; Matthijs van Soest for use of the surface mapping microscope.

REFERENCES

1. Allard P. (2010). A CO_2-rich gas trigger of explosive paroxysms at Stromboli basaltic volcano, Italy. Journal of Volcanology and Geothermal Research 189,363–374.

2. Andújar J., Costa F.,Martí J. (2010). Magma storage conditions of the last eruption of Teide volcano (Canary Islands, Spain). Bulletin of Volcanology 72,381–395.

3. Behrens H., Misiti V., Freda C., Vetere F., Botcharnikov R. E., Scarlato P. (2009).Solubility of H_2O and CO_2 in ultrapotassic melts at 1200 and 1250–C and pressure from 50 to 500 MPa. American Mineralogist 94, 105–120.

4. Burgisser A., Scaillet B. (2007). Redox evolution of a degassing magma rising to the surface. Nature 445, 194–197.

5. Burgisser A., Oppenheimer C., Alletti M., Kyle P. R., Scaillet B., Carroll M. R. (2012). Backward tracking of gas chemistry measurements at Erebus volcano.Geochemistry, Geophysics, Geosystems 13(11),

6. Caldwell D. A., Kyle P. R. (1994). Mineralogy and geochemistry of ejecta erupted from Mount Erebus, Antarctica, between 1972 and 1986. Volcanological and environmental studies of Mount Erebus, Antarctica, 147–162.

7. Carmichael I. S., Ghiorso M. S. (1990). The effect of oxygen fugacity on the redox state of natural liquids and their crystallizing phases. In: Nicholls J., Russell J. K. (eds)Modern Methods of Igneous Petrology: Understanding Magmatic Processes. Mineralogical Society of America, Reviews in Mineralogy 24, 191–212.

8. Curtis A., Kyle P. (2011). Geothermal point sources identified in a fumarolic ice cave on Erebus volcano, Antarctica using fiber optic distributed temperature sensing.Geophysical Research Letters 38(16), L16802.

9. Cioni R. (2000). Volatile content and degassing processes in the AD 79 magma chamber at Vesuvius (Italy). Contributions to Mineralogy and Petrology 140, 40–54.

10. Cocheo PA. (1993). The solubility of water in basanitic melts at low pressures,Masters thesis. Arizona State University.

11. Di Carlo I., Pichavant M., Rotolo S. G., Scaillet B. (2006). Experimental crystallization of a high-K arc basalt: the golden pumice, Stromboli volcano (Italy).Journal of Petrology 47, 1317–1343.

12. Di Carlo I., Rotolo S. G., Scaillet B., Buccheri V., Pichavant M. (2010). Phase equilibrium constraints on pre-eruptive conditions of recent felsic explosive volcanism at Pantelleria Island, Italy. Journal of Petrology 51, 2245–2276.

13. Dixon J. E., Pan V. (1995). Determination of the molar absorptivity of dissolved carbonate in basanitic glass. American Mineralogist 80(11–12), 1339–1342.

14. Dixon J. E., Stolper E. M., Holloway J. R. (1994). An experimental study of water and carbon dioxide solubilities in mid-ocean ridge basaltic liquids. Journal of Petrology36, 1607–1631.

15. Dunbar N. W., Cashman K. V., Dupré R. (1994). Crystallization processes of anorthoclase phenocrysts in the Mount Erebus magmatic system: evidence from crystal composition, crystal size distributions, and volatile contents of melt inclusions. Volcanological and environmental studies of Mount Erebus, Antarctica,129–146.

16. Feig S. T., Koepke J., Snow J. E. (2010). Effect of oxygen fugacity and water on phase equilibria of a hydrous tholeiitic basalt. Contributions to Mineralogy and Petrology 160, 551–568.

17. Fine G., Stolper E. (1986). Dissolved carbon dioxide in basaltic glasses: concentrations and speciation. Earth and Planetary Science Letters 76, 263–278.

18. Flowers G. C. (1979). Correction of Holloway's (1977) adaptation of the modified Redlich–Kwong equation of state for calculation of the fugacities of molecular species in supercritical fluids of geologic interest. Contributions to Mineralogy and Petrology 69, 315–318.

19. Freise M., Holtz F., Koepke J., Scoates J., Leyrit H. (2003). Experimental constraints on the storage conditions of phonolites from the Kerguelen Archipelago.Contributions to Mineralogy and Petrology 145, 659–672.

20. Harrison T. M., Watson E. B. (1984). The behavior of apatite during crustal anatexis: equilibrium and kinetic considerations. Geochimica et Cosmochimica Acta48, 1467–1477.

21. Holloway J. R. (1971). Internally heated pressure vessels. In: Ulmer G.C. (ed)Research Techniques for High Pressure and High Temperature. Berlin, pp. 217–258.

22. Holloway J. R. (1977). Fugacity and activity of molecular species in supercritical fluids. In: Fraser D.G. (ed) Thermodynamics in Geology. Dordrecht, pp. 161–181.

23. Iacono-Marziano G., Gaillard F., Scaillet B., Pichavant M., Chiodini G. (2009). Role of non-mantle CO_2 in the dynamics of volcano degassing: The Mount Vesuvius example. Geology 37, 319–322.

24. Iacovino K., Moore G., Roggensack K., Oppenheimer C., Kyle P. (2013). H_2O–CO_2solubility in mafic alkaline magma: applications to volatile sources and degassing behavior at Erebus volcano, Antarctica. Contributions to Mineralogy and Petrology166, 845–860.

25. Ihinger P. D., Hervig R. L., McMillan P. F. (1994). Analytical methods for volatiles in glasses. In: Carroll M. R., Holloway J. R. (eds) Volatiles in Magmas. Mineralogical Society of America, Reviews in Mineralogy 30, 67–121.

26. Kaszuba J. P., Wendlandt R. F. (2000). Effect of carbon dioxide on dehydration melting reactions and melt compositions in the lower crust and the origin of alkaline rocks. Journal of Petrology 41, 363–386.

27. Kyle P. R. (1977). Mineralogy and glass chemistry of recent volcanic ejecta from Mt Erebus, Ross Island, Antarctica. New Zealand Journal of Geology and Geophysics20, 1123–1146.

28. Kyle P. R. (1981). Mineralogy and geochemistry of a basanite to phonolite sequence at Hut Point Peninsula, Antarctica, based on core from Dry Valley Drilling Project Drillholes 1, 2 and 3. Journal of Petrology 22, 451–500.

29. Kyle P. R., Cole J. W. (1974). Structural control of volcanism in the McMurdo Volcanic Group, Antarctica. Bulletin Volcanologique 38, 16–25.

30. Kyle P. R., Moore J. A., Thirlwall M. F. (1992). Petrologic evolution of anorthoclase phonolite lavas at Mount Erebus, Ross Island, Antarctica. Journal of Petrology 33,849–875.

31. Lange R., Carmichael I. (1987). Densities of Na_2O–K_2O–CaO–MgO–FeO–Fe_2O_3–Al_2O_3–TiO_2–SiO_2 liquids—New measurements and derived partial molar properties.Geochimica et Cosmochimica Acta 51, 2931–2946.

32. Marianelli P., Sbrana A., Métrich N., Cecchetti A. (2005). The deep feeding system of Vesuvius involved in recent violent strombolian eruptions. Geophysical Research Letters 32(2).

33. Martí J., Andújar J., Costa F., Wolff J. (2008). Pre-eruptive conditions of the phonolitic magma from the El Abrigo caldera-forming eruption (Las Cañadas caldera, Tenerife, Canary Islands). In: IOP Conference Series: Earth and Environmental Science, doi:10.1088/1755-1307/3/1/012013.

34. Métrich N., Wallace P. J. (2008). Volatile abundances in basaltic magmas and their degassing paths tracked by melt inclusions. In: Putirka K. D., Tepley F. J. III. (eds)Minerals, Inclusions and Volcanic Processes. Mineralogical Society of America and Geochemical Society, Reviews in Mineralogy and Geochemistry 69, 363–402.

35. Molina I., Burgisser A., Oppenheimer C. (2012). Numerical simulations of convection in crystal-bearing magmas: A case study of the magmatic system at Erebus, Antarctica. Journal of Geophysical Research: Solid Earth (1978–2012)117(B7).

36. Moore G., Carmichael I. S. E. (1998). The hydrous phase equilibria (to 3 kbar) of an andesite and basaltic andesite from western Mexico: constraints on water content and conditions of phenocryst growth. Contributions to Mineralogy and Petrology130(3–4), 304–319.

37. Moore G., Gullikson A. (2013). Crystal/liquid thermobarometry in the rapidly erupted Peridot Mesa phonotephrite: constraints from new measurements and phase equilibria experiments [abstract]. AGU Fall Meeting Abstracts, 1(B1).

38. Morizet Y., Brooker R. A., Kohn S. C. (2002). CO_2 in haplo-phonolite melt: solubility, speciation and carbonate complexation. Geochimica et Cosmochimica Acta 66,1809–1820.

39. Moussallam Y., Oppenheimer C., Scaillet B., Kyle P. R. (2013). Experimental phase-equilibrium constraints on the phonolite magmatic system of Erebus Volcano, Antarctica. Journal of Petrology 54, 1285–1307.

40. Moussallam Y., Oppenheimer C., Scaillet B., Gaillard F., Kyle P., Peters N., DonovanA. (2014). Tracking the changing oxidation state of Erebus magmas, from mantle to surface, driven by magma ascent and degassing. Earth and Planetary Science Letters393, 200–209.

41. Nichols A. R. L., Wysoczanski R. J. (2007). Using micro-FTIR spectroscopy to measure volatile contents in small and unexposed inclusions hosted in olivine crystals. Chemical Geology 242, 371–384.

42. Ochs F. A., Lange R. A. (1999). The density of hydrous magmatic liquids. Science283(5406), 1314–1317.

43. Oppenheimer C., Kyle P. R. (2008). Probing the magma plumbing of Erebus volcano, Antarctica, by open-path FTIR spectroscopy of gas emissions. Journal of Volcanology and Geothermal Research 177, 743–754.

44. Oppenheimer C., Lomakina A. S., Kyle P. R., Kingsbury N. G., Boichu M. (2009).Pulsatory magma supply to a phonolite lava lake. Earth and Planetary Science Letters284, 392–398.

45. Oppenheimer C., Moretti R., Kyle P. R., Eschenbacher A., Lowenstern J. B., HervigR. L., Dunbar N. W. (2011). Mantle to surface degassing of alkalic magmas at Erebus volcano, Antarctica. Earth and Planetary Science Letters 306, 261–271.

46. Pandya N., Muenow D. W., Sharma S. K. (1992). The effect of bulk composition on the speciation of water in submarine volcanic glasses. Geochimica et Cosmochimica Acta 56, 1875–1883.

47. Pichavant M., Di Carlo I., Le Gac Y., Rotolo S. G., Scaillet B. (2009). Experimental constraints on the deep magma feeding system at Stromboli volcano, Italy. Journal of Petrology 50, 601–624.

48. Pichavant M., Scaillet B., Pommier A., Iacono-Marziano G., Cioni R. (2014). Nature and evolution of primitive Vesuvius magmas: an experimental study. Journal of Petrology 55, 2281–2310.

49. Popp R. K., Phillips M. W. (1995). An experimental study of phase equilibria and Fe oxy-component in kaersutitic amphibole; implications for the f_{H2} and α_{H2O} in the upper mantle. American Mineralogist 80, 534–548.

50. Popp R. K., Taras Bryndzia L. (1992). Statistical analysis of Fe^{3+}, Ti, and OH in kaersutite from alkalic igneous rocks and mafic mantle xenoliths. American Mineralogist 77, 1250–1250.

51. Prouteau G., Scaillet B. (2003). Experimental constraints on the origin of the 1991 Pinatubo dacite. Journal of Petrology 44, 2203–2241.

52. Robie R. A., Hemingway B. S. (1979). Thermodynamic properties of minerals and related substances at 298.15 K and 1 bar (105 pascals) pressure and at higher temperatures (No. 2131). USGPO; For sale by US Geological Survey, Information Services.

53. Roux J., Lefevre A. (1992). A fast-quench device for internally heated pressure vessels. European Journal of Mineralogy 4, 279–281.

54. Sack R. O., Walker D., Carmichael I. S. (1987). Experimental petrology of alkalic lavas: constraints on cotectics of multiple saturation in natural basic liquids.Contributions to Mineralogy and Petrology 96, 1–23.

55. Scaillet B., Pichavant M., Roux J., Humbert G., Lefevre A. (1992). Improvements of the Shaw membrane technique for measurement and control of fH_2 at high temperatures and pressures. American Mineralogist 77, 647–655.

56. Shaw H. R., Wones D. R. (1964). Fugacity coefficients for hydrogen gas between 0 degrees and 1000 degrees C, for pressures to 3000 atm. American Journal of Science 262, 918–929.

57. Shinohara H., Aiuppa A., Giudice G., Gurrieri S., Liuzzo M. (2008). Variation of H_2O/CO_2 and CO_2/SO_2 ratios of volcanic gases discharged by continuous degassing of Mount Etna volcano, Italy. Journal of Geophysical Research: Solid Earth (1978–2012) 113(B9), 1978–2012.

58. Spera F. J. (1984). Carbon dioxide in petrogenesis III: role of volatiles in the ascent of alkaline magma with special reference to xenolith-bearing mafic lavas.Contributions to Mineralogy and Petrology 88, 217–232.

59. Spilliaert N., Allard P., Métrich N., Sobolev A. V. (2006). Melt inclusion record of the conditions of ascent, degassing, and extrusion of volatile-rich alkali basalt during the powerful 2002 flank eruption of Mount Etna (Italy). Journal of Geophysical Research: Solid Earth (1978–2012) 111(B4), doi:10.1029/2005JB003934.

60. Sun S. S., Hanson G. N. (1975). Origin of Ross Island basanitoids and limitations upon the heterogeneity of mantle sources for alkali basalts and nephelinites.Contributions to Mineralogy and Petrology 52, 77–106.

61. Sun S. S., Hanson G. N. (1976). Rare earth element evidence for differentiation of McMurdo volcanics, Ross Island, Antarctica. Contributions to Mineralogy and Petrology 54, 139–155.

62. Tagliani S. M., Nicotra E., Viccaro M., Gianfagna A. (2012). Halogen-dominant mineralization at Mt. Calvario dome (Mt. Etna) as a response of volatile flushing into the magma plumbing system. Mineralogy and Petrology 106(1–2), 89–105.

63. Walker J. A., Roggensack K., Patino L. C., Cameron B. I., Matías O. (2003). The water and trace element contents of melt inclusions across an active subduction zone. Contributions to Mineralogy and Petrology 146, 62–77.

64. Zandomeneghi D., Aster R., Kyle P., Barclay A., Chaput J., Knox H. (2013). Internal structure of Erebus volcano, Antarctica imaged by high-resolution active-source seismic tomography and coda interferometry. Journal of Geophysical Research: Solid Earth 118, 1067–1078.

Chapter 5

The contribution of metamorphic petrology to understanding lithosphere evolution and geodynamics

Michael Brown

Laboratory for Crustal Petrology, Department of Geology, University of Maryland, College Park, MD 20742, USA

ABSTRACT

In the early 1980s, evidence that crustal rocks had reached temperatures >1000 °C at normal lower crustal pressures while others had followed low thermal gradients to record pressures characteristic of mantle conditions began to appear in the literature, and the importance of melting in the tectonic evolution of orogens and metamorphic–metasomatic reworking of the lithospheric mantle was realized. In parallel, new developments in instrumentation, the expansion of *in situ* analysis of geological materials and increases in computing power opened up new fields of investigation. The robust quantification of pressure (*P*), temperature (*T*) and time (*t*) that followed these advances has provided reliable data to benchmark geodynamic models and to investigate secular change in the thermal state of the lithosphere as registered by metamorphism through time. As a result, the last 30 years have seen significant progress in our understanding of lithospheric evolution, particularly as it relates to Precambrian geodynamics.

Eoarchean–Mesoarchean crust registers uniformly high *T/P* metamorphism that may reflect a stagnant lid regime. In contrast, two contrasting types of metamorphism, eclogite–high-pressure granulite metamorphism, with apparent thermal gradients of 350–750 °C/GPa, and granulite–ultrahigh temperature metamorphism, with apparent thermal gradients of 750–1500 °C/GPa, appeared in the Neoarchean rock record. The emergence of paired metamorphism is interpreted to register the onset of one-sided subduction, which introduced an asymmetric thermal structure at these developing convergent plate margins characterized by lower *T/P* in the subduction channel and higher *T/P* in the overriding plate. During the Paleoarchean to Paleoproterozoic the ambient mantle temperature was warmer than at present by ~300–150 °C. Although the thermal history of Earth is only poorly constrained, it is likely that prior to ca. 3.0 Ga heating from radioactive decay would have exceeded surface heat loss, whereas since ca. 2.5 Ga secular cooling has dominated the thermal history of the Earth. The advent of paired metamorphism is consistent with other changes in the geological record during the Neoarchean that are best explained as the result of a transition from a stagnant lid to subduction and a global plate tectonics regime by ca. 2.5 Ga. This interpretation is supported by results from 2-D numerical experiments of oceanic subduction that demonstrate a change to one-sided subduction is plausible as upper mantle temperature declined to <200 °C warmer than at present during the late Neoarchean–Paleoproterozoic. This is the beginning of the Proterozoic plate tectonics regime.

At 1.0 Ga the ambient mantle temperature was still ~150–100 °C warmer than at present. Continued secular cooling caused a transition to cold subduction registered in the crustal record of metamorphism by the first appearance of blueschist and high to ultrahigh pressure metamorphism during the Neoproterozoic. Results of 2-D numerical experiments of continental collision demonstrate a transition from shallow to deep slab breakoff associated with stronger crust–mantle coupling that enabled continental subduction to mantle depths as upper mantle temperature declined to <100 °C warmer than at present during the late Proterozoic. This is the beginning of the modern plate tectonics regime.

KEYWORDS

Cryogenian–Cambrian; Neoarchean–Paleoproterozoic; Plate tectonics; Rayleigh–Taylor instabilities; Stagnant lid; Subduction

1. INTRODUCTION

On contemporary Earth, regional metamorphism is associated with geodynamic settings of lower or higher than average heat flow, such as subduction zones and arc–back arc systems, or is related to tectonic processes that disrupt the steady-state thermal structure of the lithosphere by thickening or thinning, such as occurs in orogenic hinterlands and collisional mountain belts or during orogenic collapse and regional extension (Brown, 2008 and Brown, 2009). In addition, the long-term cooling of the lithosphere drives garnet-producing metamorphic reactions, which decreases the buoyancy of crustal roots and enhances lithosphere stability (Fischer, 2002). To extend this understanding of the relationship between geodynamics and metamorphism back in time, we use the forensic methods and techniques of metamorphic petrology.

Thus, metamorphic petrology is concerned with decoding the mineralogical and microstructural record of burial/heating and exhumation/cooling imprinted on pre-existing sedimentary, igneous and metamorphic rocks by processes such as subduction, accretion, trench advance or retreat, collisional orogenesis and orogenic collapse. As petrologists, we address multiple scales of activity, from the pressure (*P*)–temperature (*T*)–time (*t*) evolution of a single rock to the evolution of *P*–*T* in space and time. Both subduction/accretion and collisional orogenesis/orogenic collapse cause perturbations of steady-state geotherms on time scales and length scales constrained by the mechanical properties of the lithosphere, which vary with composition, temperature, fluid/melt presence/absence, and strain rate. Thus, by determining quantitative *P*–*T*–*t*paths petrologists make available information to parameterize subduction zone processes and collisional orogenesis. Inverting this *P*–*T*–*t* information, provides constraints on the geodynamic and metasomatic processes involved, and consequently advances understanding of lithosphere evolution and geodynamics.

Although mobile-lid plate tectonics (hereafter “plate tectonics”) has provided us with a context to understand metamorphism and its relationship to tectonics at least as far back as the dawn of the Phanerozoic

(Brown, 2008, Brown, 2010a and Brown, 2010b), during the Precambrian geodynamics may have been different (Sizova et al., 2010 and Sizova et al., 2014). For example, a strong case has been made for a stagnant-lid plate tectonics regime on the early Earth (Debaille et al., 2013, Griffin et al., 2013 and O'Neill et al., 2013; hereafter "stagnant lid"). In addition, Earth's mantle was hotter and Moho temperatures were higher in the past, and in the Archean conditions were probably appropriate for lithosphere foundering by Rayleigh–Taylor instabilities (Jull and Kelemen, 2001 and Toussaint et al., 2004), which may have played a much more important role in lithosphere evolution on the early Earth than on contemporary Earth (Johnson et al., 2014).

In the first section of this paper I review examples of major discoveries in metamorphic petrology and advances in instrumentation and methods during the last 30 years. This provides a perspective for the second section in which I review evidence from the crustal record of metamorphism and recent results from geodynamic modeling to investigate the major geodynamic transitions in Earth evolution. The first of these concerns the transition from an Archean stagnant-lid regime to a Proterozoic plate tectonics regime, as evidenced by the appearance of paired metamorphism, whereas the second concerns the Cryogenian to Cambrian transition from the Proterozoic plate tectonics regime to the modern plate tectonics regime characterized by cold subduction, as evidenced by the appearance of blueschist and ultrahigh-pressure metamorphism in the geological record.

2. THE STATE OF THE ART IN METAMORPHIC PETROLOGY

2.1. Brave new world

It is now 30 years since the discovery that some lithospheric materials record evidence of metamorphism at pressures characteristic of mantle conditions (e.g. Chopin, 1984,Chopin, 2003 and Dobrzhinetskaya et al., 2011). Just a few years earlier, persuasive evidence had appeared in the literature demonstrating that some crustal rocks registered temperatures >1000 °C over large areas at 'normal' lower crustal pressures (e.g. Ellis, 1980, Harley, 1998, Harley, 2008 and Kelsey, 2008). Although both of these statements appear to be broadly correct, it is wise to be circumspect of extreme values in case the recorded pressure was not lithostatic (Moulas et al., 2013) or where the calculated temperature approaches or exceeds the liquidus for the continental crust (~1100 °C). Also during the same period, the realization that melting and melt drainage and migration might play an important role in crustal deformation was becoming widespread (e.g.Hollister and Crawford, 1986, D'Lemos et al., 1992, Brown, 1994, Brown, 2010a and Brown, 2010b). Furthermore, it became clear that much of the underlying mantle has undergone dynamic reworking (Mercier and Nicolas, 1975), including metasomatic modifications (Thompson, 1992). This revelation opened both the mantle (Tirone et al., 2009 and Holland et al., 2013), including the deep mantle (Ganguly et al., 2009), and subducted lithospheric slabs (Hacker et al., 2003a) to study using the methods of metamorphic petrology.

These fascinating developments during the past 30 years have not only changed our perception of the field of metamorphism but also of the role that metamorphic petrology can play in understanding orogenesis. But how have these developments helped us to understand the evolution of the lithosphere and geodynamics during the Precambrian in relation to secular change?

2.2. Metamorphism and tectonics

In addressing the issue of metamorphism and tectonics, both regional and global scales of interactions are of interest. At the regional scale, we are concerned with differences from terrane to terrane that enable us to distinguish the particular combination of tectonic events responsible for the metamorphism of each terrane, and which inform us about tectonic interactions at the terrane scale. At the global scale, we are concerned with similarities in styles of metamorphism and relationship to plate tectonic setting, which relationships are likely to reflect global geodynamic processes.

The asymmetry in the thermal structure characteristic of one-sided subduction creates contrasting thermal environments—lower *T*/*P* in the subduction zone and higher *T*/*P* in the arc–backarc/orogenic hinterland—that are registered in the rock record during collisional orogenesis as two distinct types of metamorphism with different apparent thermal gradients (Brown, 2006 and Brown, 2007a). Such paired metamorphic belts are characteristic of subduction and plate tectonics processes (Brown, 2009 and Brown, 2010b). Thus, we may use the crustal record of metamorphism to identify when subduction began and by inference when Earth adopted a plate tectonics mode (Brown, 2008 and Sizova et al., 2010).

At the global scale we want to know whether these similarities in styles of metamorphism and the relationships between metamorphism and tectonics have always been the same or whether there were differences earlier in Earth history. The relationship between secular change, mantle dynamics and the metamorphic conditions retrieved from metamorphic belts has generated a succession of important contributions, such as those by de Roever, 1956 and de Roever, 1965 and Miyashiro (1961) leading up to the plate tectonics revolution, and those by Ernst, 1972, Miyashiro, 1973, England and Richardson, 1977, Watson, 1977, Grambling, 1981, Condie, 1984, England and Bickle, 1984, Galer and Mezger, 1998, Maruyama et al., 1996, Maruyama and Liou, 1998,Maruyama and Liou, 2005, Stern, 2005, Brown, 2006, Brown, 2007a, Brown, 2007b and Brown, 2008 and Anderson et al. (2012) since the plate tectonics revolution.

To address the issue of secular change in relation to understanding Precambrian lithosphere evolution and geodynamics, hypotheses about changes in geodynamics based on the crustal record of metamorphism (Brown, 2006 and Brown, 2007a) may be rigorously tested using 2-d numerical geodynamic modeling (Sizova et al., 2010, Sizova et al., 2012 and Sizova et al., 2014). Although it is now more than 50 years since the beginning of the debate about whether Phanerozoic and Precambrian metamorphic styles were different (de Roever, 1956 and de Roever, 1965), it is only during the past decade that the importance of two features of the crustal record of metamorphism—the appearance during the Neoarchean of paired metamorphism and during the Cryogenian–Cambrian of blueschist and ultrahigh pressure metamorphism—has been realized (Brown, 2006 and Brown, 2007a) and related to the underlying geodynamic causes (Sizova et al., 2010, Sizova et al., 2014 and Johnson et al., 2014).

2.3. What information is provided by metamorphic petrology?

Metamorphic rocks record information about their *P–T–t* evolution in the mineral assemblages, the chemistry and variation in composition of these minerals, and the microstructural relationships among them. It is the job of the metamorphic petrologist to invert this information to retrieve quantitative data about spatial variations in the *P–T–t*evolution of ancient metamorphic belts and to use forward models to better understand the tectonic controls over these variations in evolution.

Substantial *a posteriori* evidence indicates that mineral assemblages in rocks undergoing metamorphism equilibrate continuously on some scale throughout the prograde evolution as fluid or melt is being generated, but undergo little or no change during the retrograde evolution once the rock becomes fluid absent or after any remaining melt has reacted with peak minerals while cooling to the solidus (Powell et al., 2005). As a result, an equilibrium mineral assemblage is likely to be preserved once fluid is lost under subsolidus conditions (around the peak *T* of metamorphism) or as the last dregs of melt crystallize in residual rocks at suprasolidus conditions, this assemblage reflecting the *P–T* conditions of fluid loss or the solidus. Thus, the mineral assemblage of a metamorphic rock is the key to estimating the conditions prevailing when it formed. Furthermore, this is the basic inference underlying the concept of metamorphic facies (Fig. 1a), which are defined by sets of mineral assemblages developed in rocks of different bulk composition that recur through space and time. Thus, metamorphic facies represent the overlapping stability fields in *P–T* space for these mineral assemblages and their recurrence supports the notion of equilibrium at or soon after peak *T*. Notwithstanding the success of the facies concept, today it is preferable to think in terms of apparent thermal gradients in the classification of types of metamorphism (Fig. 1b;Brown, 2006, Brown, 2007a, Brown, 2007b, Brown, 2008 and Stüwe, 2007). The rationale for the view will be explained below.

Although the rock record stretches back to ca. 4.0 Ga, the first billion years is represented by only one millionth of the exposed continental crust and retrieving quantitative *P–T–t*information is rarely possible. However, there is a reliable record of *P–T–t* information retrieved from metamorphic belts since ca. 3.0 Ga (Brown, 2006 and Brown, 2007a), the likely beginning of a transition to one-sided subduction and steady-state plate tectonics on Earth (Brown, 2008, Sizova et al., 2010, Dhuime et al., 2012, van Hunen and Moyen, 2012 and Johnson et al., 2014).

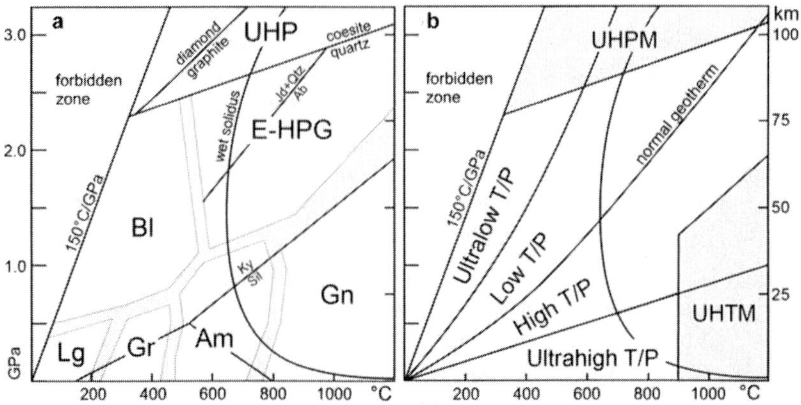

Figure 1. To show the contrasting views of *P–T* space as viewed by the petrologist (a) and the geodynamicist (b); modified after Stüwe (2007, Fig. 7.1; subject to copyright, published with kind permission of Springer Science + Business Media). The pressure to depth conversion is based in an average density of ~3000 kg m^{-3} (i.e. 3 GPa ≈ 100 km). (a) Shows the principal metamorphic facies in their simplest form. At pressures below coesite stability the facies boundaries are transitional to indicate the control of bulk composition on the change in mineral assemblages from one facies to another. Low-to-moderate pressure facies are: Lg, low-grade metamorphism, includes the zeolite facies; Gr, greenschist facies; Am, amphibolite facies; Gn, granulite facies; and, UHT, ultrahigh temperature metamorphism, which is the part of the granulite facies at $T > 900$ °C. High-to-ultrahigh pressure facies are: Bl, blueschist facies; E-HPG, eclogite–high-pressure granulite facies, which is the part of the eclogite facies where plagioclase remains stable in some bulk compositions but not in others at common *P–T* conditions; and, UHP, ultrahigh pressure metamorphism, which is the part of the eclogite facies at *P* above quartz stability. (b) *P–T* space divided in relation to a normal geotherm, where the normal geotherm is drawn to pass through 500 °C at ~33 km depth and 1200 °C at ~105 km depth. Thus, there are tectonic settings that may have either lower or higher T/P gradients than normal. On contemporary Earth, one example of where both tectonic settings occur in close proximity is along a convergent plate boundary, where the subduction zone has lower *T/P*and the arc–backarc higher *T/P*, respectively.

2.4. Advances in numerical tools, analytical instrumentation, geophysical imaging and geodynamic modeling

During the past 30 years the range of numerical tools and instrumentation available to the metamorphic petrologist has expanded beyond imagination. Examples include thermodynamic modeling and *in situ* analytical methods for U/Th–Pb isotope and trace element chemistry of minerals, and for mineral characterization.

The present level of sophistication in thermodynamic modeling of phase equilibria has its foundation in the 1980s with the introduction of appropriate internally-consistent thermodynamic datasets (Holland and Powell, 1985, Holland and Powell, 1998, Holland and Powell, 2011 and Berman, 1988) and the development and expansion of suitable activity–composition models for phases of interest in parallel with the evolution of software for quantitative phase equilibria modeling (Powell and Holland, 1985, Powell and Holland, 1988, Powell et al., 1998, White et al., 2007 and White et al., 2014, see also the THERMOCALC website at http://www.metamorph.geo.uni-mainz.de/thermocalc/index.html; Brown et al., 1988, Berman, 1988, Berman, 1991, Connolly and Kerrick, 1987,Connolly, 1990, Connolly, 2005, Connolly, 2009 and Connolly and Petrini, 2002, see also the Perple_X website http://www. perplex.ethz.ch/;de Capitani and Brown, 1987, de Capitani, 1994 and de Capitani and Petrakakis, 2010, see also the Theriak website at http:// titan.minpet.unibas.ch/minpet/theriak/theruser.html). Although the use of *in situ*U/Th–Pb geochronology (e.g. the SHRIMP) on accessory minerals was developed during the 1980s and became accepted during the 1990s (Williams, 2013), it was only with the advent of routine *in situ* trace element analysis of minerals, particularly zircon, monazite and garnet (Harley and Kelly, 2007, Harley et al., 2007 and Rubatto and Hermann, 2007), that it has become possible to link ages reliably to calculated *P–T*conditions. The recent introduction of laser ablation split-stream petrochronology (Kylander-Clark et al., 2013), in which U/Th–Pb isotope and trace element analyses are made simultaneously, provides a rapid method to obtain these data. This advance will not only enable ages to be linked to *P–T* conditions via rare earth element data from accessory minerals and garnet (e.g. Donaldson et al., 2013), but also will allow campaign-style geochronology that will provide datasets to improve our understanding of the spatial and temporal variations in *P–T* conditions in metamorphic belts (e.g. Spencer et al., 2013).

There are numerous new analytical methods that are now used routinely in metamorphic petrology—too many to review in detail here. Among spectroscopic methods employed in the characterization of geological materials (Beran and Libowtzky, 2004), Raman spectroscopy (Dubessy et al., 2012) is a highly flexible non-destructive technique used not only in the analysis of fluids, minerals and glasses but also in thermobarometry. For example, Raman spectroscopy is routinely used in the characterization of minerals such as diamond and coesite in UHPM rocks (Nasdala and Massonne, 2000 and O'Brien and Ziemann, 2008), in the assessment of peak-metamorphic conditions using carbonaceous (graphitic) materials (Beyssac et al., 2002), and in the characterization of glass (melt) and nanogranite inclusions in migmatites and granulites (Cesare et al., 2011). In addition, given a calculated or assumed *T* or *P*, Raman spectroscopy of mineral inclusions in other minerals may be used to determine either *P* or *T* (Kohn, 2014). This technique utilizes the pressure-dependence of the Raman peak positions in combination with differences in compressibility (pressure-sensitive) and thermal expansivity (temperature-sensitive) between the inclusion and host mineral, although the residual pressure retrieved depends on the elastic parameters of the host mineral (Izraeli et al., 1999, Parkinson and Katayama, 1999, Sobolev et al., 2000, Guiraud and Powell, 2006 and Enami et al., 2007).

There have also been significant advances in geophysical imaging during the past 30 years, in part driven by large National programs, such as Lithoprobe (Canada), Earthscope (USA) and SinoProbe (China), with the overall aim of exploring the composition, structure and evolution of the continental lithosphere. These advances include: establishing differences between Archean and Proterozoic lithosphere (Artemieva and Mooney, 2001); identifying melt in the deep crust under contemporary orogens and quantifying its effect on crustal rheology (Nelson et al., 1996, Li et al., 2003 and Unsworth et al., 2005), which aids in the interpretation of deeply eroded orogens; and, recognizing reaction fronts in orogenic crust (Mechie et al., 2004), identifying the water flux delivered to the deep Earth by subduction (Savage, 2012) and linking metamorphic reactions in actively subducting lithosphere to earthquake genesis (Hacker et al., 2003b and Nakajima et al., 2013).

The use of numerical experiments to investigate mantle convection (e.g. Doin et al., 1997, Tackley, 2000 and Davies, 2006) and the thermo-mechanical evolution of the lithosphere during orogenesis (e.g. Royden, 1993a,b, 1996; Willett et al., 1993;Beaumont., 1994, Beaumont et al., 1996a, Beaumont et al., 1996b, Royden, 1993a,Royden, 1993b, Royden,

1996, Pfiffner et al., 2000 and Faccenda et al., 2008) and the supercontinent cycle (e.g. Yoshida and Santosh, 2011 and Rolf et al., 2012) has expanded as computer power has increased to enable more sophisticated geodynamic modeling (e.g. Gerya and Yuen, 2003 and Gerya and Yuen, 2007). Geodynamic modeling has been used to address important processes such as slab breakoff (Gerya et al., 2004 and Duretz et al., 2011), shallow subduction (van Hunen et al., 2004 and Currie and Beaumont, 2011), ridge subduction (Groome and Thorkelson, 2009), deep crustal flow (Royden et al., 1997, Clark and Royden, 2000 and Beaumont et al., 2004), delamination (Ueda et al., 2012 and Gray and Pysklywec, 2012), intraplate orogenesis (Neil and Houseman, 1999, Pysklywec and Beaumont, 2004, Gorczyk et al., 2012, Gorczyk et al., 2013 and Gorczyk and Vogt, 2014), and exhumation of UHPM terranes (Gerya et al., 2002, Warren et al., 2008, Beaumont et al., 2009 and Sizova et al., 2012). Geodynamic modeling has also been used to evaluate crustal growth at active continental margins (Vogt et al., 2012 and Zhu et al., 2013) and to understand the links between metamorphism and tectonics (Jamieson and Beaumont, 1988, Sandiford and Powell, 1990, Jamieson et al., 2002, Jamieson et al., 2004, Gerya and Stockhert, 2006 and Li et al., 2010). Of course, geodynamic modeling has also been applied to the problem of Precambrian geodynamics (van Thienen et al., 2004, van Hunen and van den Berg, 2008, Sizova et al., 2010, Sizova et al., 2014, Gerya, 2014 and Johnson et al., 2014) and tectonics (Perchuk and Gerya, 2011), and the formation of the cratonic roots composed of strongly depleted sub-continental mantle lithosphere (Gray and Pysklywec, 2010 and Griffin et al., 2013).

3. THE EVOLUTION OF THE CONTINENTAL LITHOSPHERE

In this section I set out a frame of reference for the thermal evolution of the continental lithosphere based on my own work in metamorphic petrology and work done in collaboration with others that contributes to a better understanding of geodynamics. These contributions are twofold. First, we have used a comprehensive dataset of *P–T–t*information from Neoarchean to Cenozoic crust (Brown, 2006 and Brown, 2007a) coupled with 2-d numerical geodynamic modeling (Sizova et al., 2010, Sizova et al., 2012 and Sizova et al., 2014) to investigate secular change. This work suggests that there may have been three global-scale geodynamic regimes that were separated by two periods of change on Earth. The first of these is the initiation of subduction and the transition from a dominantly stagnant-lid regime to a plate tectonics regime in the late Archean–early Proterozoic, whereas the second is the transition to cold subduction that occurred during the Ediacaran–Cambrian. Second, we have used phase equilibria modeling of Earth's early lithosphere coupled with 2-D numerical geodynamic modeling of lithosphere stability to investigate early Archean geodynamics (Johnson et al., 2014). The results of this last study demonstrate the plausibility of a stagnant lid regime and have implications for the production of tonalite–trondhjemite–granodiorite suite magmas in the early Archean.

3.1. Early Archean (pre-Mesoarchean) lithosphere and geodynamics

3.1.1. Lithosphere

The evolution of the continental lithosphere is a response to geodynamics, which in turn is a response to the thermal evolution of Earth. There are differences in the geology and tectonic style between Archean and post-Archean crust. Gray gneisses and plutonic complexes of mainly tonalite–trondhjemite–granodiorite (TTG) suite affinity dominate the Archean geological record with volcano-sedimentary greenstone belts forming a minor component (Goodwin, 1991). Archean TTG gneisses typically have higher Na/K and Na/(Ca + Na + K) than post-Archean gneisses (e.g. Keller and Schoene, 2012), and high ratios of light to heavy rare-earth elements (L/HREE, for example, La/Yb). They are inferred to have been derived from sources of broadly basaltic composition (e.g. Foley et al., 2002, Moyen, 2011 and Ziaja et al., 2014) in which HREE depletion reflects sequestration by residual garnet. The sub-continental lithospheric mantle (SCLM) underlying Archean crust is also different from the post-Archean SCLM (e.g. Hamilton, 1998, Djomani et al., 2001 and Griffin et al., 2003). Different geology may indicate different geodynamic behavior and if this is true the differences are probably related to Earth's thermal evolution and mechanism of heat loss (Korenaga, 2006 and Labrosse and Jaupart, 2007).

Volcanic rocks of Archean greenstone belts are dominantly subaqueous (Arndt, 1999). They comprise mostly primary high-Mg non-arc basalts with 18–24 wt.% MgO produced by 30–45 vol.% partial melting of the ambient mantle together with lesser amounts of plume-related komatiites (de Wit and Ashwal, 1997), both of which indicate that the mantle was considerable hotter in the Mesoarchean–Neoarchean than today (Herzberg et al., 2007 and Herzberg et al., 2010). This is consistent with arguments based on the constancy of continental freeboard (Galer and Mezger, 1998 and Flament et al., 2008). The associated sedimentary rocks are mostly cherts and ironstones (Goodwin, 1991), and clastic sedimentary rocks only appear widely in the geological record after 3.2 Ga (Eriksson and Wilde, 2010). This is consistent with modeling that indicates the continents were mostly flooded until the Neoarchean (Flament et al., 2008 and Flament et al., 2011).

Although the thermal history of Earth is only poorly constrained (Korenaga, 2006,Korenaga, 2008, Korenaga, 2011, Korenaga, 2013, Labrosse and Jaupart, 2007, Silver and Behn, 2008, Davies, 2009 and van Hunen and Moyen, 2012), it is likely that prior to ca 3.0 Ga heating from radioactive decay would have exceeded surface heat loss, whereas since ca 2.5 Ga secular cooling has dominated the thermal history of the Earth. Ambient upper mantle temperature controls the tectonic regime and consequently the mechanism(s) for formation of primary crust and its reworking (Sizova et al., 2010,Sizova et al., 2014 and Johnson et al., 2014). Petrological data and generalizations from thermal history models suggest ambient upper-mantle temperatures in the Archean were significantly hotter than the present day, but with similar global variation in potential temperature of magma sources of ~120 °C (Herzberg et al., 2007, Herzberg et al., 2010 and van Hunen and Moyen, 2012), leading to a high percent of melting and generating thick MgO-rich primary crust underlain by highly residual mantle (Herzberg and Rudnick, 2012). However, the preserved volume of this crust is low suggesting much of it was recycled. Furthermore, since TTGs must have been sourced from broadly basaltic protoliths, as discussed above, they cannot have been generated directly from this primary MgO-rich crust.

Given the dependence of rheology on T and $P_{\text{fluid/melt}}$, the geodynamic regime in the Hadean–Archean may have been different and models for the formation of Hadean–Archean crust based on uniformitarian principles may be misleading. Geodynamic regimes proposed for the Hadean–Archean include: uniformitarian/non-uniformitarian mobile-lid plate tectonics; stagnant-lid plate tectonics, without subduction; episodic subduction events in a dominantly stagnant-lid plate tectonics regime; or, variation in regime from place to place and time to time. Further numerical geodynamic modeling, such as that recently done by Griffin et al. (2013), O'Neill et al. (2013) and Johnson et al. (2014), is required to separate plausible from improbable geodynamic regimes for this period in Earth history.

3.1.2. Geodynamics

The stability of primary crust in the early Archean may be investigated by coupling calculated phase equilibria for hydrated and anhydrous low to high MgO crust compositions and their complementary mantle residues for a range of temperatures with 2-D numerical geodynamic models to evaluate potential lithospheric instability (Johnson et al., 2014). The study by Johnson et al. (2014) showed that the density of primary crust increases more than the density of residual mantle decreases with increasing ambient mantle potential temperature. As a result, the base of MgO-rich primary crust could have become gravitationally unstable even when fully hydrated. Furthermore, Johnson et al. (2014) demonstrated that magmatically over-thickened MgO-rich crust (i.e. over-thickened with respect to the initial thickness) could have delaminated by Rayleigh–Taylor instabilities at mantle potential temperatures >1500–1550 °C. Delamination would generate a return flow of asthenospheric mantle and melting in both the dripping crust and the rising mantle. This process is illustrated in Fig. 2, which shows snapshots from a numerical experiment with 45-km-thick initial primary crust and mantle potential temperature of 1600 °C. Melting at the base of the overthickened hydrated primary crust is also plausible at the predicted temperatures.

Since most Archean primary crust would have had MgO >18 wt.%, TTG magmas in the Archean cannot have been produced directly from this crust (Johnson et al., 2014). Garnet-bearing hydrated basaltic crust with MgO <18 wt.% is required as the source of most TTGs (Johnson et al., 2014). Partial melting of the primary crust could have generated this lower MgO secondary crust; also, some lower MgO crust could have formed by crystal fractionation of sub-volcanic primary magmas. Such a scenario in which extensive reworking of the primary crust is a dominant process explains the preponderance of magmatic rocks of TTG affinity in the early Archean. Moreover, it provides an explanation for the fragmentary metamorphic record, since this early TTG crust was also subjected to extensive reworking (e.g. Friend and Nutman, 2005, Horie et al., 2010 and Nutman et al., 2013). Additionally, this scenario is consistent with production of the SCLM in a manner similar to that suggested by Spengler et al., 2006 and Spengler et al., 2008 via extensive melting in the Archean and subsequent modification, although it may be been thickened subsequently (Gray and Pysklywec, 2010).

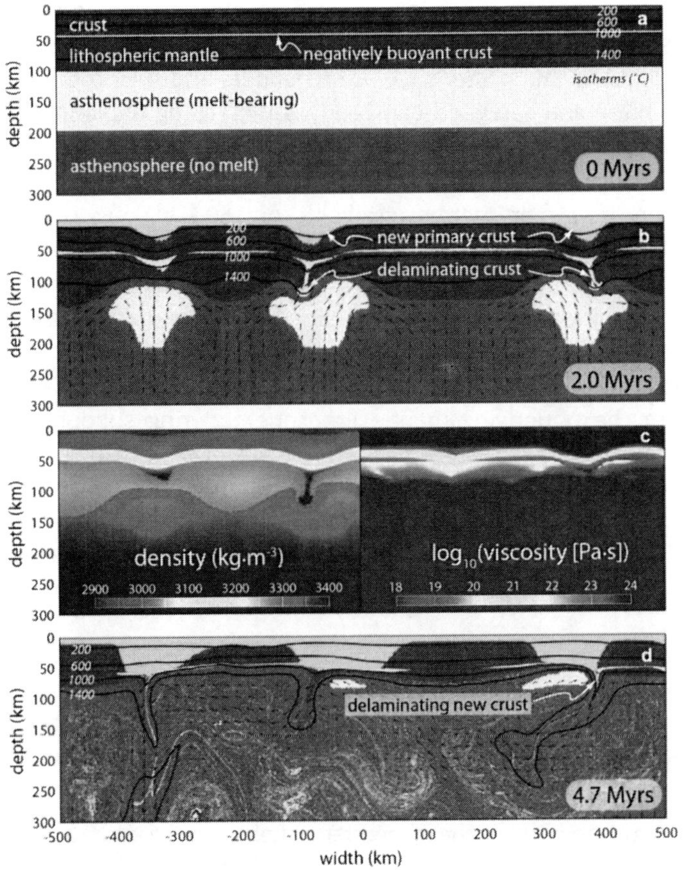

Figure 2. To show snapshots from a geodynamic model of delamination and recycling of crust caused by gravitational instabilities in a stagnant-lid plate tectonics regime. In the experiment shown, the initial primary crust was 45 km thick with a mantle potential temperature at the base of the lithosphere of 1600 °C. (a) To show the model configuration and initial conditions (dark blue, initial primary crust; light blue, negatively buoyant initial primary crust; dark pink brown, lithospheric mantle; light pink-brown, unmelted asthenosphere; yellow, melt-bearing asthenosphere). (b) To show local thickening of the initial primary crust by addition of new primary crust (green) and delamination of negatively buoyant initial primary crust. The positive buoyancy of the melt-bearing asthenosphere is the primary cause of the large-scale mantle flow; crustal drips induce smaller circulation patterns. (c) To show density (left) and viscosity (right) for the left-hand and right-hand halves of b, respectively. (d) To show local thickening and delamination of the negatively buoyant new primary crust. Figure reproduced from Johnson et al. (2014).

Unfortunately, modern quantitative metamorphic studies on crustal rocks older than Neoarchean are generally lacking. This gap in our knowledge must be filled in the future if we are to understand fully the evolution of the lithosphere and early Archean geodynamics. Using the limited available data, Brown (2009) argued that Paleoarchean–Mesoarchean crust generally registers low-to-moderate-*P*–moderate-to-high-*T*conditions, implying high but uniform apparent thermal gradients of 850–1350 °C/GPa. Ultrahigh temperature metamorphism appears to be absent from the geological record before the Neoarchean (Brown, 2009), although this may reflect a lack of recognition, and there is no imprint in the geological record of subduction of continental crust to mantle conditions prior to the Cryogenian (Brown, 2009), most likely due to decoupling of continental crust from the underlying mantle lithosphere during subduction and/or shallow slab breakoff on a hotter Earth (Sizova et al., 2014). This record is inconsistent with one-sided subduction and contemporary plate tectonics (Stern, 2005, Brown, 2006,Brown, 2010a and Brown, 2010b), as discussed above. This raises the question: at what point in the geological record of metamorphism do we find evidence of one-sided subduction and plate tectonics?

3.2. Post-Paleoarchean lithosphere and geodynamics

In contrast to the early Archean, modern quantitative metamorphic studies are widely available for crustal rocks younger than Mesoarchean. Where these studies may be linked to reliable geochronology, it is possible to interrogate the geological record to investigate the thermal history of the lithosphere.

3.2.1. Methods used to interrogate the crustal record of metamorphism

One aim of my recent metamorphic studies has been to invert the record of crustal metamorphism as a proxy for earlier thermal environments, particularly to elucidate changes in the geodynamic history of Earth, and then to test the outcome using 2-D numerical geodynamic modeling. There are two principal caveats associated with an approach based on the geological record. First, metamorphic imprints in the crust record discrete events, for example a change in plate kinematics or the evolution from subduction to terminal collision, whereas subduction is thought to be globally continuous, at least on a timescale shorter than the supercontinent cycle. Second, the record of crustal metamorphism is a function of what is preserved—if the preservation potential of some types of metamorphic rock was poor earlier in Earth history evidence of their existence may not have been retained and the geological record may be biased. Indeed, we know that the crustal record of metamorphism is generally biased in favor of preservation during periods of amalgamation of the continental lithosphere into supercontinents (Brown, 2006, Brown, 2007a, Brown, 2007b, Hawkesworth et al., 2009 and Cawood et al., 2013).

This type of analysis relies on two important features of high-temperature and high-pressure metamorphic rocks. First, that the close-to-peak mineral assemblages are robust recorders of *P* and *T*. This was discussed above in relation to metamorphic facies. In particular, at higher temperatures, prograde dehydration and melt loss produce nominally anhydrous mineral assemblages that are difficult to retrogress or overprint without fluid influx. For this reason, this study has been limited to rocks equilibrated under conditions of relatively high temperature, such as granulites and eclogites. Thus, for granulite facies metamorphism at $P < 1$ GPa a minimum temperature for inclusion was set at 700 °C, whereas for higher pressure metamorphism at $T < 700$ °C a minimum pressure of 1 GPa was applied to ensure a reasonable minimum temperature and the likelihood of an equilibrated peak mineral assemblage. Using these thresholds, the resulting *P–T* data are believed to be robust. Second, that in rocks overprinted during exhumation, close-to-peak phase assemblages are commonly preserved as inclusions in rock-forming or accessory minerals. This has proved to be particularly important for retrieving reliable information from ultrahigh-pressure metamorphic rocks.

In the discussion below, 214 *P–T–t* data, which is about a 50% increase over the size of the dataset used by Brown (2007a), have been grouped into three different types of metamorphism, as follows. (1) High-pressure–ultrahigh pressure (HP-UHP) metamorphism, characterized by lawsonite blueschists/lawsonite eclogites (27 examples) or low-temperature eclogites/UHP metamorphic rocks (59 examples), where*T* is that registered at maximum *P*, and the age of metamorphism is generally interpreted to be that around peak *P*. (2) Eclogite–high-pressure granulite (E-HPG) metamorphism (52 examples), characterized by medium-temperature eclogite or high-pressure granulite mineral assemblages, where peak *T* generally was achieved after maximum *P*, and the age of metamorphism is generally interpreted to be that around or slightly post peak *T*. (3)

Granulite–ultrahigh-temperature (G-UHT) metamorphism (76 examples), characterized by granulite or UHT mineral assemblages, where *P* is that registered at maximum *T*, and age of metamorphism is generally interpreted to be around or post–peak *T*, probably recording post-peak cooling across an elevated solidus in many cases, particularly in melt-depleted UHT metamorphic rocks.

As a general statement for the Phanerozoic, lawsonite blueschists/ lawsonite eclogites have been interpreted to record oceanic subduction, whereas low-temperature eclogites/UHP metamorphic rocks have been interpreted to record continental subduction and to define the suture formed during the transition to terminal continent–continent collision. However, with more studies during the past decade, this distinction is becoming less clear (Rebay et al., 2010 and Wei and Clarke, 2011). Thus, the development of lawsonite appears to be controlled more by H_2O-saturation versus undersaturation for a fixed bulk rock chemical composition than by tectonic setting (Rebay et al., 2010), and the development of blueschist versus eclogite seems to relate more to the bulk rock chemical composition—Ca-rich versus Ca-poor, respectively (Wei and Clarke, 2011). During the Proterozoic and the Phanerozoic, metamorphism associated with terminal continent–continent collision is generally characterized by medium-temperature eclogites and high-pressure granulites. In contrast, normal granulites and UHT metamorphic rocks were most likely generated in oceanic/continental arc–backarc settings or in orogenic hinterlands, such as the Altiplano or Tibet plateau.

I follow Miyashiro (1961) in believing that it is useful to classify metamorphic belts by type based on apparent thermal gradient in order to identify tectonic setting and understand secular change in global geodynamics. Apparent thermal gradient is the slope of the line from 0 °C/0 GPa through the peak *P–T* conditions as defined above; this is neither the metamorphic field gradient nor the piezothermic array (England and Richardson, 1977; Richardson and England, 1979) nor the transient geotherm at the metamorphic peak.

3.2.2. Results

Figure 3, Figure 4, Figure 5 and Figure 6 show plots of metamorphic belts classified by type, based on a judgment of the current best estimate of representative peak *P–T*conditions and age associated with these conditions for each metamorphic belt (Michael Brown, unpublished; based on the dataset from Brown (2007a, Tables 1–5) as updated in December 2013). From Fig. 3, it is clear that temperatures of metamorphism were uniformly high from the Neoarchean to the Paleozoic, but there is a dramatic drop in the temperature of metamorphism with the appearance of HP-UHP metamorphism in the Cryogenian–Ediacaran. This is unlikely to be related to the exclusion of data from low-temperature metamorphism at low-pressures, a conclusion supported by the data inFigs. 4 and 5. From Fig. 4, it is clear that pressures of metamorphism were uniformly low (G-UHT metamorphism) or moderate (E-HPG metamorphism) from the Neoarchean to the Paleozoic, but there is a dramatic increase in the pressure of metamorphism with the appearance of HP-UHP metamorphism.

In Fig. 5 it can be seen that although there is some variation in apparent thermal gradient within each type of metamorphism, nonetheless each falls within its own range of apparent thermal gradients, as follows. HP-UHP metamorphism mostly lies between apparent thermal gradients of 150 °C/GPa and 350 °C/GPa, with mean values of 242 °C/GPa for lawsonite blueschists/lawsonite eclogites, and 251 °C/GPa for low-temperature eclogites/UHP metamorphic rocks, respectively. E-HPG metamorphism mostly lies between apparent thermal gradients of 350 °C/GPa and 750 °C/GPa, with a mean value of 555 °C/GPa. G-UHT metamorphism mostly lies between apparent thermal gradients of 750 °C/GPa and 1500 °C/GPa, with a mean value of 1092 °C/GPa. Notwithstanding that this reflects the robustness of the metamorphic facies concept and is to be expected, it implies that each type of metamorphism is likely to have a characteristic tectonic setting with which it was associated, as discussed above. This becomes important when the temporal distribution of the different types of metamorphism is considered.

The temporal record of apparent thermal gradients retrieved from crustal rocks also provides information about secular change in thermal regimes, as shown in Fig. 6, which plots apparent thermal gradient against age. Simple linear regression suggests there is no significant change of apparent thermal gradient with age for G-UHT metamorphism. However, for E-HPG metamorphism, the apparent thermal gradient appears to decline slightly from the Neoarchean to the Paleozoic, although the slope of the linear regression is only 27 °C/GPa/Ga. Although simple linear regression suggests there is no significant change of apparent thermal gradient with age for low-temperature eclogites/UHP metamorphic rocks, the apparent thermal gradient appears to decline slightly through the Phanerozoic for lawsonite blueschists/lawsonite eclogites, although the slope of the linear regression is only 60 °C/GPa/Ga. However, based on average apparent thermal gradient, there is no significant difference between the thermal gradients recorded by low-temperature eclogites/UHP metamorphic rocks and lawsonite blueschists/lawsonite eclogites. Rather than recording "abnormally low geothermal gradients" (Tsujimori et al., 2006), both low-temperature eclogites/UHP metamorphic rocks and lawsonite blueschists/lawsonite eclogites record the characteristic thermal gradients of Phanerozoic subduction and differences in mineralogy appear to be largely a function of H_2O-saturation and bulk chemical composition rather than thermal gradient, as discussed above (Rebay et al., 2010 and Wei and Clarke, 2011).

Using these data we may investigate aspects of the thermal history of Earth, as recorded by the geological record, and the evolution of the lithosphere since the Mesoarchean. Furthermore, these data provide information that will help to clarify Precambrian geodynamics.

3.2.3. Mesoarchean to Neoarchean geodynamics

In the Mesoarchean–Neoarchean, the sporadic appearance of two types of metamorphism—E-HPG and G-UHT—with distinct apparent thermal gradients marks a transition to one-sided subduction and plate tectonics as the dominant tectonic regime (Brown, 2006, Brown, 2007a and Brown, 2008). This transition and subsequent changes in tectonic style may be investigated using 2-d numerical geodynamic modeling (van Hunen and van den Berg, 2008 and Sizova et al., 2010). In a series of experiments,Sizova et al. (2010) systematically investigated the dependence of geodynamic regime at a convergent ocean–continent plate margin on ambient upper-mantle temperature and a variety of additional parameters that have been suggested to be different in the Archean.

With increasing ambient upper-mantle temperature, the numerical experiments show unequivocally that one-sided subduction is destabilized principally by weakening of the oceanic lithosphere consequent upon increased melt flux from the underlying asthenospheric mantle. Inverting the sequence, there is a first-order transition from a no-subduction regime through a pre-subduction regime, involving limited underthrusting of oceanic lithosphere along an imposed zone of weakening at the edge of the continental lithosphere, to subduction as ambient upper-mantle temperatures decline from >200 to <175 °C warmer than present-day ambient upper-mantle. The experimental results are summarized in the series of snapshots from three numerical experiments with ambient upper mantle temperature shown in Fig. 7.

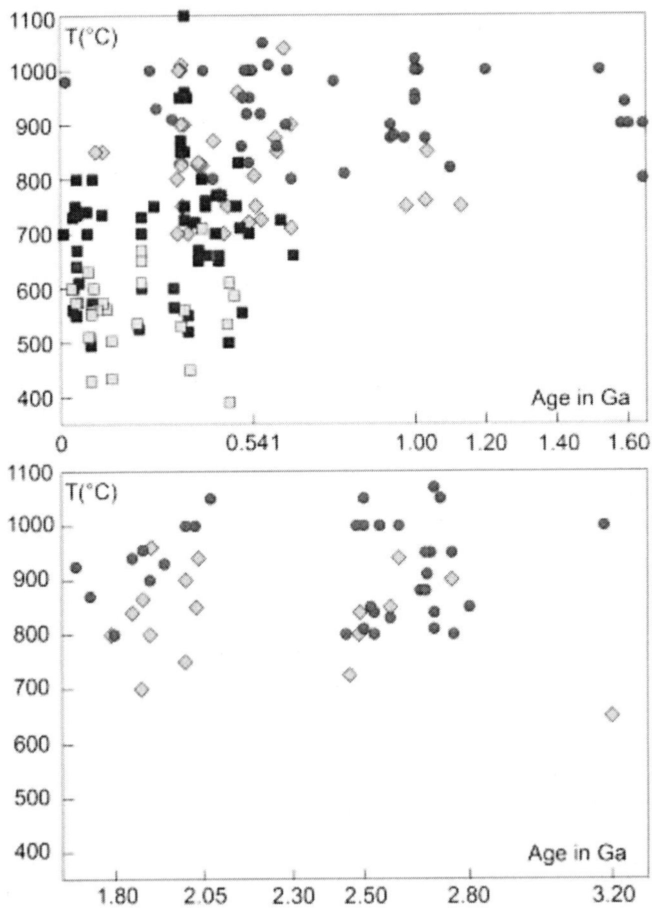

Figure 3. Peak metamorphic temperature (as defined in the text) of metamorphic belts by type plotted against age. Types of metamorphism are: granulite–ultrahigh temperature (G-UHT – orange circles) metamorphism; eclogite–high-pressure granulite metamorphism (E-HPG – light purple diamonds); and, high-pressure–ultrahigh-pressure metamorphism (HP-UHP), comprising lawsonite-bearing blueschists and eclogites (pale blue squares) and eclogites and ultrahigh-pressure metamorphic rocks (dark blue squares).

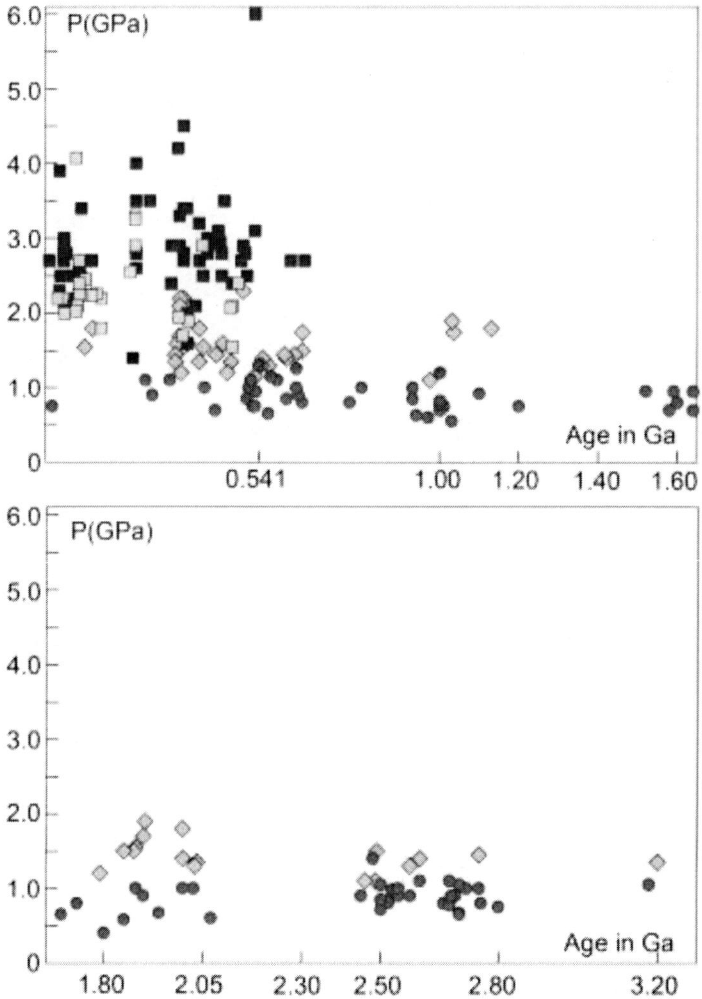

Figure 4. Peak metamorphic pressure (as defined in the text) of metamorphic belts by type plotted against age. Types of metamorphism are: granulite–ultrahigh temperature (G-UHT – orange circles) metamorphism; eclogite–high-pressure granulite metamorphism (E-HPG – light purple diamonds); and, high-pressure–ultrahigh-pressure metamorphism (HP-UHP), comprising lawsonite-bearing blueschists and eclogites (pale blue squares) and eclogites and ultrahigh-pressure metamorphic rocks (dark blue squares).

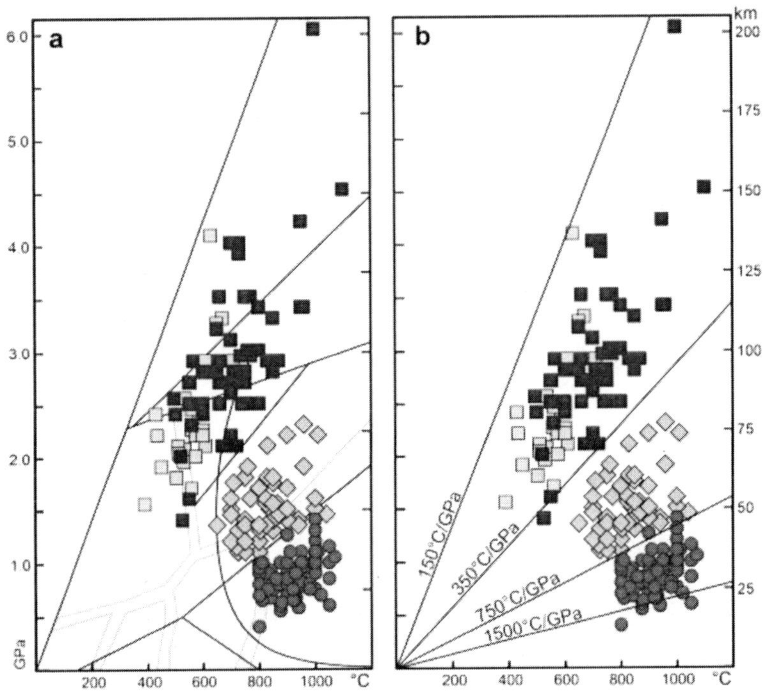

Figure 5. Peak metamorphic *P–T* conditions (as defined in the text) of metamorphic belts by type plotted on the *P–T*diagram of Fig. 1, extended to 6.0 GPa; in (b) four apparent thermal gradients are shown, as discussed in the text. Types of metamorphism are: granulite–ultrahigh temperature (G-UHT – orange circles) metamorphism; eclogite–high-pressure granulite metamorphism (E-HPG – light purple diamonds); and, high-pressure–ultrahigh-pressure metamorphism (HP-UHP), comprising lawsonite-bearing blueschists and eclogites (pale blue squares) and eclogites and ultrahigh-pressure metamorphic rocks (dark blue squares).

Neither higher radiogenic heat production nor an increased thickness of ocean crust have any significant effect on the ambient upper mantle temperature at which subduction becomes impossible. Furthermore, although increasing the density of the ocean crust via the transformation of basalt to eclogite, increasing the velocity of the ocean plate and decreasing the density of the subcontinental lithospheric mantle allow subduction to continue to ~50 °C higher ambient upper mantle temperature, subduction eventually becomes impossible at ambient upper-mantle temperatures

>200 °C warmer than the present-day ambient upper-mantle regardless of changes to these parameters (Sizova et al., 2010).

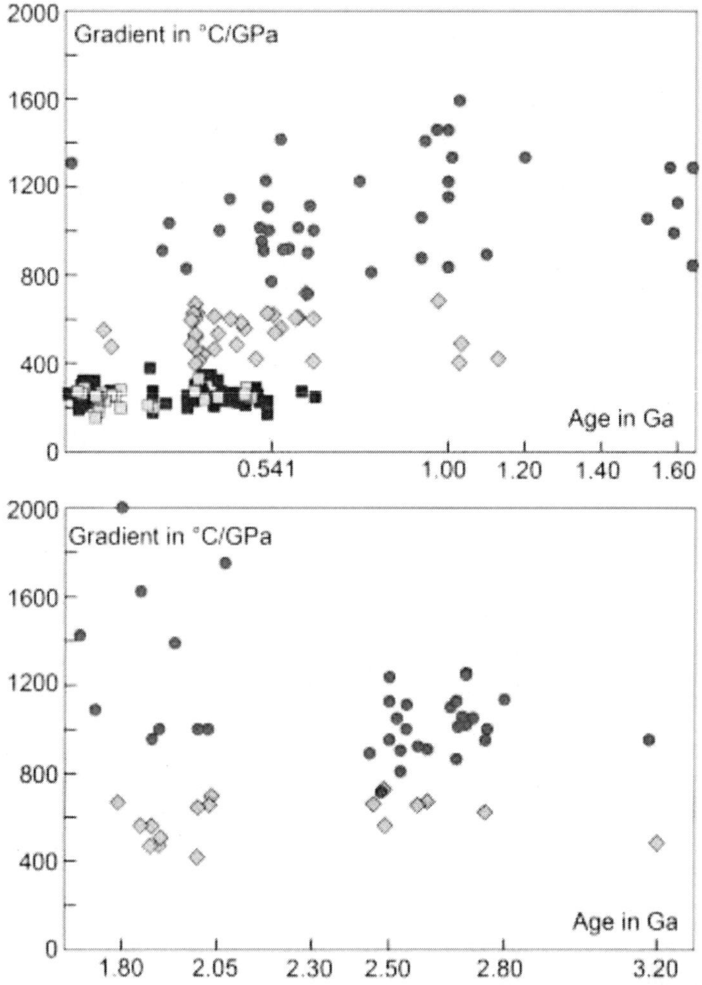

Figure 6. Apparent thermal gradients (as defined in the text) of metamorphic belts by type plotted against age. Types of metamorphism are: granulite–ultrahigh temperature (G-UHT – orange circles) metamorphism; eclogite–high-pressure granulite metamorphism (E-HPG – light purple diamonds); and, high-pressure–ultrahigh-pressure metamorphism (HP-UHP), comprising lawsonite-bearing blueschists and eclogites (pale blue squares) and eclogites and ultrahigh-pressure metamorphic rocks (dark blue squares).

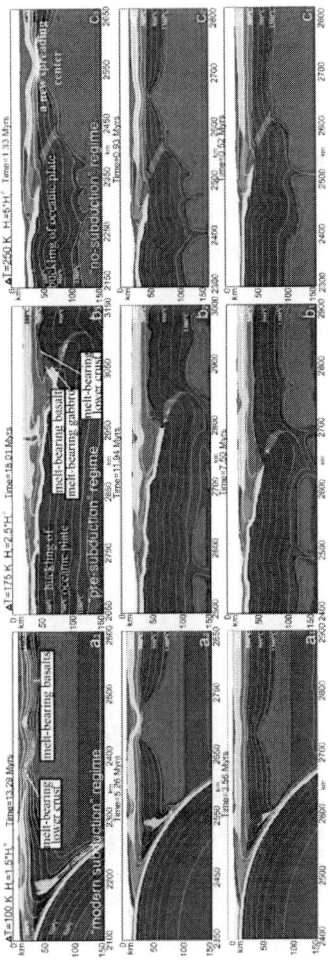

Figure 7. Snapshots from a geodynamic model used to investigate the transition from mobile-lid to stagnant-lid plate tectonics with increasing temperature of the ambient upper mantle. In the experiments, the upper-mantle temperature (ΔT) and the crustal radiogenic heat production (H) are varied independently from the present day values. The three experiments shown identify three different tectonic regimes: (a) the modern subduction regime (ΔT = 0–175 K); (b) a pre-subduction regime characterized by underthrusting but not subduction (ΔT = 175–250 K); and, (c) a no-subduction regime ($\Delta T \geq$ 250 K). The lower boundary of the melt-bearing asthenospheric mantle (red areas) in (a) and (b) is located at depths of 100–150 km, whereas in (c) this boundary extends to depths of 200 km (i.e. below the bottom of the model). Figure reproduced from Sizova et al. (2010) with permission (RightsLink License Number 3292490716639).

Taking a present-day mid-point value of 1340 °C for the mantle potential temperature of ambient mantle below oceanic ridges (Herzberg et al., 2007), the transition from a stagnant lid regime dominated by Rayleigh–Taylor instabilities to subduction and plate tectonics likely occurred at a mantle potential temperature in the range 1550–1500 °C, which corresponds to ambient mantle potential temperatures during the Neoarchean–Paleoproterozoic (Herzberg et al., 2010 and Johnson et al., 2014). Thus, the transition to dominantly steep-slab subduction occurred during the late Archean to early Proterozoic as Rayleigh–Taylor instabilities became increasingly inefficient with declining ambient mantle potential temperature (Johnson et al., 2014). This transition is consistent with the scarcity of eclogites in the orogenic rock record before 1.9 Ga, whereas eclogites recording P up to 2.0 GPa become more common through the remainder of the Proterozoic. This is the beginning of the Proterozoic plate tectonics regime.

One feature argued to be characteristic of the Archean is the reported similarity of metamorphic facies and erosion level from craton to craton (e.g. Binns et al., 1976 and Galer and Mezger, 1998). This postulate may be examined by reference toFigure 3, Figure 4 and Figure 6. From Fig. 3 it is clear that peak temperature varies widely within and overlaps considerably between G-UHT and E-HPG types of metamorphism. In contrast, from Fig. 4 we see that pressure does separate these types of metamorphism with minimal overlap, even in the Archean. Thus, in Fig. 6 these two types of metamorphism are seen to have distinct apparent thermal gradients. Thus, the postulate of similarity of metamorphic facies and erosion level from craton to craton is not supported by the data, except that the average apparent thermal gradients for the paired E-HPG and G-UHT metamorphism are similar from throughout the Neoarchean.

3.2.4. Proterozoic geodynamics

Although subduction is generally considered to be an ongoing close-to-steady-state process, it is clearly interrupted from time to time by plate collisions, such as that occurring at the present day along the northern edge of the Australian plate, and cyclically by the formation of supercontinents, although to what extent these processes change the relative lengths of the different types of plate boundaries and/or the rates of spreading and subduction is uncertain (Silver and Behn, 2008 and Korenaga, 2008). Therefore, it is no surprise to discover that the age distribution of exposed metamorphic rocks during the Proterozoic is clustered. In Fig. 6 it is apparent that there are gaps in the data in the intervals 2.45–2.05 Ga, 1.53–1.15 Ga and 0.92–0.66 Ga, which correspond in turn to supercraton breakup (Bleeker, 2003) and the supercontinent cycle from Nuna (Columbia; Zhao et al., 2004) to Rodinia (Li et al., 2008) and Gondwana–Pangea (Ernst et al., 2013). This suggests that during the Proterozoic collision-related metamorphic rocks with apparent thermal gradients of 350–750 °C/GPa were preserved as a result of orogenesis associated with supercontinent assembly (Brown, 2008 and Brown, 2009). Thus, the metamorphic record does not allow us to address the issue of whether, once established, Proterozoic plate tectonics continued without interruption or was shut down by supercontinent assembly, as argued by Silver and Behn (2008), or was episodic, driven by mantle avalanche events, as argued by O'Neill et al. (2007). Furthermore, this issue cannot be resolved by arguments based on apparent gaps in the record of magmatic activity, as argued by Condie et al. (2009), or its perceived continuity, as argued by Partin et al. (2014), since the record may be incomplete due to differential preservation of igneous rocks from tectonic setting to tectonic setting (Hawkesworth et al., 2009 and Cawood et al., 2013).

3.2.5. Proterozoic versus Phanerozoic orogenesis

There are differences in the style of collision tectonics between Proterozoic and Phanerozoic orogens, such as the widespread occurrence of G-UHTM in the former and of blueschists and UHPM in the latter. What controls these differences in style may be investigated using 2-D numerical geodynamic modeling.

To provide a starting point it is necessary to understanding how contemporary continental collisions work on Earth (Sizova et al., 2012) and then extrapolate these models back in time. To do this, Sizova et al. (2014) varied four first-order parameters in a series of numerical experiments. Parameters investigated were ambient upper-mantle temperature (increased in a series of steps up to 150 °C higher than the present-day value), radiogenic heat production in the crust (increased to one and a half times the present value), thickness of the continental lithosphere (increased from 140 km to 160 km), and the chemical density contrast between the subcontinental lithospheric mantle and the asthenospheric mantle (increased from a difference of 20 kg/m^3, corresponding to the present-day, to 50 kg/m^3, corresponding to the Proterozoic (Djomani et al., 2001)).

Results of their 2-d numerical experiments show that an increase in the mantle potential temperature to >80–100 °C above present values, corresponding to conditions of the late Proterozoic, leads to shallow slab-breakoff (Fig. 8) and modes of collision tectonics that are distinctly different from contemporary continental collision. Thus, it is the transition from shallow slab breakoff to deep slab breakoff during the Cryogenian–Cambrian and stronger crust–mantle coupling that enabled continental subduction to mantle depths as upper mantle temperature declined to <100 °C warmer than at present. This is the beginning of the modern plate tectonics regime.

The transition to the modern plate tectonics regime is shown dramatically if the data from metamorphic belts are separated into three age groups—older than the Cryogenian (>0.850 Ga), Cryogenian to Cambrian (0.850–0.485 Ga) and younger than the Cambrian (<0.485 Ga)—and replotted in *P*–*T* space, as shown in Fig. 9. The Cryogenian to Cambrian period represents the global transition from shallow to deep slab breakoff as the overall decline in ambient upper mantle temperatures allowed the lithosphere to strengthen sufficiently to maintain coherency during subduction to greater mantle depths before breakoff occurred. Deeper subduction prior to slab breakoff generates a colder environment in the subduction zone, as recorded by the appearance of HP-UHP metamorphism in the geological record (Maruyama and Liou, 1998). This regime is not only characterized by cold subduction and metamorphism of continental crust at mantle depths, but also by the decline in occurrence of ultrahigh-temperature metamorphism and its virtual absence from the geological record after the Paleozoic (Figs. 6 and 9).

Sizova et al. (2014) identified two different tectonic regimes at upper mantle temperature >80–100 °C warmer than at present, first a truncated hot collision regime, controlled by the presence of strong lower continental crust, and second a two-sided hot collision regime, controlled by the presence of weak lower continental crust. Shallow slab breakoff in both regimes precludes formation of UHPM rocks (Sizova et al., 2014). This transition to the modern collision regime would have occurred during the late Proterozoic, suggesting that collisional tectonic regimes in the Proterozoic were different compared to those in the Phanerozoic.

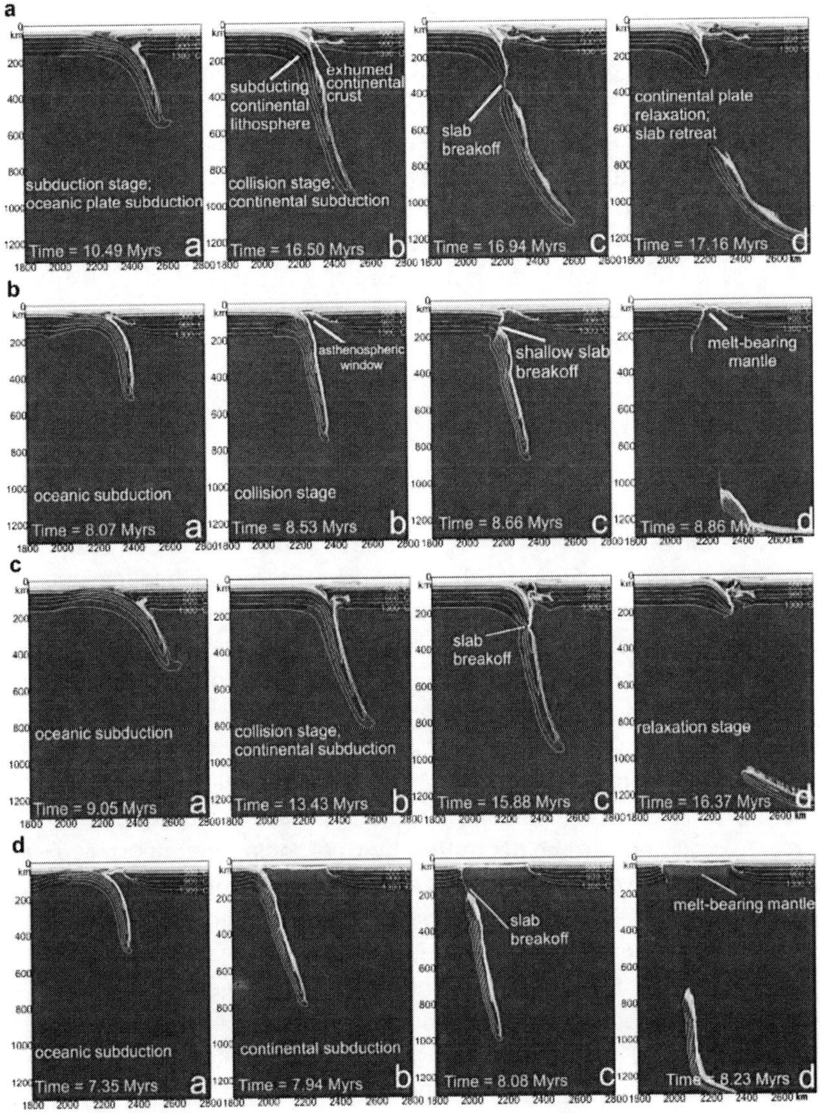

Figure 8. Snapshots from a geodynamic model used to investigate the transition from the modern collision regime to a hot collision regime with increasing temperature of the ambient upper mantle. Shown are two examples of different exhumation styles for UHPM rocks in the modern collision regime (large-scale crustal stacking and trans-lithospheric diapirism) and the consequences of a hotter mantle for the depth of slab breakoff. (a) Large-scale evolution of the large-scale crustal stacking model with a 600 km long oceanic plate and an abrupt (50 km wide) passive margin. The detailed evolution of the collision zone for this model is shown

in Sizova et al. (2014, Fig. 2). (b) Large-scale evolution of the crustal stacking model at ΔT = 100 K. A detailed snapshot of the collision zone for this model is shown in Sizova et al. (2014, Fig. 3e). (c) Large-scale evolution of the trans-lithospheric diapirism model with a 600 km long oceanic plate and a gradual (150 km wide) passive margin. The detailed evolution of the collision zone for this model is shown in Sizova et al. (2014, Fig. 6). (d) Large-scale evolution of the trans-lithospheric diapirism model at ΔT = 150 K. A detailed snapshot of the collision zone for this model is shown in Sizova et al. (2014, Fig. 7c). Figures reproduced from Sizova et al. (2014) with permission (RightsLink License Number 3292490560289).

There is a range of different styles of collisional tectonic regimes recorded by Proterozoic orogens, which may reflect the variation in mantle potential temperature globally. Some Proterozoic orogens record heating prior to thickening, expressed as CCW metamorphic*P–T–t* paths that reach G-UHT metamorphic conditions, followed by close-to-isobaric cooling; these orogens have similarities with ongoing convergence following truncated hot collision in the geodynamic models. Examples of this first style of Proterozoic orogenesis include the Paleoproterozoic Khondalite belt in the North China craton and the late Mesoproterozoic–early Neoproterozoic Eastern Ghats Province–Rayner Province of India and East Antarctica. Other Proterozoic orogens are characterized by CW looping metamorphic *P–T–t* paths and extensive granite magmatism sourced from metasedimentary and plutonic crust (syn-orogenic) and LM (post-orogenic); these orogens have similarities with two-sided hot collision in the geodynamic models. Examples of this second style of Proterozoic orogenesis include the Paleoproterozoic orogens in the southwest of the American continent (Mojave, Yavapai and Mazatzal provinces), the Paleoproterozoic Svecofennides of southern Finland, and the late Mesoproterozoic Namaqua orogen in Namibia. Finally, there are some Proterozoic orogens where the collision style was similar to large Phanerozoic orogens, but without generating HP-UHP metamorphic conditions along the future suture, perhaps identifying locations where contemporary ambient upper mantle temperature was cooler. Examples of this style of orogenesis in the Proterozoic include the Paleoproterozoic Trans–Hudson orogen of North America and the Mesoproterozoic Grenville orogen of North America.

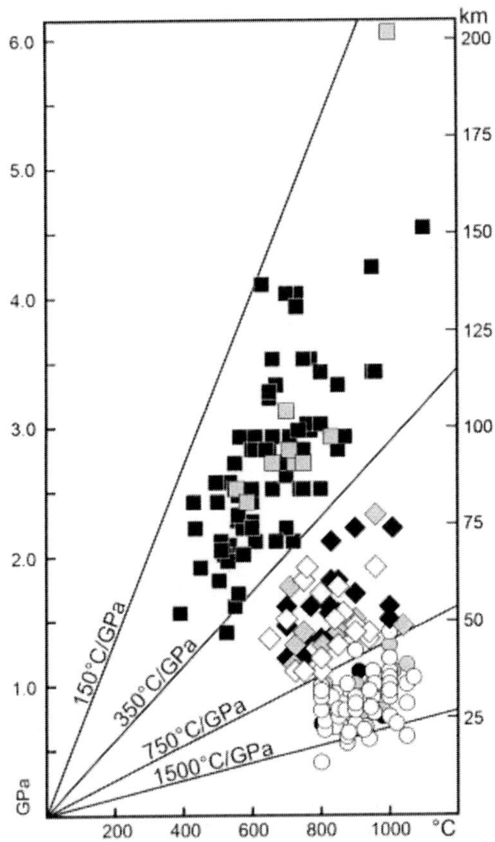

Figure 9. Metamorphic belts plotted by age in relation to thermal gradients where white symbols are belts where peak *P–T* conditions were recorded at >0.850 Ga, gray symbols are belts where peak *P–T* conditions were recorded at <0.850 but >0.485 Ga, and filled symbols are belts where peak *P–T* conditions were recorded at <0.485 Ga. Types of metamorphism are: G-UHT metamorphism – circles; E-HPG metamorphism – diamonds; and, HP-UHP metamorphism – squares.

4. CONCLUDING REMARKS

The data used in this study of secular change in the *P–T* conditions retrieved from metamorphic belts is taken from the literature, which was culled to 500 articles that were judged to provide robust estimates of *P–T* conditions and reliable age information. These articles are strongly skewed towards the twenty-first century, with 23% dating from 2010, 45% dating from 2010 to 2001, 22% dating from 2000 to 1991 and only 10% dating from before 1991. Furthermore, the dataset used in the present study is approximately 50% larger than the one used in the first iteration 8 years ago (214 *P–T–t* data in this study versus 144 *P–T–t* data used in the study of Brown, 2006 and Brown, 2007a). This is largely a reflection of the wider application of the advances discussed in the early part of this review, particularly those relating to thermodynamic modeling of phase equilibria and the combination of *in situ* geochronology with *in situ* trace element geochemistry of accessory minerals, which have led to many more precise estimates of *P–T* and a significant increase in the number of age determinations that are linked to these estimates. Consequently, this type of data compilation and analysis could not have been done reliably 30 years ago.

The trends of secular change in metamorphism identified by Brown, 2006 and Brown, 2007a are confirmed by this study. Superimposed on the secular decrease in ambient upper mantle temperature from the late Archean to the present are two changes in the style of metamorphism that are interpreted to record differences in geodynamic regime. The first change in the style of metamorphism is interpreted to register the transition from a stagnant lid to subduction and a plate tectonics geodynamic regime that occurred during the Neoarchean (Sizova et al., 2010). This transition is reflected in changes in many different elements of the Earth system during this time (Reddy and Evans, 2009,Condie and O'Neill, 2010, Bradley, 2011, Keller and Schoene, 2012, Condie and Kröner, 2013 and Young, 2013), in addition to the metamorphic record (Brown, 2006 and Brown, 2007a). The second change in the style of metamorphism is interpreted to register the transition from shallow to deep slab breakoff that occurred during the late Neoproterozoic (Sizova et al., 2014). This transition also is reflected in changes in many different elements of the Earth system during same period (Maruyama and Liou, 1998, Campbell and Squire, 2010, Bradley, 2011 and Young, 2013), in addition to the metamorphic record (Brown, 2006 and Brown, 2007a).

However, there remain some fundamental issues that must be resolved if we are to fully understand the evolution of the lithosphere in the Precambrian.

First, we need to determine how the heat for regional-scale UHT metamorphism was generated. Recent 1-D numerical thermal modeling implicates high internal concentrations of heat producing elements and low rates of erosion as essential contributors to power UHT metamorphism, as might occur, for example, in long-lived orogenic hinterlands (Clark et al., 2011). In contrast, ridge subduction and slab breakoff are currently popular in the literature as mechanisms to drive UHT metamorphism, probably in association with advected heat carried by mafic magmatism, as recently reviewed by Santosh et al. (2012). Thus, we need to determine the relative importance of enhanced radiogenic heat production and slow erosion versus ridge subduction and slab breakoff as driving forces for UHT metamorphism. In addition, alternative templates for the generation of hot orogens, such as those suggested by the 2-d numerical geodynamic modeling of Sizova et al. (2014), should be investigated in more detail.

Second, we need to determine how to interpret pressures calculated from UHP metamorphic terranes in relation to lithospheric depth. Do the recorded pressures simply reflect the lithostatic load or is there a component of the total pressure that is attributable to tectonic overpressure that was sustained for long enough to be recorded by the mineral assemblage (Moulas et al., 2013)?

Third, we need to obtain quantitative information on the *P–T–t* evolution of early Archean crust. For high-grade gneiss terrains, this will only be achieved once we have thermodynamic models for investigating suprasolidus conditions in a much wider range of protolith compositions than is possible at present, particularly for tonalitic and basaltic protoliths that are ubiquitous in these terrains. Work on these models is currently underway.

In spite of an incomplete geological record and limitations on our knowledge of geodynamic processes in the past, we are making progress in understanding lithospheric evolution, particularly in respect of our comprehension of the Precambrian. The way forward is clear.

ACKNOWLEDGMENTS

This paper was presented orally at the International Meeting on Precambrian Evolution and Deep Exploration of the Continental Lithosphere, 7–9 October 2013, Beijing, China, with support from the International Precambrian Research Centre of China. The author thanks Chris Yakymchuk for assistance with the figures, and Anonymous and Peter Cawood for their helpful review comments and suggestions.

REFERENCES

1. Anderson, J.R., Payne, J.L., Kelsey, D.E., Hand, M., Collins, W.J., Santosh, M., 2012. High-pressure granulites at the dawn of the Proterozoic. Geology 40, 431-434.

2. Arndt, N., 1999. Why was flood volcanism on submerged continental platforms so common in the Precambrian? Precambrian Researc 97, 15-164.

3. Artemieva, I.M., Mooney, W.D., 2001. Thermal thickness and evolution of Precambrian lithosphere: a global study. Journal of Geophysical Research e Solid Earth 106, 16,387-16,414.

4. Beaumont, C., Fullsack, P., Hamilton, J., 1994. Styles of crustal deformation caused by subduction of the underlying lithosphere. Tectonophysics 232, 119-132.

5. Beaumont, C., Ellis, S., Hamilton, J., Fullsack, P., 1996a. Mechanical model for subduction-collision tectonics of Alpine-type compressional orogens. Geology 24, 675-678.

6. Beaumont, C., Fullsack, P., Hamilton, J., 1996b. Styles of crustal deformation in compressional orogens caused by subduction of the underlying lithosphere. Tectonophysics 232, 119-132.

7. Beaumont, C., Jamieson, R.A., Butler, J.P., Warren, C.J., 2009. Crustal structure: a key constraint on the mechanism of ultra-high-pressure rock exhumation. Earth and Planetary Science Letters 287, 116-129.

8. Beaumont, C., Jamieson, R.A., Nguyen, M.H., Medvedev, S., 2004. Crustal channel flows: 1. Numerical models with applications to the tectonics of the HimalayaneTibetan orogen. Journal of Geophysical Research e Solid Earth 109, B06406.

9. Beran, A., Libowtzky, E., 2004. Spectroscopic methods in Mineralogy. EMU Notes in Mineralogy 6, 655 pp.

10. Berman, R.G., 1988. Internally-consistent thermodynamic data for minerals in the system Na2OeK2OeCaOeMgOeFeOeFe2O3eAl2O3 eSiO 2eTiO2eH2OeCO2. Journal of Petrology 29, 445-522.

11. Berman, R.G., 1991. Thermobarometry using multi-equilibrium calculations e a new technique, with petrological applications. Canadian Mineralogist 29, 833-855.

12. Beyssac, O., Goffe, B., Chopin, C., Rouzaud, J.N., 2002. Raman spectra of carbonaceous material in metasediments: a new geothermometer. Journal of Metamorphic Geology 20, 859-871.

13. Bleeker, W., 2003. The late Archean record: a puzzle in ca. 35 pieces. Lithos 71, 99-134.

14. Binns, R.A., Gunthorpe, R.J., Groves, D.I., 1976. Metamorphic Patterns and Development of Greenstone Belts in the Eastern Yilgarn Block, Western Australia. In: Windley, B.F. (Ed.), The Early History of the Earth. Wiley, New York, N.Y., pp. 303-316.

15. Bradley, D.C., 2011. Secular trends in the geologic record and the supercontinent cycle. Earth-Science Reviews 108, 16-33.

16. Brown, M., 1994. The generation, segregation, ascent and emplacement of granite magma: the migmatite-to-crustally-derived granite connection in thickened orogens. Earth-Science Reviews 36, 83-130.

17. Brown, M., 2006. Duality of thermal regimes is the distinctive characteristic of plate tectonics since the Neoarchean. Geology 34, 961-964.

18. Brown, M., 2007a. Metamorphic conditions in orogenic belts: a record of secular change. International Geology Review 49, 193-234.

19. Brown, M., 2007b. Metamorphism, plate tectonics, and the supercontinent cycle. Earth Science Frontiers 14, 1-18.

20. Brown, M., 2008. Characteristic thermal regimes of plate tectonics and their metamorphic imprint throughout Earth history: when did Earth first adopt a plate tectonics mode of behavior? In: Condie, K., Pease, V. (Eds.), When Did Plate Tectonics Begin?, Geological Society of America Special Paper, vol. 440, pp. 97-128.

21. Brown, M., 2009. Metamorphic patterns in orogenic systems and the geological record. In: Cawood, P.A., Kröner, A. (Eds.), Accretionary

Orogens in Space and Time, Geological Society, London, Special Publications, vol. 318, pp. 37-74.

22. Brown, M., 2010a. The spatial and temporal patterning of the deep crust and implications for the process of melt extraction. Philosophical Transactions of the Royal Society, Series A 368, 11-51.

23. Brown, M., 2010b. Paired metamorphic belts revisited. Gondwana Research 18, 46-59.

24. Brown, T.H., Berman, R.G., Perkins, E.H., 1988. Ge0eCalc e Software package for calculation and display of pressure-etemperatureecomposition phase-diagrams using an IBM or compatible personal-computer. Computers & Geosciences 14, 279-289.

25. Campbell, I.H., Squire, R.J., 2010. The mountains that triggered the late neoproterozoic increase in oxygen: the Second Great Oxidation Event. Geochimica et Cosmochimica Acta 74, 4187-4206.

26. Cawood, P.A., Hawkesworth, C.J., Dhuime, B., 2013. The continental record and the generation of continental crust. Geological Society of America Bulletin 125, 14-32.

27. Cesare, B., Acosta-Vigil, A., Ferrero, S., Bartoli, O., 2011. Melt inclusions in migmatites and granulites. Journal of the Virtual Explorer (Electronic Edition) 40 paper 2. Chopin, C., 1984. Coesite and pure pyrope in high-grade blueschists of the Western Alps e a 1st record and some consequences. Contributions to Mineralogy and Petrology 86, 107-118.

28. Chopin, C., 2003. Ultrahigh-pressure metamorphism: tracing continental crust into the mantle. Earth and Planetary Science Letters 212, 1-14.

29. Clark, C., Fitzsimons, I.C.W., Healy, D., Harley, S.L., 2011. How does the continental crust get really hot? Elements 7, 235-240.

30. Clark, M.K., Royden, L.H., 2000. Topographic ooze: building the eastern margin of Tibet by lower crustal flow. Geology 28, 703-706.

31. Condie, K., 1984. Archean geotherms and supracrustal assemblages. Tectonophysics 105, 29-41.

32. Condie, K.C., O'Neill, C., Aster, R.C., 2009. Evidence and implications for a widespread magmatic shutdown for 250 My on Earth. Earth and Planetary Science Letters 282, 294-298.

33. Condie, K.C., Kröner, A., 2013. The building blocks of continental crust: evidence for a major change in the tectonic setting of continental growth at the end of the Archean. Gondwana Research 23, 394-402.

34. Condie, K.C., O'Neill, C., 2010. The Archean-Proterozoic boundary: 500 My of tectonic transition in Earth history. American Journal of Science 310, 775-790.

35. Connolly, J.A.D., 1990. Multivariable phase-diagrams e an algorithm based on generalized thermodynamics. American Journal of Science 290, 666-718.

36. Connolly, J.A.D., 2005. Computation of phase equilibria by linear programming: a tool for geodynamic modeling and its application to subduction zone decarbonation. Earth and Planetary Science Letters 236, 524-541.

37. Connolly, J.A.D., 2009. The geodynamic equation of state: what and how. Geochemistry Geophysics Geosystems 10, Q10014

38. Connolly, J.A.D., Kerrick, D.M., 1987. An algorithm and computer-program for calculating composition phase-diagrams. CALPHAD 11, 1-55.

39. Connolly, J.A.D., Petrini, K., 2002. An automated strategy for calculation of phase diagram sections and retrieval of rock properties as a function of physical conditions. Journal of Metamorphic Geology 20, 697-708.

40. Currie, C.A., Beaumont, C., 2011. Are diamond-bearing Cretaceous kimberlites related to low-angle subduction beneath western North America? Earth and Planetary Science Letters 303, 59-70.

41. Davies, G.F., 2006. Gravitational depletion of the early Earth's upper mantle and the viability of early plate tectonics. Earth and Planetary Science Letters 243, 376-382.

42. Davies, G.F., 2009. Effect of plate bending on the Urey ratio and the thermal evolution of the mantle. Earth and Planetary Science Letters 287, 513-518.

43. de Capitani, C., 1994. Gleichgewichts-Phasendiagramme: Theorie und Software. Beihefte zum European Journal of Mineralogy 72, 48. Jahrestagung der Deutschen Mineralogischen Gesellschaft 6.

44. de Capitani, C., Brown, T.H., 1987. The computation of chemical-equilibrium in complex-systems containing nonideal solutions. Geochimica et Cosmochimica Acta 51, 2,639-2,652.

45. de Capitani, C., Petrakakis, K., 2010. The computation of equilibrium assemblage diagrams with Theriak/Domino software. American Mineralogist 95, 1,006-1,016.

46. de Roever, W.P., 1956. Some differences between post-paleozoic and older regional metamorphism. Geologie en Mijnbouw 18, 123-127.

47. de Roever, W.P., 1965. On the cause of the preferential distribution of certain metamorphic minerals in orogenic belts of different age. Geologische Rundschau 54, 933-943.

48. de Wit, M.J., Ashwal, L.D., 1997. Greenstone Belts. Clarendon Press, Oxford, 840 pp. Debaille, V., O'Neill, C., Brandon, A.D., Haenecour, P., Yin, Q.Z., Mattielli, N., Treiman, A.H., 2013. Stagnant-lid tectonics in early Earth revealed by 142Nd variations in late Archean rocks. Earth and Planetary Science Letters 373, 83-92.

49. Dhuime, B., Hawkesworth, C.J., Cawood, P.A., Storey, C.D., 2012. A change in the geodynamics of continental growth 3 billion years ago. Science 335, 1334-1336.

50. Djomani, Y.H.P., O'Reilly, S.Y., Griffin, W.L., Morgan, P., 2001. The density structure of subcontinental lithosphere through time. Earth and Planetary Science Letters 184, 605-621.

51. D'Lemos, R.S., Brown, M., Strachan, R.A., 1992. Granite magma generation, ascent and emplacement within a transpressional orogen. Journal of the Geological Society, London 149, 487-490.

52. Dobrzhinetskaya, L.F., Faryad, S.W., Wallis, S., Cuthbert, S., 2011. Ultrahigh-Pressure Metamorphism 25 Years after the Discovery of Coesite and Diamond. Elsevier, 670 pp.

53. Donaldson, D.G., Webb, A.A.G., Menold, C.R., Kylander-Clark, A.R.C., Hacker, B.R., 2013. Petrochronology of Himalayan ultrahigh-pressure eclogite. Geology 41, 835-838.

54. Doin, M.P., Fleitout, L., Christensen, U., 1997. Mantle convection and stability of depleted and undepleted continental lithosphere. Journal of Geophysical Research e Solid Earth 102, 2771-2787.

55. Dubessy, J., Caumon, M.-C., Rull, F., 2012. Raman spectroscopy applied to earth sciences and cultural heritage. EMU Notes in Mineralogy 12, 438 pp.

56. Duretz, T., Gerya, T.V., May, D.A., 2011. Numerical modelling of spontaneous slab breakoff and subsequent topographic response. Tectonophysics 502, 244-256.

57. Ellis, D.J., 1980. Osumiliteesapphirineequartz granulites from Enderby Land, Antarctica e PeT conditions of metamorphism, implications for garnetecordierite equilibria and the evolution of the deep crust. Contributions to Mineralogy and Petrology 74, 201-210.

58. Enami, M., Nishtyama, T., Mouri, T., 2007. Laser Raman microspectrometry of metamorphic quartz: a simple method for comparison of metamorphic pressures. American Mineralogist 92, 1303-1315.

59. England, P., Bickle, M., 1984. Continental thermal and tectonic regimes during the Archean. Journal of Geology 92, 353-367.

60. England, P.C., Richardson, S.W., 1977. The influence of erosion upon the mineral facies of rocks from different metamorphic environments. Journal of the Geological Society, London vol. 134, 201-213.

61. Eriksson, K.A., Wilde, S.A., 2010. Palaeoenvironmental analysis of Archaean siliciclastic sedimentary rocks in the westecentral Jack Hills, Western Australia with new constraints on ages and correlations. Journal of the Geological Society, London 167, 827-840.

62. Ernst, R.E., Bleeker, W., Soderlund, U., Kerr, A.C., 2013. Large Igneous Provinces and supercontinents: toward completing the plate tectonic revolution. Lithos 174, 1-14.

63. Ernst, W.G., 1972. Occurrence and mineralogic evolution of blueschist belts with time. American Journal of Science 272, 657-668.

64. Faccenda, M., Gerya, T.V., Chakraborty, S., 2008. Styles of post-subduction collisional orogeny: influence of convergence velocity, crustal rheology and radiogenic heat production. Lithos 103, 257-287.

65. Fischer, K.M., 2002. Waning buoyancy in the crustal roots of old mountains. Nature 417, 933-936.

66. Flament, N., Coltice, N., Rey, P.F., 2008. A case for late-Archaean continental emergence from thermal evolution models and hypsometry. Earth and Planetary Science Letters 275, 326-336.

67. Flament, N., Rey, P.F., Coltice, N., Dromart, G., Olivier, N., 2011. Lower crustal flow kept Archean continental flood basalts at sea level. Geology 39, 1159-1162.

68. Foley, S., Tiepolo, M., Vannucci, R., 2002. Growth of early continental crust controlled by melting of amphibolite in subduction zones. Nature 417, 837-840.

69. Friend, C.R.L., Nutman, A.P., 2005. Complex 3670-3500 Ma orogenic episodes superimposed on juvenile crust accreted between 3850 and 3690 Ma, Itsaq Gneiss Complex, Southern West Greenland. Journal of Geology 113, 375-397.

70. Galer, S.J.G., Mezger, K., 1998. Metamorphism, denudation and sea level in the Archean and cooling of the Earth. Precambrian Research 92, 389-412.

71. Ganguly, J., Freed, A.M., Saxena, S.K., 2009. Density profiles of oceanic slabs and surrounding mantle: integrated thermodynamic and thermal modeling, and implications for the fate of slabs at the 660 km discontinuity. Physics of the Earth and Planetary Interiors 172, 257-267.

72. Gerya, T.V., 2014. Precambrian geodynamics: concepts and models. Gondwana Research.

73. Gerya, T., Stockhert, B., 2006. Two-dimensional numerical modeling of tectonic and metamorphic histories at active continental margins. International Journal of Earth Sciences 95, 250-274.

74. Gerya, T.V., Yuen, D.A., 2003. Characteristics-based marker-in-cell method with conservative finite-differences schemes for modeling geological flows with strongly variable transport properties. Physics of the Earth and Planetary Interiors 140, 295-320.

75. Gerya, T.V., Yuen, D.A., 2007. Robust characteristics method for modelling multiphase visco-elasto-plastic thermo-mechanical problems. Physics of the Earth and Planetary Interiors 163, 83-105.

76. Gerya, T.V., Stockhert, B., Perchuk, A.L., 2002. Exhumation of high-pressure metamorphic rocks in a subduction channel: a numerical simulation. Tectonics 21.

77. Article Number: 1056. Gerya, T.V., Yuen, D.A., Maresch, W.V., 2004. Thermomechanical modelling of slab detachment. Earth and Planetary Science Letters 226, 101-116.

78. Guiraud, M., Powell, R., 2006. PeVeT relationships and mineral equilibria in inclusions in minerals. Earth and Planetary Science Letters 244, 683-694.

79. Goodwin, A.M., 1991. Precambrian Geology. Academic Press, London, 666 pp. Gorczyk, W., Hobbs, B., Gerya, T., 2012. Initiation of RayleigheTaylor instabilities in intra-cratonic settings. Tectonophysics 514e517, 146-155.

80. Gorczyk, W., Vogt, K., 2014. Tectonics and melting in intra-continental settings. Gondwana Research.

81. Gorczyk, W., Hobbs, B., Gessner, K., Gerya, T., 2013. Tectonics and melting in intracontinental settings. Gondwana Research 24, 838-848.

82. Grambling, J.A., 1981. Pressures and temperatures in Precambrian metamorphic rocks. Earth and Planetary Science Letters 53, 63e68. Gray, R., Pysklywec, R.N., 2010. Geodynamic models of Archean continental collision and the formation of mantle lithosphere keels. Geophysical Research Letters 37.

83. Article Number: L19301. Gray, R., Pysklywec, R.N., 2012. Geodynamic models of mature continental collision: evolution of an orogen from lithospheric subduction to continental retreat/ delamination. Journal of Geophysical Research e Solid Earth 117. Article Number: B03408.

84. Griffin, W.L., O'Reilly, S.Y., Abe, N., Aulbach, S., Davies, R.M., Pearson, N.J., Doyle, B.J., Kivi, K., 2003. The origin and evolution of Archean lithospheric mantle. Precambrian Research 127, 19-41.

85. Griffin, W.L., Belousova, E.A., O'Neill, C., Reilly, S.Y., Malkovets, V., Pearson, N.J., Spetsius, S., Wilde, S.A., 2013. The world turns over: HadeaneArchean crustmantle evolution. Lithos 189, 2-15.

86. Groome, W.G., Thorkelson, D.J., 2009. The three-dimensional thermo-mechanical signature of ridge subduction and slab window migration. Tectonophysics 464, 70-83.

87. Hacker, B.R., Abers, G.A., Peacock, S.M., 2003a. Subduction factory 1. Theoretical mineralogy, densities, seismic wave speeds, and H2O contents. Journal of Geophysical Research 108, 2029,.

88. Hacker, B.R., Peacock, S.M., Abers, G.A., Holloway, S.D., 2003b. Subduction factory 2. Are intermediate-depth earthquakes in subducting slabs linked to metamorphic dehydration reactions? Journal of Geophysical Research 108, 2030.

89. Hamilton, W.B., 1998. Archean magmatism and deformation were not products of plate tectonics. Precambrian Research 91, 143-179.

90. Harley, S.L., 1998. On the occurrence and characterisation of ultrahigh-temperature (UHT) crustal metamorphism. In: Treloar, P.J., O'Brien, P.J. (Eds.), What Drives Metamorphism and Metamorphic Reactions?, Geological Society, London, Special Publication, vol. 138, pp. 75-101.

91. Harley, S.L., 2008. Refining the PeT records of UHT crustal metamorphism. Journal of Metamorphic Geology 26, 125-154.

92. Harley, S.L., Kelly, N.M., 2007. Zircon: tiny but timely. Elements 3, 13-18.

93. Harley, S.L., Kelly, N.M., Möller, A., 2007. Zircon behaviour and the thermal histories of mountain chains. Elements 3, 25-30.

94. Hawkesworth, C., Cawood, P., Kemp, A., Storey, C., Dhuime, B., 2009. A matter of preservation. Science 323, 49-50.

95. Herzberg, C., Condie, K., Korenaga, J., 2010. Thermal history of the Earth and its petrological expression. Earth and Planetary Science Letters 292, 79-88.

96. Herzberg, C., Rudnick, R., 2012. Formation of cratonic lithosphere: an integrated thermal and petrological model. Lithos 149, 4-15.

97. Herzberg, C., Asimow, P.D., Arndt, N., Niu, Y., Lesher, C.M., Fitton, J.G., Cheadle, M.J., Saunders, A.D., 2007. Temperatures in ambient mantle and plumes: constraints from basalts, picrites and komatiites. Geochemistry Geophysics Geosystems 8, Q02006.

98. Holland, T.J.B., Powell, R., 2011. An improved and extended internally consistent thermodynamic dataset for phases of petrological interest, involving a new equation of state for solids. Journal of Metamorphic Geology 29, 333-383.

99. Holland, T.J.B., Hudson, N.F.C., Powell, R., Harte, B., 2013. New thermodynamic models and calculated phase equilibria in NCFMAS for basic and ultrabasic compositions through the transition zone into the uppermost lower mantle. Journal of Petrology 54, 1901-1920.

100. Hollister, L.S., Crawford, M.L., 1986. Melt-enhanced deformation e a major tectonic process. Geology 14, 558-561.

101. Holland, T.J.B., Powell, R., 1985. An internally-consistent thermodynamic dataset with uncertainties and correlations. 2. Data and results. Journal of Metamorphic Geology 3, 343-370.

102. Holland, T.J.B., Powell, R., 1998. An internally-consistent thermodynamic dataset for phases of petrological interest. Journal of Metamorphic Geology 16, 309-344.

103. Holland, T.J.B., Powell, R., 2011. An improved and extended internally consistent thermodynamic dataset for phases of petrological interest, involving a new equation of state for solids. Journal of Metamorphic Geology 29, 333-383.

104. Holland, T.J.B., Hudson, N.F.C., Powell, R., Harte, B., 2013. New thermodynamic models and calculated phase equilibria in NCFMAS for basic and ultrabasic compositions through the transition zone into the uppermost lower mantle. Journal of Petrology 54, 1901-1920.

105. Hollister, L.S., Crawford, M.L., 1986. Melt-enhanced deformation e a major tectonic process. Geology 14, 558-561.

106. Horie, K., Nutman, A.P., Friend, C.R.L., Hidaka, H., 2010. The complex age of orthogneiss protoliths exemplified by the Eoarchaean Itsaq Gneiss Complex (Greenland): SHRIMP and old rocks. Precambrian Research 183, 25-43.

107. Izraeli, E.S., Harris, J.W., Navon, O., 1999. Raman barometry of diamond formation. Earth and Planetary Science Letters 173, 351-360.

108. Jamieson, R.A., Beaumont, C., 1988. Orogeny and metamorphism e a model for deformation and pressureetemperatureetime paths with applications to the central and southern Appalachians. Tectonics 7, 417-445.

109. Jamieson, R.A., Beaumont, C., Medvedev, S., Nguyen, M.H., 2004. Crustal channel flows: 2. Numerical models with implications for metamorphism in the HimalayaneTibetan orogen. Journal of Geophysical Research e Solid Earth 109.

110. Jamieson, R.A., Beaumont, C., Nguyen, M.H., Lee, B., 2002. Interaction of metamorphism, deformation and exhumation in large convergent orogens. Journal of Metamorphic Geology 20, 9-24.

111. Johnson, T.E., Brown, M., Kaus, B., VanTongeren, J.A., 2014. Delamination and recycling of Archaean crust caused by gravitational instabilities. Nature Geoscience 7, 47-52.

112. Jull, M., Kelemen, P.B., 2001. On the conditions for lower crustal convective instability. Journal of Geophysical Research e Solid Earth 106, 6,423-6,446.

113. Keller, B., Schoene, B., 2012. Statistical geochemistry reveals disruption in secular lithospheric evolution about 2.5 Gyr ago. Nature 485, 490-493.

114. Kelsey, D.E., 2008. On ultrahigh-temperature crustal metamorphism. Gondwana Research 13, 1-29.

115. Kohn, M.J., 2014. "Thermoba-Raman-try": calibration of spectroscopic barometers and thermometers for mineral inclusions. Earth and Planetary Science Letters 388, 187-196.

116. Korenaga, J., 2006. Archean geodynamics and the thermal evolution of Earth. Geophysical Monograph Series 164, 7-32.

117. Korenaga, J., 2008. Urey ratio and the structure and evolution of Earth's mantle. Reviews of Geophysics 46. Article Number RG2007.

118. Korenaga, J., 2011. Thermal evolution with a hydrating mantle and the initiation of plate tectonics in the early Earth. Journal of Geophysical Research e Solid Earth 116. Article Number B12403.

119. Korenaga, J., 2013. Initiation and evolution of plate tectonics on Earth: theories and observations. Annual Review of Earth and Planetary Sciences 41, 117-151.

120. Kylander-Clark, A.R.C., Hacker, B.R., Cottle, J.M., 2013. Laser-ablation split-stream ICP petrochronology. Chemical Geology 345, 99-112.

121. Labrosse, S., Jaupart, C., 2007. Thermal evolution of the Earth: secular changes and fluctuations of plate characteristics. Earth and Planetary Science Letters 260, 465-481.

122. Li, Z.X., Bogdanova, S.V., Collins, A.S., Davidson, A., De Waele, B., Ernst, R.E., Fitzsimons, I.C.W., Fuck, R.A., Gladkochub, D.P., Jacobs, J., Karlstrom, K.E., Lu, S., Natapov, L.M., Pease, V., Pisarevsky, S.A., Thrane, K., Vernikovsky, V., 2008. Assembly, configuration, and break-up history of Rodinia: a synthesis. Precambrian Research 160, 179-210.

123. Li, Z.H., Gerya, T.V., Burg, J.-P., 2010. Influence of tectonic overpressure on PeT paths of HPeUHP rocks in continental collision zones: thermomechanical modeling. Journal of Metamorphic Geology 28, 227-247.

124. Li, S.H., Unsworth, M.J., Booker, J.R., Wei, W.B., Tan, H.D., Jones, A.G., 2003. Partial melt or aqueous fluid in the mid-crust of Southern Tibet? Constraints from INDEPTH magnetotelluric data. Geophysical Journal International 153, 289-304.

125. Maruyama, S., Liou, J.G., 1998. Initiation of ultrahigh-pressure metamorphism and its significance on the ProterozoicePhanerozoic boundary. The Island Arc 7, 6e35. Maruyama, S., Liou, J.G., 2005. From snowball to phanerozoic earth. International Geology Review 47, 775-791.

126. Maruyama, S., Liou, J.G., Terabayashi, M., 1996. Blueschists and eclogites of the world and their exhumation. International Geology Review 38, 490-596.

127. Mechie, J., Sobolev, S.V., Ratschbacher, L., Babeyko, A.Y., Bock, G., Jones, A.G., Nelson, K.D., Solon, K.D., Brown, L.D., Zhao, W., 2004. Precise temperature estimation in the Tibetan crust from seismic detection of the alpha-beta quartz transition. Geology 32, 601-604.

128. Mercier, J.C.C., Nicolas, A., 1975. Textures and fabrics of upper-mantle peridotites as illustrated by xenoliths from basalts. Journal of Petrology 16, 454-487.

129. Miyashiro, A., 1961. Evolution of metamorphic belts. Journal of Petrology 2, 277-311.

130. Miyashiro, A., 1973. Paired and unpaired metamorphic belts. Tectonophysics 17, 241-254.

131. Moulas, E., Podladchikov, Y.Y., Aranovich, L.Ya, Kostopoulos, D., 2013. The problem of depth in geology: when pressure does not translate into depth. Petrology 21, 527-538.

132. Moyen, J.-F., 2011. The composite Archaean grey gneisses: petrological significance, and evidence for a non-unique tectonic setting for Archaean crustal growth. Lithos 123, 21-36.

133. Nakajima, J., Uchida, N., Shiina, T., Hasegawa, A., Hacker, B.R., Kirby, S.H., 2013. Intermediate-depth earthquakes facilitated by eclogitization-related stresses. Geology 41, 659-662.

134. Nasdala, L., Massonne, H.-J., 2000. Microdiamonds from the Saxonian Erzgebirge, Germany: in situ micro-Raman characterization. European Journal of Mineralogy 12, 495-498.

135. Neil, E.A., Houseman, G.A., 1999. Rayleigh-Taylor instability of the upper mantle and its role in intraplate orogeny. Geophysical Journal International 138, 89-107.

136. Nelson, K.D., Zhao, W.J., Brown, L.D., Kuo, J., Che, J.K., Liu, X.W., Klemperer, S.L., Makovsky, Y., Meissner, R., Mechie, J., Kind, R., Wenzel, F., Ni, J., Nabelek, J., Chen, L.S., Tan, H.D., Wei, W.B., Jones, A.G., Booker, J., Unsworth, M., Kidd, W.S.F., Hauck, M., Alsdorf, D.,

Ross, A., Cogan, M., Wu, C.D., Sandvol, E., Edwards, M., 1996. Partially molten middle crust beneath southern Tibet: synthesis of project INDEPTH results. Science 274, 1684-1688.

137. Nutman, A.P., Bennett, V.C., Friend, C.R.L., Hidaka, H., Yi, K., Lee, S., Kamiichi, T., 2013. Episodic 3920-3660 Ma juvenile crust formation and 3660e3600 Ma recycling in the Itsaq Gneiss Complex of southern West Greenland. American Journal of Science 313, 877-911.

138. O'Brien, P.J., Ziemann, M.A., 2008. Preservation of coesite in exhumed eclogite: insights from Raman mapping. European Journal of Mineralogy 20, 827-834.

139. O'Neill, C., Debaille, V., Griffin, W., 2013. Deep earth recycling in the Hadean and constraints on surface tectonics. American Journal of Science 313, 912-932.

140. O'Neill, C., Lenardic, A., Moresi, L., Torsvik, T.H., Lee, C.-T.A., 2007. Episodic Precambrian subduction. Earth and Planetary Science Letters 262, 552-562.

141. Parkinson, C.D., Katayama, I., 1999. Present-day ultrahigh pressure conditions of coesite inclusion in zircon and garnet: evidence from laser Raman microspectroscopy. Geology 27, 979-982.

142. Partin, C.A., Bekker, A., Sylvester, P.J.,Wodicka, N., Stern, R.A., Chacko, T., Heaman, L.M., 2014. Filling in the juvenile magmatic gap: evidence for uninterrupted Paleoproterozoic plate tectonics. Earth and Planetary Science Letters 388, 123-133.

143. Perchuk, L.L., Gerya, T.V., 2011. Formation and evolution of Precambrian granulite terranes: a gravitational redistribution model. Geological Society of America Memoirs 207, 289-310.

144. Pfiffner, O.A., Ellis, S., Beaumont, C., 2000. Collision tectonics in the Swiss Alps: insight from geodynamic modeling. Tectonics 19, 1065-1094.

145. Powell, R., Holland, T.J.B., 1985. An internally consistent dataset with uncertainties and correlations. 1. Methods and a worked example. Journal of Metamorphic Geology 3, 327-342.

146. Powell, R., Holland, T.J.B., 1988. An internally consistent dataset with uncertainties and correlations. 3. Applications to geobarometry, worked examples and a computer-program. Journal of Metamorphic Geology 6, 173-204.

147. Powell, R., Holland, T.J.B., Worley, B., 1998. Calculating phase diagrams involving solid solutions via non-linear equations, with examples using Thermocalc. Journal of Metamorphic Geology 16, 577-588.

148. Powell, R., Guiraud, M., White, R.W., 2005. Truth and beauty in metamorphic phaseequilibria: conjugate variables and phase diagrams. The Canadian Mineralogist 43, 21-33.

149. Pysklywec, R.N., Beaumont, C., 2004. Intraplate tectonics: feedback between radioactive thermal weakening and crustal deformation driven by mantle lithosphere instabilities. Earth and Planetary Science Letters 221, 275-292.

150. Rebay, G., Powell, R., Diener, J.F.A., 2010. Calculated phase equilibria for a MORB composition in a PeT range, 450-650 C and 18-28

151. kbar: the stability of eclogite. Journal of Metamorphic Geology 28, 635e645. Reddy, S.M., Evans, D.A.D., 2009. Palaeoproterozoic supercontinents and global evolution: correlations from core to atmosphere. Geological Society, London, Special Publications 323, 1-26.

152. Richardson, S.W., England, P.C., 1979. Metamorphic consequences of crustal eclogite production in overthrust orogenic zones. Earth and Planetary Science Letters 42, 183-190.

153. Rolf, T., Coltice, N., Tackley, P.J., 2012. Linking continental drift, plate tectonics and the thermal state of the Earth's mantle. Earth and Planetary Science Letters 351, 134-146.

154. Royden, L.H., 1993a. The tectonic expression of slab pull at continental convergent boundaries. Tectonics 12, 303-325.

155. Royden, L.H., 1993b. Evolution of retreating subduction boundaries formed during continental collision. Tectonics 12, 629-638.

156. Royden, L.H., 1996. Coupling and decoupling of crust and mantle in convergent orogens: implications for strain partitioning in the crust. Journal of Geophysical Research e Solid Earth 101, 17,679-17,705.

157. Royden, L.H., Burchfiel, B.C., King, R.W., Wang, E., Chin, Z.L., Shen, F., Liu, Y.P., 1997. Surface deformation and lower crustal flow in eastern Tibet. Science 276, 788-790.

158. Rubatto, J., Hermann, J., 2007. Zircon behavior in deeply subducted rocks. Elements 3, 31-35.

159. Sandiford, M., Powell, R., 1990. Some isostatic and thermal consequences of the vertical strain geometry in convergent orogens. Earth and Planetary Science Letters 98, 154-165.

160. Santosh, M., Liu, S.J., Tsunogae, T., Li, J.H., 2012. Paleoproterozoic ultrahightemperature granulites in the North China Craton: implications for tectonic models on extreme crustal metamorphism. Precambrian Research 222-223, 77-106.

161. Savage, B., 2012. Seismic constraints of the water flux delivered to the deep earth by subduction. Geology 40, 235-238.

162. Silver, P.G., Behn, M.D., 2008. Intermittent plate tectonics? Science 319, 85e88. Sizova, E., Gerya, T., Brown, M., Perchuk, L., 2010. Subduction styles in the Precambrian: insight from numerical experiments. Lithos 116, 209-229.

163. Sizova, E., Gerya, T., Brown, M., 2012. Exhumation mechanisms of melt-bearing ultrahigh pressure crustal rocks during collision of spontaneously moving plates. Journal of Metamorphic Geology 30, 927-955.

164. Sizova, E., Gerya, T., Brown, M., 2014. Contrasting styles of Phanerozoic and Precambrian continental collision. Gondwana Research.

165. Sobolev, N.V., Fursenko, B.A., Goryainov, S.V., Shu, J.F., Hemley, R.J., Mao, H.K., Boyd, F.R., 2000. Fossilized high pressure from the Earth's deep interior: the coesite-in-diamond barometer. Proceedings of the National Academy of Sciences of the United States of America 97, 11,875-11,879.

166. Spencer, K.J., Hacker, B.R., Kylander-Clark, A.R.C., Andersen, T.B., Cottle, J.M., Stearns, M.A., Poletti, J.E., Seward, G.G.E., 2013. Campaign-style titanite UePb dating by laser-ablation ICP: implications for crustal flow, phase transformations and titanite closure. Chemical Geology 341, 84-101.

167. Spengler, D., van Roermund, H.L.M., Drury, M.R., Ottolini, L., Mason, P.R.D., Davies, G.R., 2006. Deep origin and hot melting of an Archaean orogenic peridotite massif in Norway. Nature 440, 913-917.

168. Spengler, D., Davies, G., Pearson, G., Mason, P., Van Roermund, H., Drury, M., Vukmanovic, Z., Wiersma, E., 2008. Petrogenesis of Archaean sub-lithospheric mantle preserved in the Otroy peridotite massif, Norway. Geochimica et Cosmochimica Acta 72 (12 Suppl. 1) pp. A888-A888.

169. Stern, R.J., 2005. Evidence from ophiolites, blueschists, and ultrahigh-pressure metamorphic terranes that the modern episode of subduction tectonics began in Neoproterozoic time. Geology 33, 557-560.

170. Stüwe, K., 2007. Geodynamics of the Lithosphere. SpringereVerlag, Berlin Heidelberg, 493 pp. Tackley, P.J., 2000. Mantle convection and plate tectonics: toward an integrated physical and chemical theory. Science 288, 2002-2007.

171. Thompson, A.B., 1992. Water in the Earth's upper mantle. Nature 358, 295e302. Tirone, M., Ganguly, J., Morgan, J.P., 2009. Modeling petrological geodynamics in the Earth's mantle. Geochemistry Geophysics Geosystems 10. Article Number: Q04012.

172. Toussaint, G., Burov, E., Jolivet, L., 2004. Continental plate collision: unstable vs. stable slab dynamics. Geology 32, 33-36.

173. Tsujimori, T., Sisson, V.B., Liou, J.G., Harlow, G.E., Sorensen, S.S., 2006. Very lowtemperature record in subduction process: a review of worldwide lawsonite eclogites. Lithos 92, 609-624.

174. Ueda, K., Gerya, T.V., Burg, J.-P., 2012. Delamination in collisional orogens: thermomechanical modeling. Journal of Geophysical Research e Solid Earth 117. Article Number: B08202.

175. Unsworth, M.J., Jones, A.G., Wei, W., Marquis, G., Gokarn, S., Spratt, J., 2005. Crustal rheology of the Himalaya and Southern Tibet inferred from magnetotelluric data. Nature 438, 78-81.

176. van Hunen, J., Moyen, J.-F., 2012. Archean subduction: fact or fiction? Annual Review of Earth and Planetary Sciences 40, 195-219.

177. van Hunen, J., van den Berg, A.P., 2008. Plate tectonics on the early Earth: limitations imposed by strength and buoyancy of subducted lithosphere. Lithos 103, 217-235.

178. van Hunen, J., van den Berg, A.P., Vlaar, N.J., 2004. Various mechanisms to induce present-day shallow flat subduction and implications for the younger Earth: a numerical parameter study. Physics of the Earth and Planetary Interiors 146, 179-194.

179. van Thienen, P., van den Berg, A.P., Vlaar, N.J., 2004. On the formation of continental silicic melts in thermochemical mantle convection models: implications for early Earth. Tectonophysics 394, 111-124.

180. Vogt, K., Gerya, T.V., Castro, A., 2012. Crustal growth at active continental margins: numerical modeling. Physics of the Earth and Planetary Interiors 192, 1-20.

181. Warren, C.J., Beaumont, C., Jamieson, R.A., 2008. Modelling tectonic styles and ultrahigh pressure (UHP) rock exhumation during the transition from oceanic subduction to continental collision. Earth and Planetary Science Letters 267, 129-145.

182. Watson, J.V., 1977. Precambrian thermal regimes. Philosophical Transactions of the Royal Society of London A 288, 431-440.

183. Wei, C.J., Clarke, G.L., 2011. Calculated phase equilibria for MORB compositions: a reappraisal of the metamorphic evolution of lawsonite eclogite. Journal of Metamorphic Geology 29, 939-952.

184. White, R.W., Powell, R., Holland, T.J.B., 2007. Progress relating to calculation of partial melting equilibria for metapelites. Journal of Metamorphic Geology 25, 511-527.

185. White, R.W., Powell, R., Holland, T.J.B., Johnson Green, 2014. Journal of Metamorphic Geology 32, 261-286.

186. Willett, S., Beaumont, C., Fullsack, P., 1993. Mechanical model for the tectonics of doubly vergent compressional orogens. Geology 21, 371-374.

187. Williams, I.S., 2013. 30 Years of Progress in Applying Microdating Techniques to Understanding Precambrian Crustal Evolution. Precambrian Evolution and Deep Exploration of the Continental Lithosphere, 7-9 October 2013, Beijing, China, Abstract Volume, IAGR Conference Series 15, pp. 110-112.

188. Yoshida, M., Santosh, M., 2011. Supercontinents, mantle dynamics and plate tectonics: a perspective based on conceptual vs. numerical models. Earth-Science Reviews 105, 1-24.

189. Young, G.M., 2013. Precambrian supercontinents, glaciations, atmospheric oxygenation, metazoan evolution and an impact that may have changed the second half of Earth history. Geoscience Frontiers 4, 247-261.

190. Ziaja, K., Foley, S.F., White, R.W., Buhre, S., 2014. Metamorphism and melting of picritic crust in the early Earth. Lithos 189, 173-184.

191. Zhao, G.C., Sun, M., Wilde, S.A., Li, S.Z., 2004. A Paleo-Mesoproterozoic supercontinent: assembly, growth and breakup. Earth-Science Reviews 67, 91-123.

192. Zhu, G.Z., Gerya, T.V., Tackley, P.J., Kissling, E., 2013. Four-dimensional numerical modeling of crustal growth at active continental margins. Journal of Geophysical Research e Solid Earth 118, 4682-4698.

Chapter 6

Petrology of the Volcanic Rocks from Bioko Island ("Cameroon Hot Line")

Fadimatou Ngounouno Yamgouot[1*], Bernard Déruelle[2], Isaac Bertrand Gambie Mbowou[3], Ismaïla Ngounouno[3]

[1]Département des Sciences de la Terre, Faculté des Sciences, Université de Ngaoundéré, Ngaoundéré, Cameroun

[2]Laboratoire de Magmatologie et Géochimie Inorganique et Expérimentale (MAGIE), Institut de Physique du Globe de Paris, Université Pierre et Marie Curie, Paris, France

[3]Département des Mines et de la Géologie, Ecole de Géologie et d'Exploitation Minière (EGEM), Université de Ngaoundéré, Meiganga, Cameroun

ABSTRACT

Bioko island (3008 m a.s.l.), is composed of the alkaline basaltic lavas (basalts and hawaiites) with xenoliths. These lavas have a microlitic texture and consisted of euhedral to subhedral phenocrysts (>1 mm) of olivine (0.83 < Mg# < 0.87) and diopside ± Ti-augite. Plagioclase ($An_{62-67}Ab_{35-32}Or_{3-1}$) phenocrysts are present only in hawaiites. In Harker diagrams, SiO_2, Al_2O_3, Na_2O and K_2O contents of the lavas increase and Fe_2O_3, MgO and CaO decrease with increasing differentiation from basalts to hawaiites. The compatible elements Ni, Cr and V decrease strongly in basalts and remain at low levels in hawaiites. Basaltic lavas from Bioko lavas have low Hf contents (4.2 - 9.2 ppm) and consequently higher Hf/Zr ratios (50 - 90) than those (<50) of similar lavas from other volcanoes of the Cameroon Hot Line. Such high ratios are commonly observed in alkaline basaltic lavas associated with carbonatitic and/or nephelinitic magmatism. The Sr and Nd isotopic compositions point to a slightly depleted mantle source.

KEYWORDS

Petrology, Geochemistry, Lavas, Bioko, Cameroon Hot Line

1. INTRODUCTION

Bioko Island (formerly Fernando Poo) is the northernmost island of the oceanic sector of the Cameroon Hot Line (CHL, inset Figure 1); it lies upon the border of the continental plateau, at about 35 km to the SW of the African continent shore. The island is composed of three adjacent large strato-volcanoes (Figure 1): Pico Santa Isabel which culminates at 3008 m a.s.l., Pico Biao characterized by a 6 km diameter caldera, and San Carlos. The published ages of these volcanoes are all younger than 1.33 ± 0.07 Ma [1] -[3] . Paleomagnetic data yield ages younger than 0.7 Ma [1] . It is probable that the Bioko ages are similar to those of Mt Cameroon [4] . Mt Cameroon is younger than nephelinites from Mt Etinde that have been dated between 1.10 and 0.65 Ma [3] . Historical eruptions occurred on Bioko in 1898, 1903 and 1923 [5] .

The objectives of this paper are to present new mineralogical and geochemical data for volcanic rocks from Bioko Island, in order to compare it to others volcanic rocks from the Cameroon Hot Line.

2. PETROGRAPHY AND MINERALOGY

2.1. The Lavas

All the samples of Bioko Island belong to the alkaline series. Only basaltic lavas (basalts and hawaiites) have been found (Figure 2) whilst felsic lavas (trachytes and phonolites) are present in the other islands of CHL

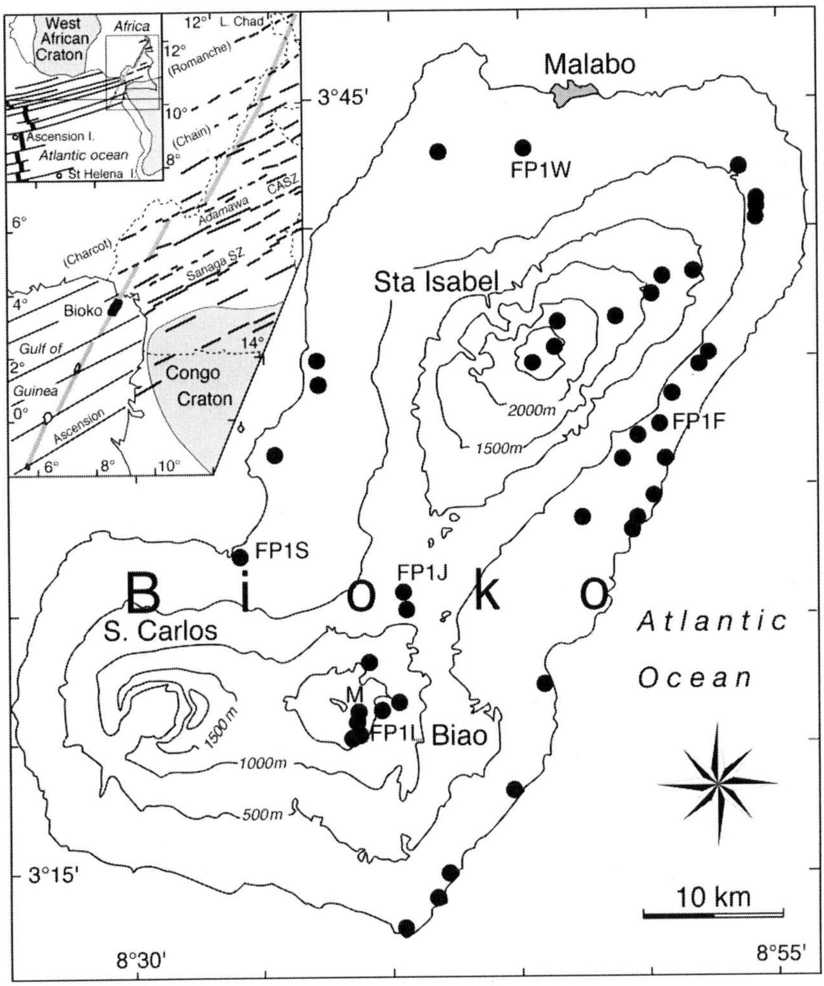

Figure 1. Sample location on Bioko Island. Upper inset: propagation of the N70°E continental shear zones into the fracture zones of the Atlantic Ocean and the N30°E Cameroon Hot Line. Lower inset: major volcanoes of the Cameroon Hot Line (black) from Pagalu Island to Lake Chad. Other volcanic areas (Adamawa and Biu plateaus) are shown in grey.

(Pagalu, Principe and Sao Tome). A phonolite has been described by [6] , but it has been sampled as a loosed block on the sea shore and could originate from Sao-Tome or Principe islands from which it could have been carried as ship ballast: phonolite has never been found anymore on Bioko Island. Basaltic lavas only have also been sampled on the neighbouring continental Mt Cameroon volcano (Figure 2). The nomenclature used in this paper is based upon the differentiation index (D.I., [7]) and on peculiar mineralogical distribution. Basalts have D.I. < 35 (42 samples) and hawaiites have 35 < D.I. < 50 (15 samples). Sample FP1K which has hornblende phenocrysts, high SiO_2 content (>53.0 wt%) and normative quartz (>4.0 wt%) has a D.I. of 49.2 (value obtained on a recalculated anhydrous analysis): it is classified as a mugearite.

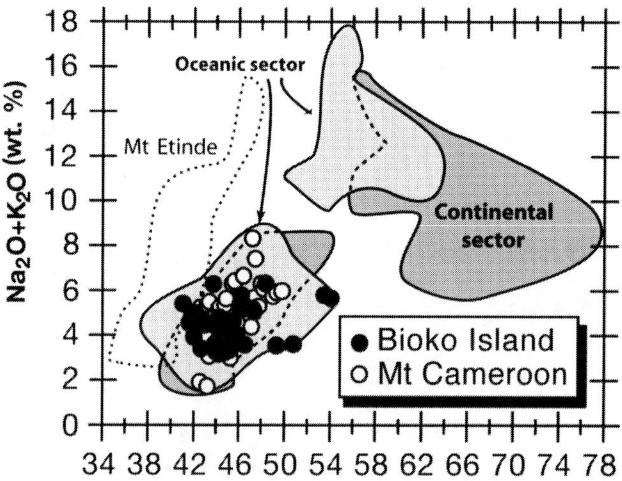

Figure 2. SiO_2 vs (Na_2O + K_2O) diagram for Bioko lavas. All the Bioko lavas plot inside the field of the lavas from the CHL oceanic sector. The data from Mt Cameroon are given for comparison (data after [17]).

The lavas have a microlitic texture. Basalts contain euhedral to subhedral phenocrysts (>1 mm) of olivine (0.83 < Mg# < 0.87) and clinopyroxene. Plagioclase ($An_{62-67}Ab_{35-32}Or_{3-1}$) is also present in hawaiites. The matrix contains microlites of clinopyroxene, plagioclase and magnetite and microphenocrysts of olivine. Diopside and Ti-augite occur respectively as euhedral to subhedral microphenocrysts (<2.5 mm) and as phenocrysts of variable size (2.5 to 6.0 mm). One lava sample (B17) of basaltic affinity contains skeletal hypersthene phenocrysts, embayed olivine phenocrysts and euhedral cordierite microphenocryts with opacified core. Cordierite has already been described in hypersthene-bearing lava from St. Helena Island [8] .

Crystal size distribution (CSD) measurements have been performed on clinopyroxene crystals from 12 basaltic lavas using CSD Corrections software of [9] . The data were plotted following general convention [10] as linear crystal size vs population density—ln(n) (mm^{-4}) vs l (mm) (Figure 3). Most of the CSDs appear curved with nearly straight distribution for small (<2 mm) grains and curved patterns for larger (>2 mm) crystals (Fig- ure 4). The small grains are distributed along a steep slope with intercepts between 8.5 and 12.0 mm^{-4} while the larger grains plot along a more gentle slope with lower intercepts (4 to 6 mm^{-4}). Following [11] , [9] and [12] , these curved CSDs can be interpreted as mixing of two populations of crystals with contrasted crystallization histories. The population of large crystals could represent clinopyroxene that has crystallized under different physical condition. To strengthen this hypothesis, the crystallization pressure of the clinopyroxenes was estimated using the [10] structural barometer of [13] . This barometer is suitable for anhydrous alkaline melts as the Bioko basaltic melt. The small phenocrysts of Ti-augite composition were equilibrated at low P (0 - 2 kbar) while the few larger ones (of diopside composition) crystallized at significantly higher pressure (~6 kbar). In few samples, the clinopyroxene phenocrysts are zoned: the diopside cores crystallized at high pressure and the Ti-augite rims at low pressure.

CSD pattern and P estimate suggest a polybaric crystallization of Bioko basaltic magma. Similar multi- chamber magmatic evolution is common in alkaline [14] and peralkaline [15] volcanic rocks.

2.2. Xenoliths

Some basaltic flows and tephra deposits contain abundant ultramafic and mafic xenoliths. Xenoliths of gabbro are composed of clinopyroxene and cloudy plagioclase (due to the presence of numerous scattered small green spinel crystals) sometimes with olivine. The variety of ultramafic xenoliths is quite large. In a basalt lava flow at the NE flank of Santa Isabel volcano, harzburgites, lherzolites and dunites occur together scattered in a restricted area of few m^2. In fact, xenoliths are mainly harzburgites with a protogranular texture and lherzolites; dunites are rare. Olivine is Mg-rich (Fo ≈ 90.3 ± 0.5) as well as clinopyroxene and orthopyroxene (Mg#: 0.92 - 0.95 and 0.91 - 0.92, respectively). Clinopyroxene is also Cr-rich (0.13 < Cr a.p.f.u. < 0.31) when compared to other crystals in xenoliths from the CHL (Cr a.p.f.u. < 0.18) ([16] and references therein, [17]). Harzburgites contain holly-leaf and vermicular spinel. This spinel has a composition similar to that of crystals in xenoliths from the Kapsiki plateau but is quite disctinct from xenoliths in other basalts all along the CHL. Rare phlogopite and very scarce plagioclase crystals are present in veinlets between olivine and orthopyroxene grains. Amphibole occurs in veinlets in the dunites. High fO_2 (above the FMQ buffer) characterize dunites. Similar xenoliths have also been found at Mt Cameroon, Mt Bambouto, Mt Oku and Nyos (see [16]) in the continental sector of the CHL. Equilibrium temperatures for pyroxene pairs calculated after [9] are between 889˚C and 1033˚C (mean 943; 2 sigmas: 38˚C).

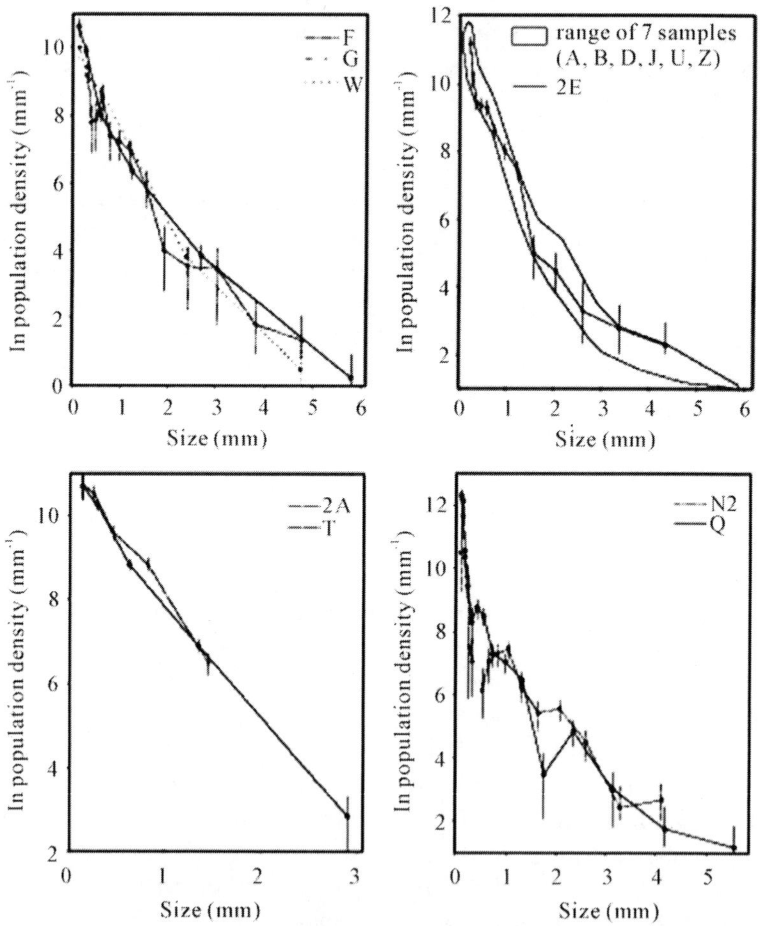

Figure 3. Different types of crystal size distribution (CSD) patterns observed for the clinopyroxene phenocrysts of selected Bioko basaltic lavas.

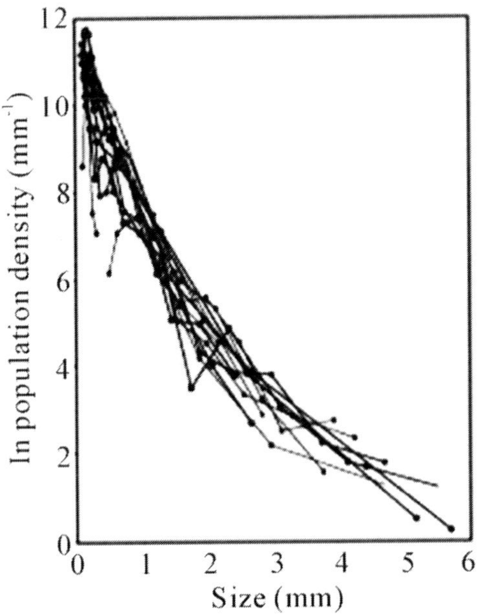

Figure 4. Crystal size distribution of clinopyroxenes from the Bioko lavas show typical curved patterns.

3. GEOCHEMISTRY

In Harker diagrams (Figure 5), the samples are somewhat scattered due to the limited range of D.I. variations (from 12 to 49). Samples from Santa Isabel and Pico Biao volcanoes roughly plot along the same broad evolution trends (one sample from San Carlos only has been analysed). Nevertheless, increase in SiO_2, Al_2O_3, Na_2O and K_2O and decrease in Fe_2O_3, MgO and CaO are note worthy with increasing differentiation. TiO_2 and P_2O_5 contents do not vary significantly.

The trace element contents of Bioko lavas (Figure 6) are plotted against Th content (incompatible element used as geochemical differentiation index) and compared to data from Mt Cameroon ([18] and unpublished). The compatible elements Ni, Cr and V decrease strongly in basalts and remain at low levels in hawaiites. The incompatible LIL elements (Rb, Sr, Ba, Zr, and Ta) and the REEs increase with increasing Th contents. The limited Bioko differentiation series as a whole and the Mt Cameroon data are very similar except for Ba and Zr that are higher in Bioko lavas. Zr/Hf ratios are higher at Bioko (50 - 90) than at Mt Cameroon (18, unpubl. data).

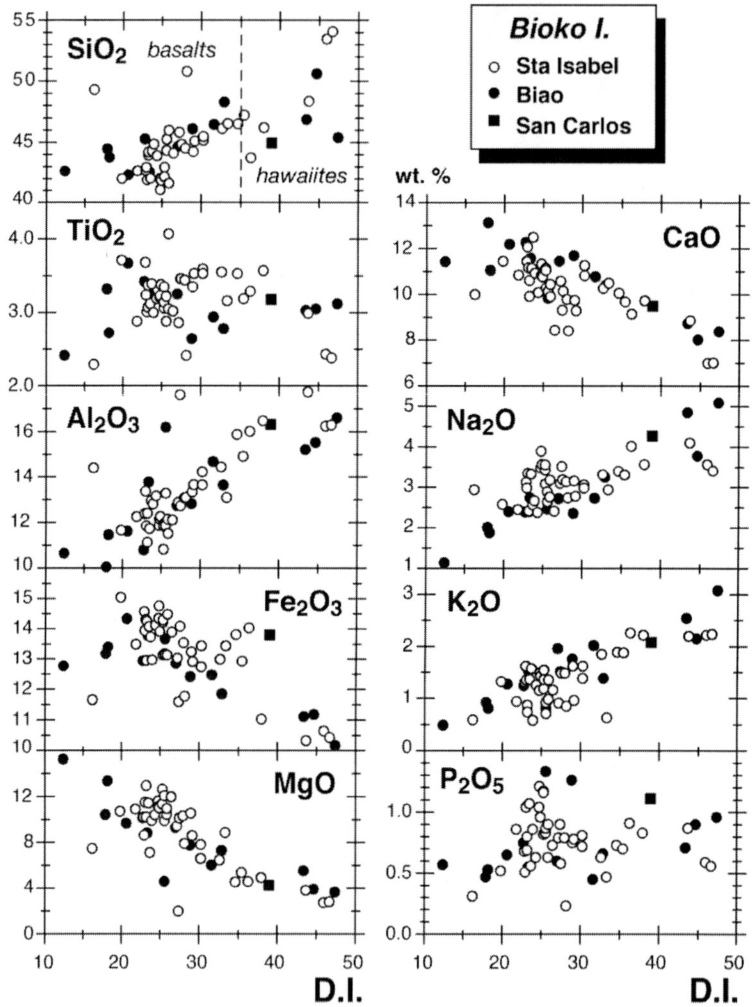

Figure 5. Major element (in wt%) versus I.D. (used as a geochemical differentiation index) diagrams for Bioko Island.

In primitive mantle normalized multi-element diagram (Figure 7), the patterns of Bioko lavas have the classical shape of alkali basaltic lavas; they have high LIL incompatible element abundances (50 to 100× chondrites) and lower HFSE and HREE abundances (<20× chondrites). They also show pronounced negative anomaly for

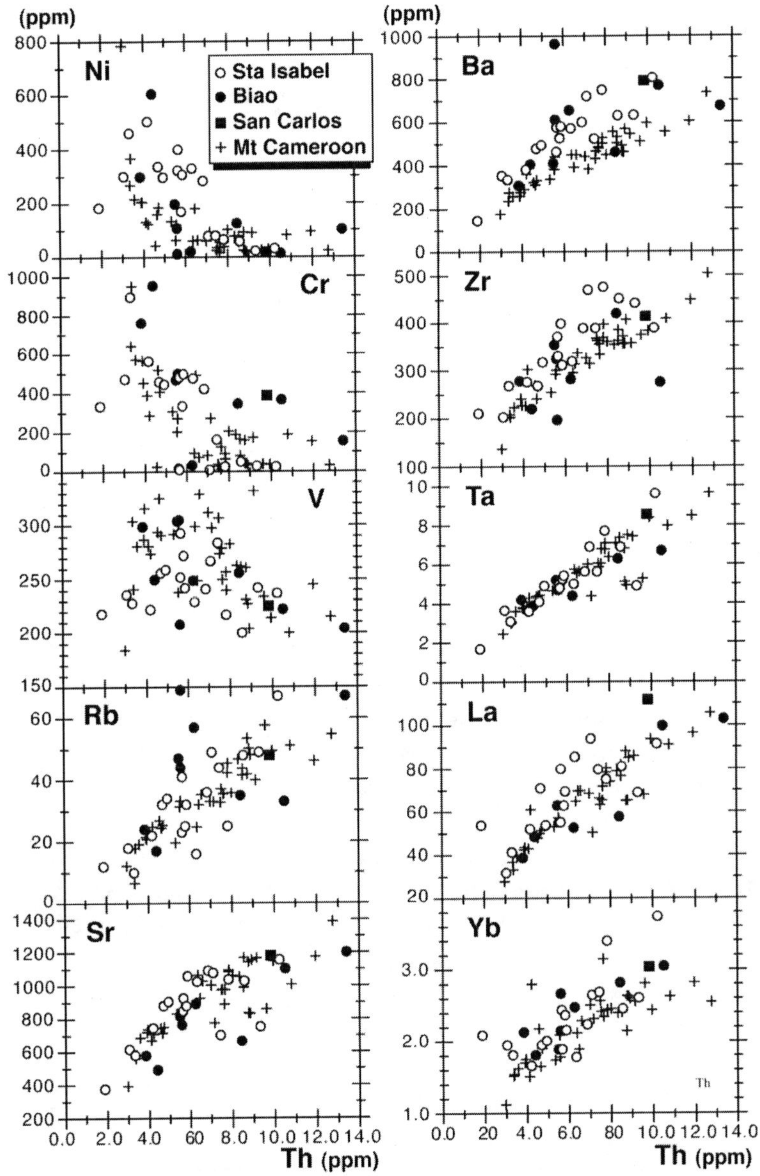

Figure 6. Trace element variations (in ppm) versus Th (used as a geochemical differentiation index) diagrams for Bioko lavas compared to Mt Cameroon.

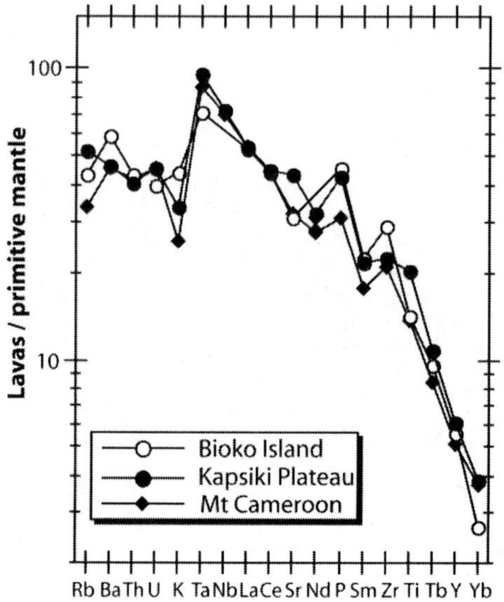

Figure 7. Spider diagram for representative lavas from Bioko compared with basalts from Mt Cameroon and St Kapsiki Plateau ([16] and ref. therein). Normalisation values from [22] .

K, a less important one for Hf and positive anomalies for Ba, U, Ta, P and Zr. The LaN/CeN ratios (1.4 - 1.6) are roughly similar for all the lavas from the three Bioko volcanoes but are significantly higher than those from the neighbouring continental volcanoes of Mt Etinde (1.25 - 1.27, [19]). The range of measured $^{87}Sr/^{86}Sr$ isotope compositions (0.70319 - 0.70343; n = 4) by [20] and our unpublished data (0.70321 - 0.70352; n = 4) are largely overlapping and in the range of data for the basaltic lavas of the CHL (0.70299 - 0.70343, compilation of [17]). The Nd isotopic compositions (ε_{Nd}: + 2.6 to + 4.0; same ref as Sr) point to a slightly depleted upper mantle source.

4. DISCUSSION

4.1. Basaltic Lavas and Scattering of Data

The three volcanoes of Bioko Island only emitted basaltic lavas (basalts and hawaiites); this is similarly to what is observed on the nearby Mt

Cameroon. These volcanoes are the only ones of the CHL that have only basaltic lavas. Moreover, apart the phreatic eruptions of Monoun and Nyos lakes, Bioko and Mt Cameroon are the only volcanoes that had eruptions during the last century. Mt Etinde which is located at the very shore of the continent between Bioko and Mt Cameroon only consists of nephelinitic lavas which are completely unknown on the other volcanoes.

4.2. Fractional Crystallization Modelling

Evolution by crystal fractionation has been modeled for the short differentiation series of Bioko. Despite the restricted compositional range observed for the data set, the mineralogical (progressive evolution of mineral compositions with increasing differentiation) and geochemical (regular and progressive major and trace element evolution) data suggest that fractional crystallization had a major role in the differentiation of the lava series. Mass balance modelling for major elements gave results that are consistent with the distribution of mineralogical phases present as phenocrysts in the lavas. For example, the evolution from basalt (D) to hawaiite (W) can be modeled by fractional crystallization of ol + cpx + pl + ox in the proportions 6.9 + 0.7 + 0.3 + 0.1 ($\Sigma r^2 < 0.6$) and to hawaiite FP1F by fractionation of ol + cpx + ox in the proportions 5.3 + 14.2 + 2.0 ($\Sigma r^2 < 0.5$). The limited number of mineral phases involved in the process lead to quite large Σr^2.

Basaltic lavas from Bioko lavas have low Hf contents (4.2 - 9.2 ppm) and consequently higher Zr/Hf ratios (50 - 90) than those (<50) of similar lavas from other volcanoes of the CHL. Such high ratios are commonly observed in alkaline basaltic lavas associated with carbonatitic and/or nephelinitic magmatism. It is interesting to remind that Bioko is situated quite near the nephelinitic Mt Etinde volcano.

4.3. Metasomatism

The occurrence of amphibole (pargasite) in ultramafic xenoliths of Bioko and of many other volcanoes of the CHL (see review in [17]) suggests that the lithospheric mantle beneath the CHL has been metasomatized. Metasomatized peridotites are likely to be representative of enriched lithosphere, the commonly accepted metasomatic agent being partial melts that form a network of thin veins lacing mantle peridotite beneath zones of recent volcanism and that induce the crystallization of K-rich amphibole [21] . The low Al_2O_3 (<4 wt%) and high CaO (21 - 24 wt%) contents of clinopyroxenes and also the relatively high fO_2 (above the FMQ buffer) in dunites are suggestive of a metasomatic influence by carbonatitic liquids.

5. CONCLUSIONS

Bioko Island (formerly Fernando Poo) belongs to the "Cameroon Hot Line" (CHL), a major geological structure of Central Africa. It lies on the continental plateau, close (~35 km) to the continent shore and consists of three large strato-volcanoes (Pico Santa Isabel culminating at more than 3000 m).

The differentiation series is incomplete: only mafic lavas—basalts and hawaiites—have been found as on the huge Mount Cameroon volcano. Ultramafic and mafic xenoliths are locally abundant; they contain pargasite and rare phlogopite pointing to a metasomatic mantle beneath CHL. A polybaric differentiation process is inferred from the curved CSD patterns of clinopyroxene phenocrysts and from pressure estimates: the large diopside phenocrysts were equilibrated at ~6 kbar while the small Ti-augite crystals equilibrated at low (<2 kbar) pressure.

Fadimatou Ngounouno Yamgouot, Bernard Déruelle, Isaac Bertrand Gambie Mbowou, Ismaïla Ngounouno (>50) are significantly higher than those of most basalts from CHL (<50); these high values could be related to the presence of the nearby Mt Etinde nephelinites. The Sr and Nd isotopic compositions point to a slightly depleted mantle source.

REFERENCES

1. Aka, F.T., Nagao, K., Kusakabe, M., Sumino, H., Tanyileke, G., Ateba, B. and Hell, J. (2004) Symmetrical Helium Isotope Distribution on the Cameroon Volcanic Line, West Africa. Chemical Geology, 203, 205-223.

2. Chauvel, C., Dia, A.N. Bulourde, M., Chabaux, F., Durand, S., Ildefonse, P., Gérard, M., Déruelle, B. and Ngounouno, I. (2005) Do Decades of Tropical Rainfall Affect the Chemical Compositions of Basaltic Lava

Flows in Mt Cameroon? Journal of Volcanology and Geothermal Research, 141, 195-223.

3. Hedberg, J.D. (1969) A geological Analysis of the Cameroon Trend. Ph.D. Thesis, Princeton University, Princeton, 188 p.

4. Piper, J.D.A. and Richardson, A. (1972) The Palaeomagnetism of the Gulf of Guinea Volcanic Province, West Africa. Geophysical Journal of Royal Astronomical Society, 29, 147-171.

5. Enciclopedia Universal Ilustrada 1924. Fernando Poo, Espasa-Calpe SA Hijos de J. Espasa, Madrid 23, 833.

6. Boëse, W. (1912) Petrographische Untersuchungen an jungvulkanischen Ergussgesteinen von São Thomé und Fernando Poo. Neues Jahrbrbuch für Geologie und Paläontologie Abhandlungen, 34, 253-320.

7. Thornton, C.P. and Tuttle, O.F. (1960) Chemistry of Igneous Rocks—[Part] 1 I. Differentiation Index. American Jour- nal of Science, 258, 664-684.

8. Ridley, W.I. and Baker, I. (1973) The Petrochemistry of a Unique Cordierite-Bearing Lava from St. Helena Island, South Atlantic. American Mineralogist, 58, 813-818.

9. Higgins, M.D. (2006) Verification of Ideal Semi-Logarithmic, Lognormal or Fractal Crystal Size Distributions from 2D Datasets. Journal of Volcanology and Geothermal Research, 154, 8-16.

10. Marsh, B.D. (1988) Crystal Size Distribution (CSD) in Rocks and the Kinetics and Dynamics of Crystallization. 1. Theory. Contribributions to Mineralogy and Petroogy, 99, 277-291.

11. Higgins, M.D. (1996) Magma Dynamics beneath Kameni Volcano, Thera, Greece, as Revealed by Crystal Size and Shape Measurements. Journal of Volcanology and Geothermal Research, 70, 37-48.

12. Marsh, B.D. (1998) On the Interpretation of Crystal Size Distribution in Magmatic Systems. Journal of Petrology, 39, 553-599.

13. Nimis, P. and Ulmer, P. (1998) Clinopyroxene Geobarometry of Magmatic Rocks Part 1: An Expanded Structural Geobarometer for

Anhydrous and Hydrous, Basic and Ultrabasic Systems. Contributions to Mineralogy and Petrology, 133, 122-135.

14. Woodland, A.B. and Jugo, P.J. (2007) A Complex Magmatic System beneath the Devès Volcanic Field, Massif Central, France: Evidence from Clinopyroxene Megacrysts. Contributions to Mineralogy and Petrology, 153, 719-731.

15. Berger, J., Ennih, N., Liégeois, J.P., Nkono, C., Mercier, J.C.C. and Demaiffe, D. (2008) A Complex Multi-Chamber Magmatic System beneath a Late Cenozoic Volcanic Field: Evidence from CSDs and Thermobarometry of Clinopyroxene from a Single Nephelinite Flow (Djbel Saghro, Morocco). Geological Society, London, Special Publications, 297, 509-524.

16. Déruelle, B., Ngounouno, I. and Demaiffe, D. (2007) The "Cameroon Hot Line" (CHL): A Unique Example of Active Alkaline Intraplate Structure in Both Oceanic and Continental Lithospheres. Comptes Rendus Geoscience, 339, 589- 600.

17. Lee, D.C., Halliday, A.N., Davies, G.R., Essene, E.J., Fitton, J.G. and Temdjim, R. (1996) Melt Enrichment of Shallow Depleted Mantle: A Detailed Petrological, Trace Element and Isotopic Study of Mantle-Derived Xenoliths and Mega- crysts from the Cameroon Line. Journal of Petrology, 37, 415-441.

18. Déruelle, B., Bardintzeff, J.M., Cheminée, J.L., Ngounouno, I., Lissom, J., Nkoumbou, C., Etamé, J., Hell, J.V., Tanyileke, G., N'ni, J., Ateba, B., Ntepe, N., Nono, A., Wandji, P., Fosso, J. and Nkouathio, D.G. (2000) Éruptions simultanées de basalte alcalin et de hawaiite au mont Cameroun (28 mars-17 avril 1999). Comptes Rendus de l'Académie des Sciences Paris, 331, 525-531.

19. Nkoumbou, C., Déruelle, B. and Velde, D. (1995) Petrology of Mt Etinde Nephelinite Series. Journal of Petrology, 36, 373-395.

20. Halliday, A.N., Davidson, J.P., Holden, P., DeWolf, C., Lee, D.C. and Fitton, J.G. (1990) Trace-Element Fractionation in Plumes and the Origin of HIMU Mantle beneath the Cameroon Line. Nature, 347, 523-528.

21. Ngounouno, I. and Déruelle, B. (2007) Pétrologie des xénolithes de wehrlites et clinopyroxénites du Mont Cameroun: Evidence d'un

métasomatisme mantellique. Journal of Cameroon Academics Science, 7, 35-56.

22. Hofmann, A.W. (1988) Chemical Differentiation of the Earth: The Relationship between Mantle, Continental Crust, Oceanic Crust. Earth and Planetary Science Letters, 90, 297-314.

Chapter 7

Coal petrology and genesis of Jurassic coal in the Ordos Basin, China

Weihua Ao [a], Wenhui Huang [a, *], Chengmin Weng [a], Xiuling Xiao [a], Dameng Liu [a], Xiuyi Tang [b], Ping Chen [b], Zhigen Zhao [b], Huan Wan [a], Robert B. Finkelman [c]

[a] School of Energy Resources, China University of Geosciences, Beijing 100083, China

[b] Anhui University of Science and Technology, Huainan, Anhui 232001, China

[c] The University of Texas at Dallas, Richardson, TX 75083-0688, USA

ABSTRACT

Multiple sets of thick coal beds characterized by simple structure and shallow burial depth were developed in the Early and Middle Jurassic strata of the Ordos Basin, northwestern China. The huge reserves of this high quality coal have a high commercial value. We studied the coal's petrologic characteristics and its maceral distribution to determine the maceral's contribution to generation of oil and gas. The results show that the Jurassic coals in the Ordos Basin have special petrological features because of the Basin's unique depositional environment which was mainly a series of high-stand swamps in the upper fluvial system. These petrographic features are a result of the development of typical inland lakes where some sand bodies were formed by migrating rivers. After burial, the peat continued to undergo oxidizing conditions, this process generated extensive higher inertinite contents in the coals and the vitrinite components were altered to semi-vitrinite. The macroscopic petrographic types of these Jurassic coals are mainly semi-dull coal, dull coal, semilustrous and lustrous coal. The proportions of semi-dull coal and dull coal are higher in the basin margins, especially in the area near the northern margin. The numbers of semilustrous and lustrous coals increase southwards and towards the central basin. This situation indicates that different coal-forming swamp environments have major controlling effects on the coal components. Another observation is that in the Ordos' coal sequences, especially in the lower part, some sandstone beds are thick, up to 20 m with a coarse grain size. The higher fusinite content in the macerals accompanies a higher semi-vitrinite content with more complete and regular plant cell structure. The fusinite structure is clear and well preserved. After burial, the lithology of the roof and floor rocks can continue to affect the evolution of coal petrology. The sand bodies in the roof and floor exhibit good physical conditions so that pore water can maintain a long-term state of oxidation, circulation and connection to the coal. So coal components remain in an oxidation environment for a long time. Conversely, in the basin center, lacustrine facies developed and peat was rapidly covered by mudstone after burial and subsequent coal beds rapidly entered a reducing environment. As a result, abundant gelatification occurred and the vitrinite content increased. Exinite often accumulated in a specific position in the coal bed. Although the average exinite content is not high on the whole, it does significantly contribute to the total hydrocarbon generation. The exinite content has been underestimated, especially the amorphous bituminous fluid and its importance is emphasized here. The reason is that the fluid flows easily into fusinite which has strong rigidity, or flows into some fissures, where it is commonly neglected.

KEYWORDS

Lower-Middle Jurassic; Coal-forming environment; Coal maceral; Ordos Basin

1. INTRODUCTION

Huge reserves approaching 1500 Gt of Jurassic coal are present in NW China. The coal rank is low, but because of its special properties, the coal can be used in several ways. No matter what application is considered, a basic understanding of the maceral and mineral content of the coals is necessary. For example, coal properties are important for efficient direct coal liquefaction; these properties include high hydrogen content with the H/C atomic ratio greater than 0.8, and low inertinite content, generally lower than 20%. A direct coal liquefaction technique has been developed by the Shenhua Company, the biggest coal mining company in China (Zhang, 2006). It has been proven successful because it directly exploits transitional semi-vitrinite and active semi-inertinite components in coal (the macerals are described in detail in section 4.1).

Some of the Jurassic coals in NW China have oil-generating potential. For example, coal-generated oil occurs in the Tuha Basin, where coals from the Middle Jurassic Xishanyao Formation provide one of the major oil source rocks. The main concern is to determine which macerals are the major oil-generation materials in the Tuha Basin. Suberinite was once considered the primary oil-generation component. Degradinite, containing ultramicro-lipid components, was also regarded as another major oil-generation contributor. Zhao et al. (1994) have conducted many studies in the Tuha, Junggar, Tarim and other basins. They consider that the primary hydrocarbon-generation components are exinite (including algainite and bitumenite) and fluorescent degradinite. However, Liu et al. (2000) considered most of northwestern China's Jurassic coal to be poor in exinite. As with other areas in NW China, the Jurassic coals in the Ordos Basin possess high inertinite content. However, there are large differences in the proportion of each maceral and some coal seams are rich in exinite. Coal-forming environments control changes in the maceral content, therefore, research on the paleogeographic framework of the Ordos Basin and the sedimentary environments of each of its major coal beds will bring benefits to the prediction of maceral contents within them and the potential uses of the coals.

2. REGIONAL GEOLOGIC SETTING

The Ordos Basin is a huge basin in China rich in coal, oil and gas resources. It is almost 400,000 km^2 in area and is located on the western margin of the

North China Platform (Fig. 1). The basin developed on a stable craton with a stratigraphic thickness generally over 4000 m, and up to 6000 m in the central basin. During its tectonic evolution, the basin underwent a Paleozoic platform phase, a foreland basin flexural phase in the Middle and Late Triassic similar, and a relatively simple subsidence phase during the Jurassic. The long-term development of this superimposed basin led to the occurrence of Permo-Carboniferous sequences, the Wayaobao coal sequences in the Triassic and the Yan'an Formation coal sequences in the Jurassic. The coal sequences coexist in some area, but also have different areal development. From the standpoint of coal resources, the Jurassic coals with burial depths of less than 2000 m are estimated to be 14,878 Gt, accounting for 75% of the total resources (Zhang et al., 2005).

The major coal-bearing seams in the Ordos Basin are located in Early and Middle Jurassic strata and are widely distributed throughout the basin. The petrography of the Jurassic coals has been well studied (Qian et al., 1987, Huang et al., 1995, Dai et al., 2007 and Dai et al., 2008), and indicates that Ordos's coals are very different in their properties and genetic types in different parts of the basin. Coal formation in Ordos Basin can be divided into four temporal phases, namely, Early Jurassic, Middle Jurassic, Late Jurassic, and Early Cretaceous. The Early Jurassic is an important coal-forming period with a total thickness of coal-bearing beds of about 300 m (Wang and Zhang, 1999). The lower part of the sequence is named the Fu County Formation and the upper part is the Yan'an Formation (Fig. 1). Recent surveys have found that the Middle Jurassic Zhiluo Formation also has recoverable coal beds (Zhang and Jian, 2007), but the Yan'an Formation, with the most extensive distribution and simplest structure of recoverable coal beds, is still the major coal strata in the Ordos Basin.

Because of the influences of tectonism and sedimentation, configuration and centers of basinal subsidence are different in every phase (Figure 2 and Figure 3). As a result, coal-accumulation zones display differences in the number of coal seams, their thicknesses and their lateral extents in different areas of the Ordos Basin. The Jurassic coal-rich belt in the Ordos Basin has a uniform distribution, the center of the coal-accumulation zone and depositional center are nonoverlapping but parallel. The distribution of the thick coal seam is a broad gentle syncline and its axis is located at N35°E in the line from the Wushenqi to Huan County. This trend parallels the distribution direction of the coal-rich belt. The coal-accumulation center in the Ordos Basin is located near Wushenqi (Fig. 4). The maximum thickness is over 40 m and gets thinner towards the periphery. Local coal accumulations occur in small synclinal basin near Pingliang and Huating in the southwestern Ordos Basin and in northeastern areas (Zhang et al., 1995). Coal sequences in the Ordos Basin show a thinning trend from north to south; coal seams in the northern basin are always more than 20 m-thick but in the southern basin they decrease to 10 m. Moreover, the coal seams also become non-continuous.

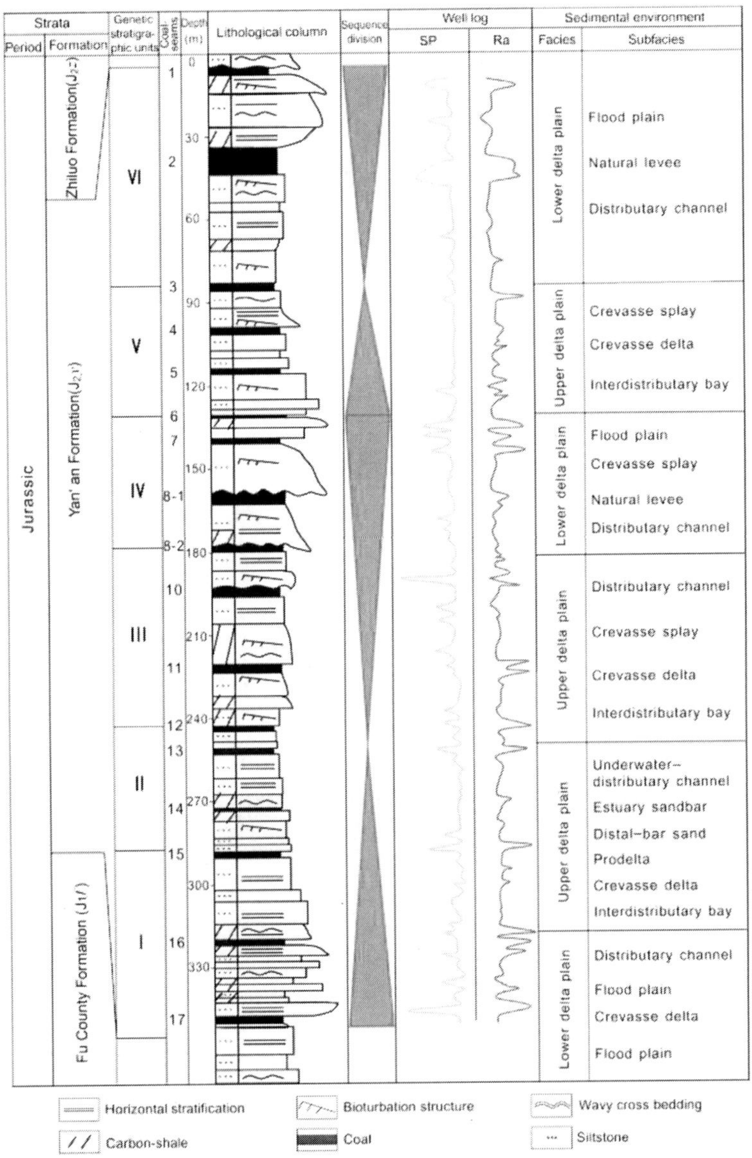

Figure 1. Diagram of stratigraphy and coal seams in the Yan'an Formation of Suishi mine, Lingwu coalfield, Ordos Basin.

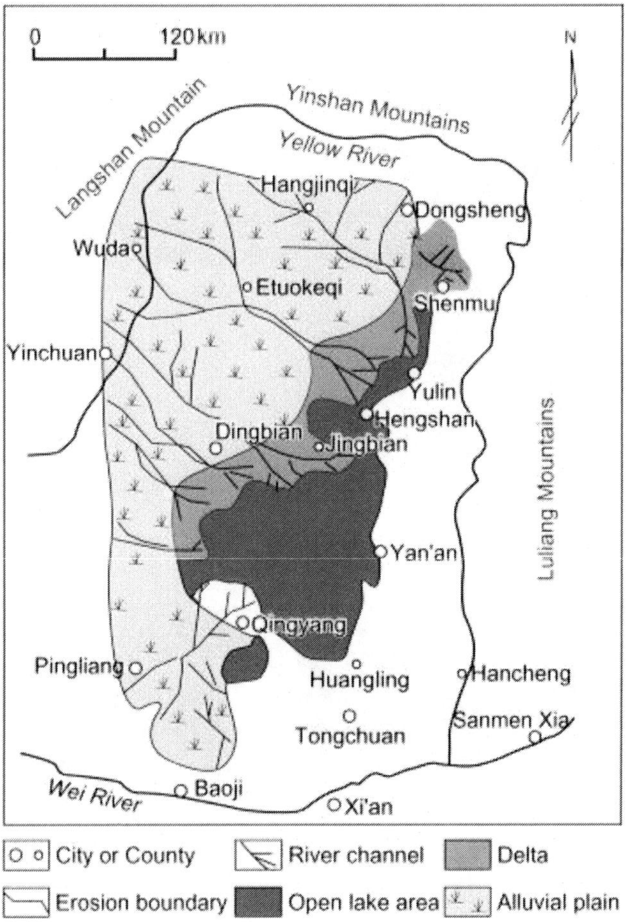

Figure 2. Distribution of sedimentary environments of the Early–Middle Jurassic Yan'an Formation (after Wang and Zhang, 1999).

The distribution of the Jurassic coal-rich belt has a specific zoning property. From the basin's western margin to its eastern area, coal sequences show a gradual thinning trend. Although at the western margin, coal sequences are thick, individual coal seams are thinner and the phenomena of branching off and merging are common. The development of the coal-forming swamp was near the basin's western sedimentary provenance, a depositional setting not favorable for the formation of thick individual coal beds. For this reason the economic coal-rich belt is only developed somewhat east of the depositional center.

According to the coal seams' thickness and the numbers, the Ordos Basin has two coal-rich districts. One is in the northern area located along the line of Xinzhao, Wushenqi and Dongsheng. The Lingwu district, in western margin of northern basin, has a great number of coal seams with large thickness, more than 22 m (Figure 1 and Figure 5). The number of coal seams gradually declines and they thin southwards to Huan County and Qingyang City. Some coal-rich belts occur in the southern margin with coal seams more than 20 m-thick appearing locally in Pingliang, Huating, Tongchuan and Huangling. There are no coal deposits in the Dali River to the south, Hulu River to the north and Wuqi area to the east of the basin (Fig. 4). Based on our observation, the sedimentary environment and depositional microfacies are important in controlling the coal's development.

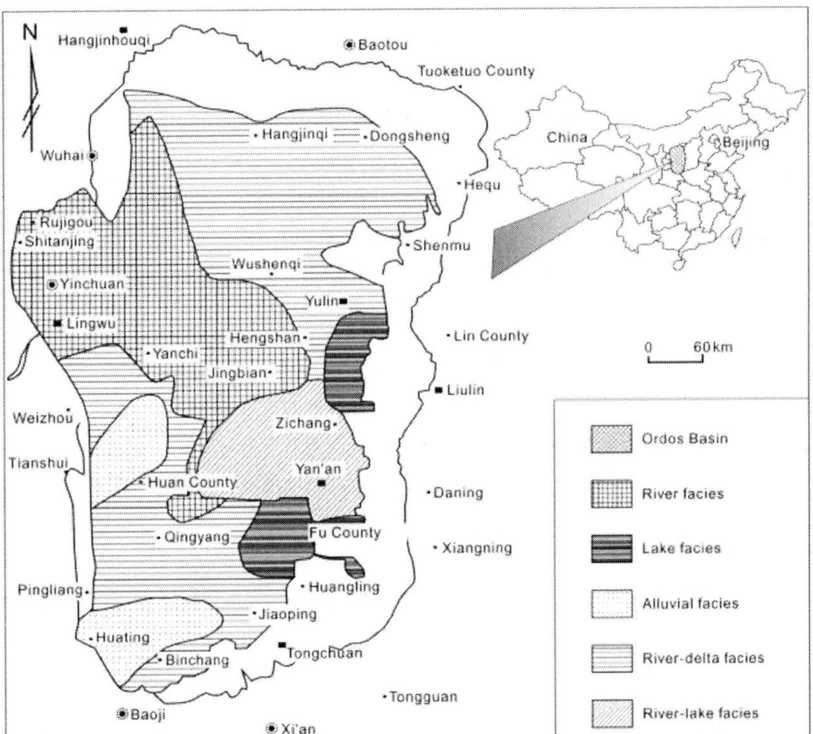

Figure 3. Sedimentary facies distribution of Jurassic rocks in the Ordos Basin (after Yang et al., 2005).

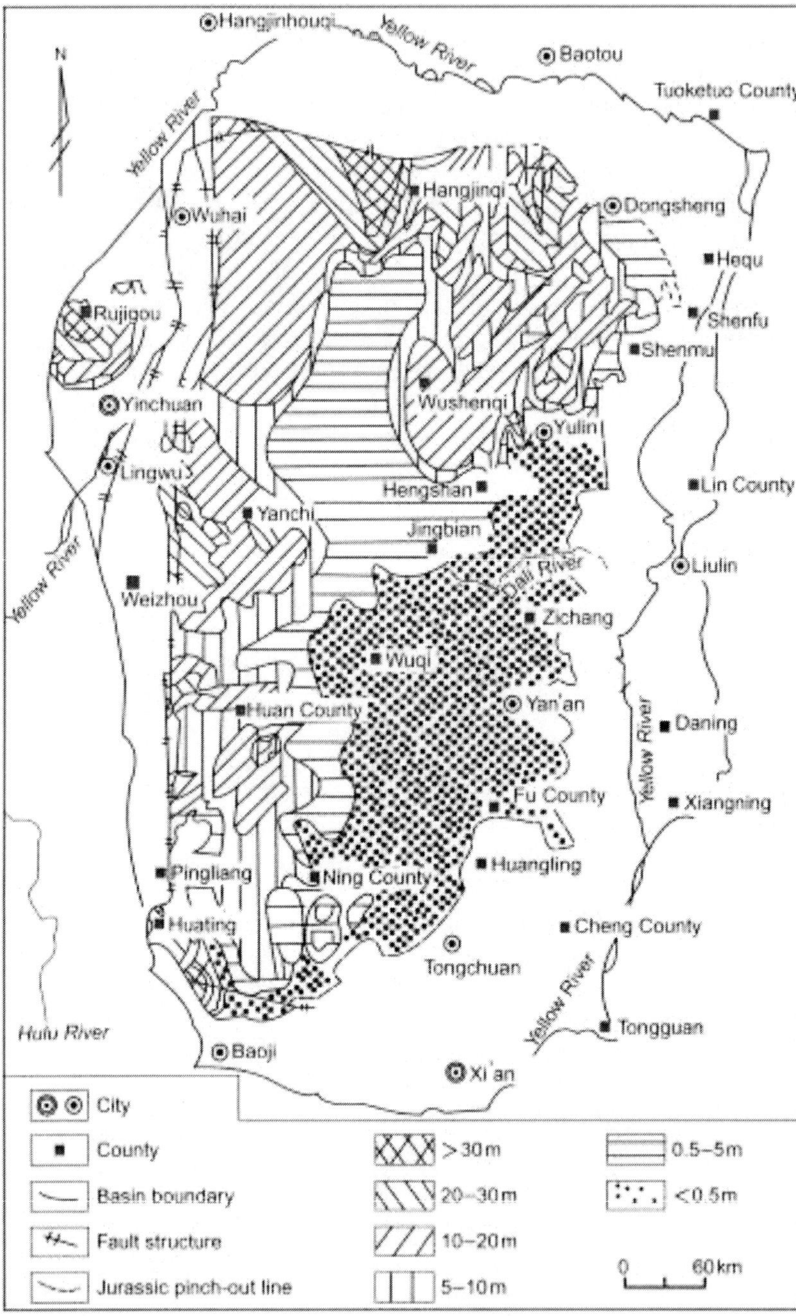

Figure 4. Distribution of Jurassic coal's thickness in the Ordos Basin.

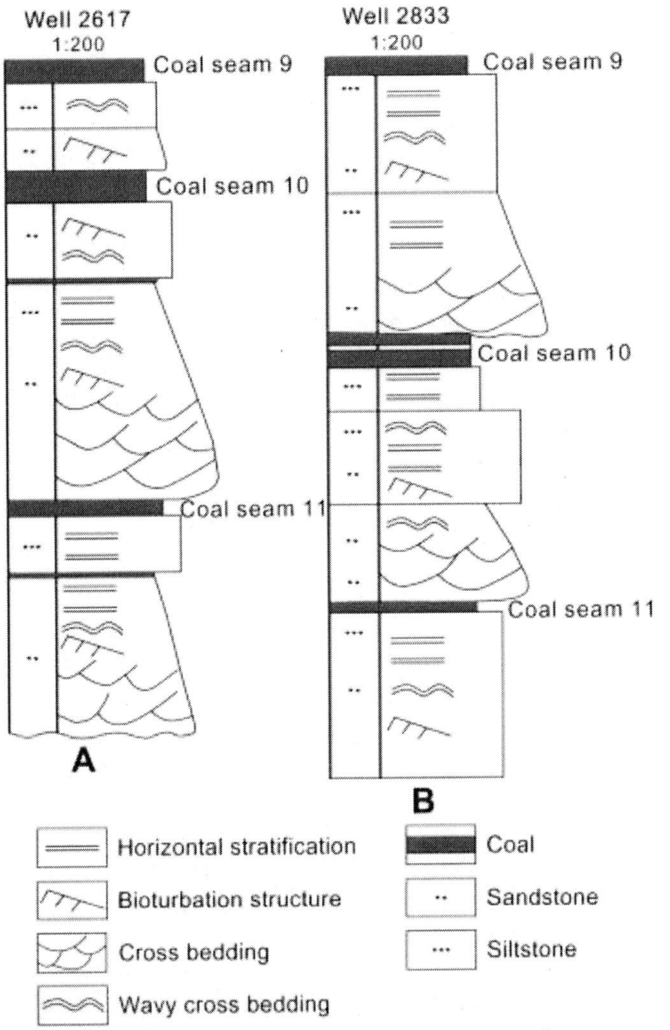

Figure 5. Diagrammatic sections of the rock sequence with coal seams in the Lingwu coalfield area.

3. CHARACTERISTICS OF COAL PETROLOGY

3.1. Macroscopic types of coals

The coals of the Jurassic Yan'an Formation form in two general depositional environments. One is high-stand swamp oxidation, which mainly develops in swamp environments in upper fluvial systems near the sediment provenance. Coarse sand bodies develop in the lower coal beds and a thin mudstone layer always occurred in the bottom of the sand bodies. This type of environment mainly develops semi-dull coals, dull coals with lesser clarite, and bright coals. It is mainly distributed in the northeastern Dongsheng area. The proportion of dull coal is higher from the basin's south edge to the north. Commonly, the combined percentage of semi-dull coal and dull coal can account for 60% to more than 75% of high-stand swamp coals with a gradually declining trend from the basin north edge to the south. It should be pointed out that in previous studies (Yang et al., 2005), much bright coal and clarite, approximately 90% or more, was observed in the Ruqigou area. These coals were considered to be a special type. Actually, the metamorphic degree of these coals is so high that they are anthracitic and even their liptinite and inertinite cannot be distinguished by common petrographic methods. As a result, the type of swamp environment cannot be determined by coal petrology or the types of macerals in such highly metamorphosed coals. Semi-dull coal and dull coal are dominant in the mining areas with a moderate degree of coalification (R^0_{max}=0.5% − 0.8%). Judging from our observations, semi-bright and bright coals always occurred in the bottom of the coal seams. The other type of Ordos's coal environment is called a low-stand swamp. It mainly develops in swamp environments in the upper fluvial system or in lakes (lacustrine facies). In this environment, fine sandstone bodies occurred in the lower, poorly developed, coal seams. The depositional layers are very fine-grained muddy shale and coaly mudstone, which are well developed in the floor and the roof. The precursor peat quickly becomes anoxic after burial and undergoes major gelatification. Therefore, the vitrinite content is high. The macroscopic types of coal are mainly bright coal and semi-bright coal, such that the combined proportion of dull coal and semi-dull coal is less than 50%. This type of coal occurs in Shenfu and Yulin-Hengshan in Shaanxi Province, in the central and southern basin and in Fu County in the southeastern basin. Although semi-bright coal and bright coal are dominant in this environment, some semi-dull coal and dull coal is well developed locally, indicating that the sedimentary microfacies still controls the distribution of the coal types.

In a single coal sequence, oxidation generally characterizes the lower coal beds, in contrast to reducing conditions in the upper sequence. This phenomenon results from microfacies changes. In general, oxidizing swamp waters were in circulation in early periods; later backwater environmental conditions were reductive. The whole sedimentary cycle presents a normal graded bed fining-upward sequence. If overlying water conditions are stagnant and reductive enough, no matter whether in peat swamps, upper fluvial or upper lacustrine systems, and the gelatification process is adequate, the vitrinite content will be high. This phenomenon is common in the Ordos Basin, but not all coal-bearing cycles obey the same law.

The macroscopic types of coal display specific differences (Table 1). However, it should be pointed out that the reason why the types of coals in Ruqigou are mainly bright coal is that the coal rank is anthracite. This metamorphic degree makes it difficult to distinguish exinite and vitrinite and this is the main reason for statistical compositional errors.

Table 1. Macro-petrologic types as percentages (%) of Jurassic coal in the Ordos Basin.

Mining area	Bright coal	Semi-bright coal	Semi-dull coal	Dull coal
Lingxin	3	19	66	12
Tianshuibao	1	26	14	59
Diantou	6	19	44	31
Dongsheng	8	28	47	18
Rujigou	45	46	4	5
Wayaobao	10	20	47	23
Majiatan	13	16	38	33
Suishijing	8	24	25	43
Shenfu	18	29	32	21
Yulin-Hengshan	12	29	33	26
Jiaoping	5	26	34	35
Huangling	7	19	33	41
Binchang	15	15	34	36

The low-stand type of coal environment cannot develop coals abundantly for two reasons. One is that some areas do not possess sufficient water to cover the swamp. The other is that the timing of the occurrence of this kind of environment is commonly too short.

In many coal sequences, individual sequences display the dominant dull coal and semi-dull coal in their lower part, more and more semi-bright coal and bright coal in their middle part and then increasing semi-dull coal and dull coal contents upwards. The reason for this is that the upper swamp environment with sufficient water to cover the swamp was rapidly interrupted and then replaced by an oxidizing swamp environment. In short, in many incomplete coal-bearing sequences, the coal beds display changes from dull coal to bright coal, then to dull coal, according to the cycle development.

It has also been discovered in this study that in the Huangling mining area in the SE part of the basin, the No. 4 coal bed in the lower Yan'an Formation contains sapropelic and cannel coals. These oily coals are black with black streak and a dark silky luster. Depending on the vitrinite content, the luster appears glassy or earthy. The cannel coal density is low; it is flexible and has a flat, conchoidal or conchoidal-like fractured surface. Under the microscope, the sapropelic coal can be distinguished from humic coal because it has no internal structure or layers. Other properties include a unique texture and high strength (Zhang et al., 1996). The Huangling area had been a sag in the SE basin in the depositional period of the coal-bearing sequences. At that time, the lake was deep, which was not favorable for the development of coal-forming swamps, but sapropelic coal can be formed in deep and semi-deep lacustrine environments. The areal extent of the sapropelic coal corresponds exactly to the extent of the deep lake in the Early–Middle Jurassic sedimentary basin.

3.2. Characteristics of macerals

Many studies have been done on the characteristics of macerals in mining areas in the Ordos Basin (Wu et al., 1995, Shu et al., 1996, Zeng, 2000, Bai and Li, 2001, Bai et al., 2002, Duan et al., 2004, Chen et al., 2006 and Wang and Tang, 2007). In the comprehensive research of the whole basin, vitrinite was absolutely the predominant maceral group in the Permo-Carboniferous coals, inertinite was next in abundance and exinite was minor. These abundances are basically similar with those in other areas in North China. In the Jurassic, large-scale inland lacustrine and fluvial systems became the major environment with accompanying changes in climate and plant types. These changes resulted in an increase of inertinite in the Jurassic coals of the Ordos, with lesser vitrinite and still lesser exinite compositions.

The maceral types in Jurassic coals are mainly vitrinite and inertinite, with inertinite most abundant. The most common vitrinite components are anthraxylon, homocollinite and degradinite. The most common inertinite components are fusinite, fusovitrite and macrinite, and the most common exinite components are microsporinite, cutinite and amorphous bitumen. Some resinite, cortical tissue can be found, but no macrosporinite. We noted that exinite can be accumulated locally in the middle part of thick coal beds, which is very important for strong hydrocarbon generating and reserve capacity. Fusinite often acts as the matrix material. In most situations, the fusinite and vitrinite coexist in bent, irregular, or lenticular shapes and is distributed in blocky forms or in a dispersed state. Some fusinite is distributed as complicated blocky forms or in a dispersed state. Fusinite has a high rigidity so that cavities and pores are developed, which is favorable for the migration and the storage of hydrocarbons. So it is apparent that fusinite acts to enhance the reserve capacity of the coal beds. Vitrinite and exinite have strong hydrocarbon generation capacities without many vugs and pores. Therefore, further research on the distribution and allocation of macerals in the thick coal seams of the Ordos Basin will be useful to evaluate the hydrocarbon generation capacity of the coal sequences and to make an assessment of the reasonable utility of various components in the coal beds.

4. MACERAL DISTRIBUTION FEATURES

The macerals in Yan'an Formation coals in different mining areas of the Ordos Basin (Fig. 3) are mainly vitrinite and liptite; the vitrinite content is lower than that of liptite (Table 2); and there is rare exinite and little mineral matter. The vitrinite content generally ranges from 32% to 60%, except in the Rujigou mining area. The distribution trend of the macerals is such that the vitrinite content increases from the basin's periphery to the center; the fusinite content has an opposite trend. The vitrinite content is lowest in the Dongsheng area on the northern periphery of the basin, ranging from 27.5% to 40%. This is because the water cover in the coal-forming swamp became well-oxygenated near the northern part. The vitrinite content is higher in the Yulin-Hengshan mining area, ranging from 60% to 70% because of the high coal rank in Rujigou, Shitanjing, Huating and other mining areas near the western margins of the basin; in other areas the vitrinite value is about 40%. From old strata to young strata, the vitrinite content in the Yan'an Formation coals changes from low to high, then from high to low. The changes of liptinite content are exactly opposite.

Table 2. Maceral statistic data (in %) and mean reflectance of some Jurassic Yan'an Formation coal samples from the Ordos Basin.

Mining area	Vitrinite	Liptite	Exinite	Mineral	R^0_{max}
Dongsheng	56	38	3	3	0.46
Rujigou	84	9	3	4	0.92
Ciyaobao	30	57	4	9	0.55
Majiatan	40	50	2	8	0.70
Suishijing	44	46	3	7	0.63
Shenfu	50	42	4	4	0.50
Yulin-Hengshan	60	30	3	7	0.64
Jiaoping	34	59	7	3	0.71
Huangling	37	52	7	4	0.79
Binchang	35	55	5	5	0.64
Huating	84	6	7	3	0.49

Note: R^0_{max} = mean vitrinite reflectance.

Microscopic observations of the macerals show that the vitrinite is mainly anthraxylon, homocollinite or degradinite. Liptite contains mainly fusinite, with a little micrinite and sclerotinite. The main components in exinite are sporophytes and cutinite with rare resinite. Overall, fusinite (liptite) content is high and the content of mineral materials is low.

4.1. Vitrinite: telinite, collinite, degradinite, corpocollinite

4.1.1. Telinite

The cell walls of telinite are reddish-brown to brownish-red in transmitted light. The cell structure is well preserved with a regular arrangement that can be observed clearly. Some cell walls expand and then the cell cavities (lumen) shrink gradually; some of them are empty (Plate 1a, b). Others are filled in with collinite, resinite, microsome or clay. Extrusion and deformation of the structure of the cell cavities are common and occasionally the telinite cell cavities are filled in by cutinite fragments.

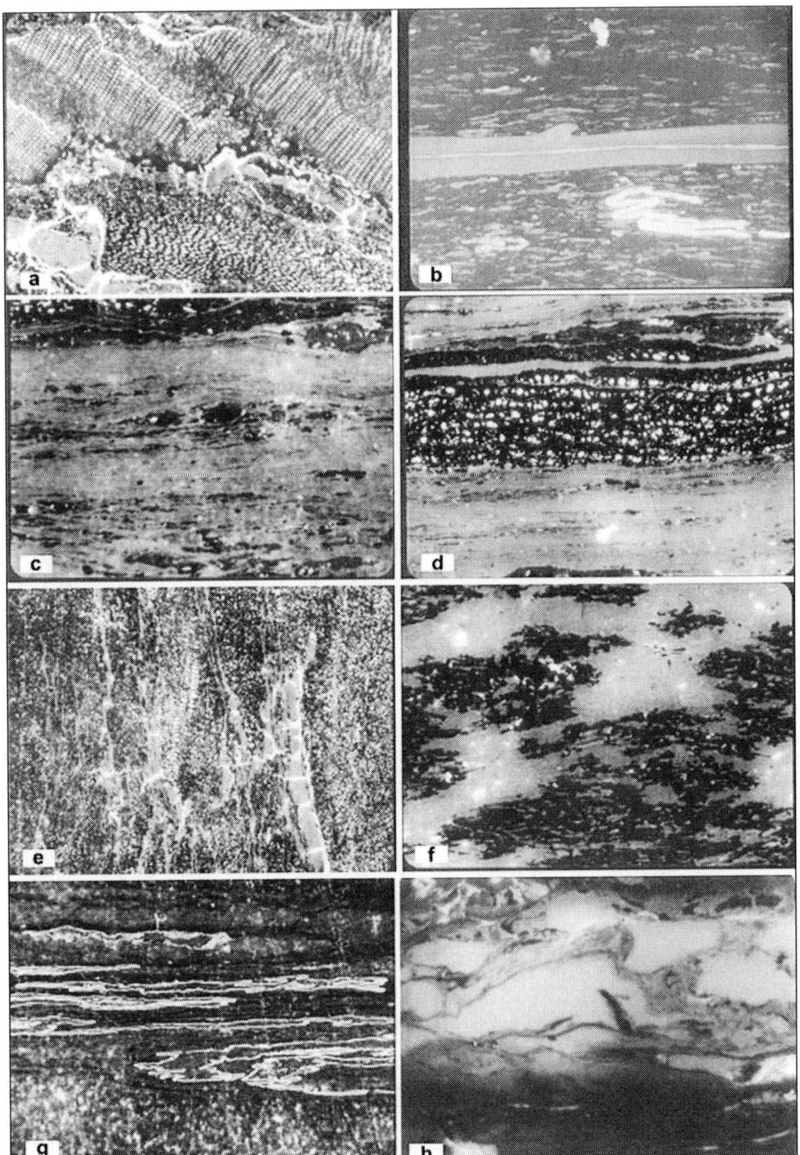

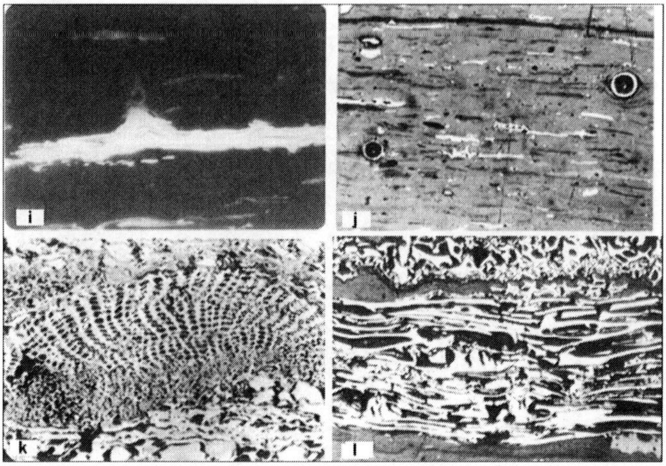

Plate 1. Petrographical micrographs of Jurassic coal samples from the Ordos Basin. a: anthraxylon, orange-red, clear and regular arrangement of cell texture, void cell cavity, transforming into semi-vitrinite and fusinite and surrounded by cutinite. Micro-fissures are developed in the network shape. Dongsheng, $J_{1-2}y$, transmitted light, 50×, [orange-yellow components shown in Plate 1a are homocollinites]; b: mainly cutinite, macrosporinite and microsporinite, in amorphous degradinite, the latter dark brown in reflected blue light. The macrosporinite (banded cement in the middle) is yellow and shows strong fluorescent light; the microsporinite and cutinite are yellow and show weak or dark brown fluorescent light. Yuhebao, Yulin, $J_{1-2}y$, blue reflected light, 50×; c: degradinite, medium-degree degradation, amorphous matrix, cemented macrinite. Shenmu, $J_{1-2}y$, transmitted light, 130×; d: fusinite, brownish-black with clear cell texture between degradinite bands with sharp boundaries and without any transformational phenomenon. Yuhebao, Yulin, $J_{1-2}y$, transmitted light, 50×; e: transition from anthraxylon, semi-fusinite to fusinite (varying from left to right); the rate of change is slower on the left and quicker on the right. There are intercalated vitrinite bands; several horizontal partings in the bands are the result of shrinkage of vitrinite gelatination. Dongsheng, $J_{1-2}y$, transmitted light, 50×; f: macrinite, black, distributed irregularly in amorphous degradinite, the latter orange-red in mixed distribution, but still can be connected. The macrinite may result from the residue of separation of vitrinites during hydrocarbon evolution. Yuhebao, Yulin, $J_{1-2}y$, transmitted light, 125×; g: cutinite, thin cell wall, fine and bent as a wire, distributed in degradinite. Shenmu, $J_{1-2}y$, transmitted light, 130×; h: phlobaphinite, yellow fluorescent light, the arrangement of the cell texture of suberitoid is irregular with the separation of bitumen. Dongsheng, $J_{1-2}y$, blue reflected light, 600×; i: seepage bitumen, strong yellow fluorescent

light, infilling bedding fissures, showing flowing phenomenon. Well Yi-23, Etuokeqi, $J_{1-2}y$, blue reflected light, 780×; j: sclerotinite (emphasized by white rings), white, void rings, sclerotium, in degradinite. Well Yi-23, Etuokeqi, $J_{1-2}y$, oil-immersed reflected light, 310×; k: fusinite, lined, very regular and complete arrangement of cell texture, the product of early-stage fusainization, void cell cavity. Dongsheng, $J_{1-2}y$, oil-immersed reflected light, 390×; l: burnt fusinite, easily broken. The cell wall is compressed and broken, extending in one direction, showing that well-textured burnt fusinite was broken in-situ by compression creating better permeability of the coal bed. Well Yi-23, Etuokeqi, $J_{1-2}y$, oil-immersed reflected light, 490×.

4.1.2. Collinite

Based on its shape and genesis, collinite sometimes can be subdivided. In the Ordos Basin, it contains homocollinite and degradinite with a little corpocollinite. Collinite can evolve from gelatinous substance, which could have been wood fibers or other components in plants. Under the microscope, the homocollinite appears uniform but shows different degrees of gray color in oil immersion and reflected light. The color is reddish orange to brown in transmitted light; it is distributed uniformly as uneven bands and lenses with locally orange-yellow color. Vertical innate fissures can be observed occasionally (Plate 1a), with cutinite fringes locally or scattered orange-yellow cuticularized aggregates.

4.1.3. Degradinite

Degradinite is one of the primary macerals in this area presenting amounts up to 30% in most coal samples. In transmitted light, degradinite appears as a sheet or "cement" (Plate 1b). Its color is sometimes reddish orange to orange-yellow without a stable shape or cell structure. In addition, some degradinites can cement macrinite and clay minerals (Plate 1c).

4.1.4. Corpocollinite

Corpocollinite occurs singly or as cell-filling material; it is mainly derived from phlobaphinite and is seldom found in this area. Its color is brownish red-maroon and is distributed as oval, circular and lenticular.

The vitrinite reflectance of Jurassic coals generally ranges from 0.5% to 0.8%, which is in the range of low metamorphic bituminous coal. Bai et al. (2005) considered that the volatile yield of such low metamorphic degree coals does not correlate well with the maceral composition and vitrinite reflectance; it is probably caused by different degrees of reduction.

4.2. Inertinite: fusinite, semi-fusinite, macrinite, micrinite, sclerotinite

4.2.1. Fusinite, semi-fusinite, macrinite

Fusinite has clear cell structure. It appears black and brownish-black in transmitted light (Plate 1d). It is distributed in bands, lenses, strips and irregular shapes. The cell wall is white in oil immersion and reflected light (Plate 1k, l). Semi-fusinite often occurs in the Ordos Basin and generally appears in a transformed state with anthraxylon. The color ranges from reddish-black to brownish-black in reflected light (Plate 1e). The reflectance of semi-fusinite is lower than that of fusinite, but higher than vitrinite. Its cell structure is not preserved as well as that of fusinite. Macrinites are observed as different-size circular and oval grains without original cell structure. It also appears as a mixture with degradinite in the matrix (Plate 1f).

4.2.2. Micrinite, sclerotinite

Micrinite occurs as white aggregates observed in oil immersion and reflected light. It often appears in bands or cell-cavity fillings. Minor sclerotinite is also present in some Ordos coals (Plate 1j); it looks like a circular white vein in oil immersion and reflected light. It often occurs in degradinite with black sclerotium.

After their study on low grade Jurassic coals in NW China, Bai et al. (2005) considered that the high content of inertinite and transmitted light components has some connection with a weak reducing environment. The Jurassic coal-forming swamp was mainly of the high-stand type. Because of drought and forest fires, burnt fusinite occasionally appears as intercalated layers in the coal. The Jurassic coals in NW China are generally non-caking coals and weakly caking coals (Han et al., 1996). Because of weak reduction the coal has low volatile yield, low hydrogen content, but high oxygen content.

After his study on inertinite-rich coal in the Shendong (Shenfu-Dongsheng) mining area, a famous coal mine near Yulin City, Li (2005) considered that the main conditions for formation of this coal were dry weather with weak oxidation conditions on the peat surface. In addition, the plant remains underwent the influence of fungal activity, which has been indicated as an important mechanism for forming inertinite (Stach et al., 1982). This study (see Plate 1j) has not shown that the fungus coexists with fusinite which has a complete cell structure. However, the fungus always exists with semi-vitrinite, fine grain cutinite and microsporinite in dull coal or semi-dull coal belts.

We have also found that in these coals the inertinite content varies depending on the coal-forming environment and swamp types. Adjacent to the northern basin the inertinite content is the highest, exceeding 80%; the content often reaches more than 60% towards Jiaoping, Huangling and Binchang mining area in the southern part of the basin. Fire-formed fusinite is minor, but the content of semi-fusinite is high. The features of clear cell structure and fracturing of fire-formed fusinite are quite different from the structure of semi-fusinite or semi-vitrinite (Plate 1l). It is noted that in previous studies (Chandra, 1965, Stach et al., 1982, Dai et al., 2003, Tang and Huang, 2004 and Zeng et al., 2005), the semi-fusinite or semi-vitrinite, is always categorized as inertinite, and basically still remains active. Thus semi-fusinite and semi-vitrinite should not be classified as inertinite but should be reassessed according to their percentage in the whole maceral groups or its hydrocarbon ratio. The high content of these transitional components in the middle part of Ordos Basin should be given consideration.

4.3. Exinite

4.3.1. Sporophyte

Sporophyte is sclerine of pollen or sporinite. Microsporinite generally exhibits strong yellow fluorescence under ultraviolet light. It has compressed elongated shapes and occurs in degradinite or amorphous bitumen in groups. It is widely distributed, but in low concentrations.

Macrosporinite is rare in this area and when present, has weak fluorescence (Plate 1b).

This study discovered that exinite often accumulated locally in the middle part of coal beds. Its average content is not high in the whole coal bed, but the presence increases the capacity of liquid hydrocarbon generation. In previous research (Stach et al., 1982), the exinite content is often underestimated because exinite has fine grain size, occurs in other components, and it is always difficult to analyze accurately.

4.3.2. Cutinite

Cutinite is from plant's leaves, branch or skin secretion, and it is grayish black in reflected light and oil immersion and is often filled with vitrinite (Plate 1g). It occurs in single bodies or layers. In fluorescent light, it appears yellowish green to brownish yellow with smooth or very clear sawtooth boundaries in reflected light, it is yellow or orange-yellow. In transmitted light, it is bright yellowish white and distributed in banding, limbate and flake shape. Cutinite is widely distributed, but present in low concentration in the study area.

4.3.3. Resinite

Resinite was the plant's resin or oil secretion, in the coal beds it generally occurs in oblong, oval, kidney shape particles, with thin and thick banding or dispersed and locally as fillings in cell cavities. Occasionally the resinite appears compressed and deformed. In transmitted light, it appears homogeneous and bright yellow and in reflected light it is dark grayish black. In the excitation of blue reflected light, the color of fluorescence varies from brownish yellow to yellowish green.

4.3.4. Phlobaphinite

Phlobaphinite (cortical tissue) is one of the more important macerals in the Ordos Basin coals. Its cell structure is well preserved. The regular degree of the cell structure arrangement is uneven, mostly imbricated, and elongated along the beds. Clear cortical structure can be observed. The color is dark grayish light gray in reflected light and oil immersion (Plate 1h). The fluorescence is yellow, brownish yellow and yellowish green in blue reflected light excitation.

4.3.5. Seepage amorphous bitumen

Seepage amorphous bitumen is very common. It has features of low reflectance and mostly weak light brown fluorescent light. In blue reflected light, it shows dark-gray or yellow fluorescent light. The bitumen fills in fissures and pores; it sometimes appears to have a flowing structure (Plate 1i).

4.3.6. Minerals

The mineral content of the Jurassic coal in the study area is mainly kaolinitic clay distributed in disseminated or small lenticular shape. The sulfur content is low, which indicates that an acidic terrigenous environment was predominant. Pyrite crystals occur in some coal beds, but framboidal pyrite is rare. In addition, a kind of matrix mineral bitumen that doesn't belong to any maceral can be distinguished in plain light. This component is very common in bulk rock samples and has not been identified.

5. CONCLUSIONS

In this study, we found that the coal petrological features of Jurassic coal in the Ordos Basin in NW China are its transitional components, semi-vitrinite and semi-fusinite. The coal's semi-fusinite content is significantly high. Therefore, the chemical characteristics of these coals and their technical properties are different from other coals with the same metamorphic degree in other areas. In-depth research on the genesis of these transitional components will be helpful for better predictions on the varying trend of the technical properties of these Jurassic coals.

Coal petrology research on the Jurassic coal sequences in the Ordos Basin shows that macroscopically the coals are mainly semi-dark and dark with a little semi-bright and bright coal. The coals are mainly a two-component petrographic system: vitrinite and fusinite. The coal's maceral composition has close relationship with the depositional environment evolution, no matter whether in horizontal or stratigraphic distribution it mainly shows the features of vitrinite content of Jurassic coal. On plane distribution, vitrinite content gradually increases from the basin periphery to the center, but the fusinite content exhibits a reverse trend. It indicates that the oxidation and fusinization are stronger on the basin periphery than in the center. When the covering water deepens in the basin and forms a stable and closed water body, the reduction strengthens, which lead to stronger gelatification and higher vitrinite content in basin centers. In stratigraphic distribution, the lower vitrinite content in Jurassic coal at the two edges but higher in the middle, indicates the evolution of the depositional environment. In other words, the basin underwent a process of advance–retreat–advance with the lake in a state of expansion–shrinkage–expansion, and the paleoclimatic environment was in a damp–drought–damp cycle.

Exinite is often observed in middle part of thick coal beds in the Ordos Basin. Although its average content is not high, it contributes a lot on the total hydrocarbon generation capacity of coal beds. The hydrocarbon generated from coal seam can flow into and storage in layers, which are rich in fusinite. In Ordos Basin, the exinite content increases in the south. Its content is lower than other components in the coal beds, but if it rises a little, the whole hydrocarbon generation capacity strengthens. And further research should put more attention to the observation of thin sections, the precise description of the distribution state of exinite in coal beds, and an objective estimation of its content, especially fluid bitumen. The reason is that it can flow easily into hard fusinite or fissures and then is always neglected.

ACKNOWLEDGMENTS

This paper is supported by Chinese Ministry of Science and Technology Project (No. 2006CB202202), and the National Natural Science Foundation of China (No. 40972106and No. 40772095). The authors are very grateful for these financial supports. We would like to thank Prof. Gregory A. Davis, and two anonymous reviewers for critical and constructive review of the manuscript.

REFERENCES

1. Bai, X.F., Li, W.H., 2001. Coal characteristics and governed factors for coal properties of Datong Jurassic 10-11 coal. Coal Conversion 24 (4), 21e24 (in Chinese with English abstract).

2. Bai, X.F., Li, W.H., Chen, W.M., Ma, W.W., 2005. Study on distribution and characteristics of coals with weak reductive degree in West China. Journal of China Coal Society 30 (4), 502e506 (in Chinese with English abstract).

3. Bai, X.F., Li, W.H., Zhang, S.L., Zhao, Y.X., 2002. Study on properties of Shendong Jurassic 2-2# coal. Coal Conversion 25 (3), 87-88

4. (in Chinese with English abstract). Chandra, D., 1965. Reflectance of Indian coals. Quarterly Journal of the Geological Mining and Metallurgical Society of India 37, 37-39.

5. Chen, H.B., Li, W.H., Jiang, Y., Bai, X.F., 2006. Macerals and structure characteristic parameters of Shendong coals. Coal Conversion 29 (1), 6-10 (in Chinese with English abstract).

6. Dai, S.F., Li, D., Chou, C.-L., Zhao, L., Zhang, Y., Ren, D.Y., Ma, Y.W., Sun, Y.Y., 2008. Mineralogy and geochemistry of boehmite-rich coals: new insights from the Haerwusu Surface Mine, Jungar Coalfield, Inner Mongolia, China. International Journal of Coal Geology 74, 185-202.

7. Dai, S.F., Ren, D.Y., Hou, X.Q., Shao, L.Y., 2003. Geochemical and mineralogical anomalies of the late Permian coal in the Zhijin coalfield of southwest China and their volcanic origin. International Journal of Coal Geology 55, 117-138.

8. Dai, S.F., Ren, D.Y., Li, S.S., Zhao, L., Zhang, Y., 2007. Coal facies evolution of the main minable coal-bed in the Heidaigou Mine, Jungar Coalfield, Inner Mongolia, northern China. Science in China (Series D) 50 (Suppl. II), 144-152.

9. Duan, X.Q., Wang, Z.N., Qu, J.W., 2004. Study on structural property of inertinite and vitrinite of Shengfu coal. Coal Science and Technology 32 (2), 19-23 (in Chinese with English abstract).

10. Han, D.X., Ren, D.Y., Wang, T.B., 1996. China Coal Petrology. China University of Mining and Technology Press, Beijing, 251 pp (in Chinese).

11. Huang, D.F., Qin, K.Y., Wang, T.G., 1995. The Formation of Coal into Oil and its Hydrocarbon Generation Mechanism. Petroleum Industry Press, Beijing, pp. 313-369 (in Chinese).

12. Li, X.Y., 2005. Conditions of inertinite-rich coal generation, Shendong mining area: significance of fungal alternating origin of inerts. Coal Geology and Exploration 33 (5), 1e4 (in Chinese with English abstract).

13. Liu, G.X., Pan, W.L., Shao, H.N., 2000. Evaluation of hydrocarbon generating potential of the middle and lower Jurassic coal in Northwest China. Experimental Petroleum Geology 22 (3), 232-235 (in Chinese with English abstract).

14. Qian, L.J., Bai, Q.Z., Xiong, C.W., 1987. The Accumulating Features of Jurassic Coal-bearing Strata in Northern Shanxin Province. Northwest University Press, Xi'an, pp. 128-167 (in Chinese).

15. Shu, X.Q., Wang, Z.N., Xu, J.Q., Ge, L.M., 1996. Structural characteristics and differences among lithotypes. Journal of Fuel Chemistry and Technology 24 (5), 426-433 (in Chinese with English abstract).

16. Stach, E., Mackowsky, M.-Th., Teichm€uller, M., Taylor, G.H., Chandra, D., Teichm€uller, R., 1982. Textbook of Coal Petrology. Gebr€uder Borntraeger, Berlin, pp. 177-197.

17. Tang, X.Y., Huang, W.H., 2004. Trace Elements in Chinese Coal. The Commercial Press, Beijing, pp. 3-11 (in Chinese).

18. Wang, S.M., Zhang, Y.P., 1999. Study on the formation, evolution and coalaccumulating regularity of the Jurassic Ordos Basin. Earth Science Frontiers 6 (Suppl.), 147-154 (in Chinese with English abstract).

19. Wang, S.Q., Tang, Y.G., 2007. Coal petrological characteristics and coal facies in Shendong mining area. Coal Geology of China 19 (5), 4-7 (in Chinese with English abstract).

20. Wu, C.R., Zhang, H., Li, Y.L., 1995. Research on the Petrography, Quality and Metamorphism of EarlyeMiddle Jurassic Coal in Northwest China. China Coal Industry Publishing House, Beijing, pp. 19-64 (in Chinese).

21. Yang, Q., Liu, D.M., Huang, W.H., 2005. Comprehensive Evaluation on the Geology and the Resources of Coalbed Methane in Northwest China. Geological Publishing House, Beijing, pp. 32-41 (in Chinese).

22. Zeng, F.G., 2000. The petrographic feature of Daliuta 2-2 coalbed in Shenfu. Coal Geology and Exploration 28 (3), 25-27 (in Chinese with English abstract).

23. Zeng, R.S., Zhuang, X.G., Koukouzas, N., Xu, W.D., 2005. Characterization of trace elements in sulphur-rich Late Permian coals in the Heshan coal field, Guangxi, South China. International Journal of Coal Geology 61, 87-95.

24. Zhang, H., Bai, Q.Z., Zhang, X.W., Gao, X.Z., He, Z.L., Li, H.T., Lv, Z.F., 1995. Formation of the Ordos Basin and its tectonic environment. Coal Geology and Exploration 23 (3), 1e9 (in Chinese with English abstract).

25. Zhang, H., He, Z.L., Jin, X.L., 2005. The Tectonic Evolution and the Incoalation of Ordos Basin e Abbreviation of 1:500,000 Geognostic Map of Coal-bearing Ordos Basin. Geological Publishing House, Beijing, 68 pp (in Chinese).

26. Zhang, J.G., Jian, S.G., 2007. Coal forming environment of Zhiluo Formation in Erdaoling Mine of Lanshan coal field, Inner Mongolia. Northwestern Geology 40 (1), 89-93 (in Chinese with English abstract).

27. Zhang, X.Y., Tong, Y.Z., Xiao, D.X., Zhang, C.R., 1996. Study on origin of cannel coal in Huangling mining district. Coal Geology of China (8 Suppl.), 10-16 (in Chinese with English abstract).

28. Zhang, Y.Z., 2006. The advance on Shenhua direct liquefaction. Science and Technology of China 2, 32-35 (in Chinese).

29. Zhao, C.Y., He, Z.H., Chen, K.M., 1994. The desmocollinite's hydrocarbon generation potential and characteristics of coals in Tuha Basin. Chinese Science Bulletin 39 (21), 1979-1981 (in Chinese with English abstract).

Chapter 8

Constraints on the Nature and Evolution of the Magma Plumbing System of Mt. Etna Volcano (1991–2008) from a Combined Thermodynamic and Kinetic Modelling of the Compositional Record of Minerals

Maren Kahl[1],[2] Sumit Chakraborty[2], Massimo Pompilio[3] and Fidel Costa[4]

[1]School of Earth and Environment, University of Leeds, Leeds LS2 9JT, UK

[2]Institut für Geologie, Mineralogie und Geophysik, Ruhr-Universität Bochum, Bochum 44780, Germany

[3]Istituto Nazionale di Geofisica e Vulcanologia–Sezione di Pisa, Pisa 56126, Italy

[4]Earth Observatory of Singapore, Nanyang Technological University, Singapore 639798, Singapore

1. OLIVINE COMPOSITIONS

Olivines with forsterite contents greater than Fo84 were reported by Molloet *al.* (2015) in their Supplementary Data, but were not found by Kahl *et al.*(2015) in their large body of chemical analyses of olivines from the

1991–2008 eruptive products of Mt. Etna, or among the compositions produced in their MELTS calculations. For clarification the experiments of Mollo *et al.*(2015) did not crystallize olivine. The experimental olivine compositions reported in their Supplementary Data were from experiments conducted by Métrich & Rutherford (1998). The forsterite contents of these experimental olivines range from Fo69 to Fo83. The high-Fo olivine compositions included in the Supplementary Data of Mollo *et al.* (2015)are actually analyses of natural olivines reported by Nicotra & Viccaro (2012). Inspection of these analyses by Kahl *et al.* (2015) revealed that they showed poor stoichiometry (cation sums much less than 3·0 on a four-oxygen basis) and therefore the reported high Mg# values are likely to be analytical artefacts.

2. CLINOPYROXENE COMPOSITIONS

Mollo *et al.* (2015) reported a bimodal distribution of natural clinopyroxene compositions (Mg# 54–74 and Mg# 75–85) from the 2001–2012 Etna activity, based on literature data. Kahl *et al.* (2015), however, found only cpx with Mg# > 70 in their analyses of natural clinopyroxenes, as well as in their MELTS calculations. The results of experiments reported by Mollo *et al.* (2015) under water-saturated [data of Métrich & Rutherford (1998)] as well as anhydrous conditions yielded clinopyroxene with Mg# of 70–80, whereas water-undersaturated conditions resulted in cpx with Mg# of 52–69. The data from water-saturated and anhydrous experiments are consistent with the conclusions of Kahl *et al.* (2015).

REFERENCES

1. Kahl M., Chakraborty S., Pompilio M., Costa F. (2015). Constraints on the nature and evolution of the magma plumbing system of Mt. Etna Volcano (1991–2008) from a combined thermodynamic and kinetic modelling of the compositional record of minerals. Journal of Petrology 56, 2025–2068, doi:10.1093/petrology/egv063.

2. Métrich N., Rutherford M. J. (1998). Low pressure crystallization paths of H_2O-saturated basaltic–hawaiitic melts from Mt Etna: implications for open-system degassing of basaltic volcanoes. Geochimica et Cosmochimica Acta 62, 1195–1205.

3. Mollo S.,Giacomoni P. P., Coltorti M., Ferlito C., Iezzi G., Scarlato P. (2015).Reconstruction of magmatic variables governing recent Etnean eruptions: Constraints from mineral chemistry and *P*–*T*–fO_2–H_2O modelling. Lithos 212–215,311–320.

4. Nicotra E., Viccaro M. (2012). Transient uprise of gas and gas-rich magma batches fed the pulsating behavior of the 2006 eruptive episodes at Mt. Etna volcano. Journal of Volcanology and Geothermal Research 227–228, 102–118.

Chapter 9

Petrology and Geochemistry of Basement Complex Rocks in Okom-Ita Area, Oban Massif, Southeastern Nigeria

Kelechi Denis Opara*, Young E. Obioha, Samuel Okechukwu Onyekuru*, Chikwendu Okereke, Sabinus Ikechukwu Ibeneme

Department of Geosciences, Federal University of Technology, Owerri, Nigeria

ABSTRACT

The determination of the mineralogical composition of rocks that underlie Okom-Ita area, in Oban Massif became necessary for the interpretation of the petrogenesis and protoliths of rocks in the area. Twelve representative rock samples were selected for thin section petrography and bulk rock geochemical analysis. The results reveal that the dominant intrusive rocks in Okom-Ita area are pegmatites, quartz veins and dolerites emplaced within gneisses, schists and phyllites host rocks. The quartz veins and pegmatites are leucocratic consisting predominantly of quartz, plagioclase and subordinate amount of Muscovites. The dolerite is dark grey, fine to medium grained and texturally ophitic and consists dominantly of pyroxenes, olivine and opaques. Three varieties of gneisses were recognized: feldspathic, banded and biotite gneisses. The gneisses, schists, and phyllites are dominated by SiO_2 in the range of (60% - 75%). The rocks exhibit higher molecular concentration {Al_2O_3 wang##Brack$_e$t# (C$_a$O + Na2O + K2O)}, high alkali concentr$_a$tion$_w$ith Na2O w$_a$n$_g$##Bracket# K2O, high Al2O3 to alkali ratios and low TiO2, CaO and MgO concentrations. These distribution trends suggest a compensation for the high silica and alumina concentrations and support a granitic protolith for the pegmatites and their host gneisses and schists. The dolerites and amphibolites are impoverished in SiO2, but relatively enriched in lime and alkali oxi$_d$es. Pl$_o$t$_s$ in the AFM, T$_i$O2-K2O-$_P$2$_{O5}$ and Na2O/Al2O3-K2O/Al2O3 diagrams show that the pegmatite was derived from calc-alkaline magmatic source, while the dolerite was derived from tholeiitic basalt magma.

KEYWORDS

Okom-Ita, Oban Massif, Petrology, Geochemistry, Petrogenesis, Protolith, Tectonics

1. INTRODUCTION

Nigeria lies in the Pan-African mobile belt, being sandwiched between the West African Craton to the west, eastern Saharan Craton to the northeast and the Gabon-Congo Craton to the southeast [1] . It is underlain by crystalline Basement Complex and sedimentary rocks. The Precambrian Basement Complex rocks cover about 50% of Nigeria's land area and extend into neighboring countries like Benin to the west and Cameroun Republic to the east. The Nigerian Precambrian rocks have been grouped into two based on age and composition: the pre-Pan African rocks (2.8 - 1.3 Ga) which are mainly migmatite-gneiss complexes and ancient granitic rocks and the Pan-African rocks (600 ± 1500 Ma) comprising of the Older Granites and Metavolcano sedimentary series [2] . The Oban Massif area in the southeastern part of Nigeria is underlain by some of these Basement Complex rocks. The area is a part of the giant spur that represents the western prolongation of the Cameroun Mountains into the Cross River plains of southeastern Nigeria (Figure 1). In fact the Oban Massif is a deeply dissected plateau which in places attains a height of 1125 m above mean sea level [3] . Okom-Ita area, in Akamkpa Local Government Area of Cross River State lies in the southwestern end of the Oban Massif. The area is defined by latitudes 5°19'N and 5°23'N and longitudes 8°19'E and 8°23'E, covering an approximate areal extent of about 42 km^2 (Figure 1).

The determination of the composition of rocks in Okom-Ita area using X-Ray fluorescence spectrometry and thin section petrography became necessary to obtain more comprehensive geological information of the study area. And combining the geological information with a detailed chemical classification of rocks in the area using dominant mineralogical compositions, the prediction of the petrogenesis and the protoliths of the rocks in Okom-Ita area would be achieved.

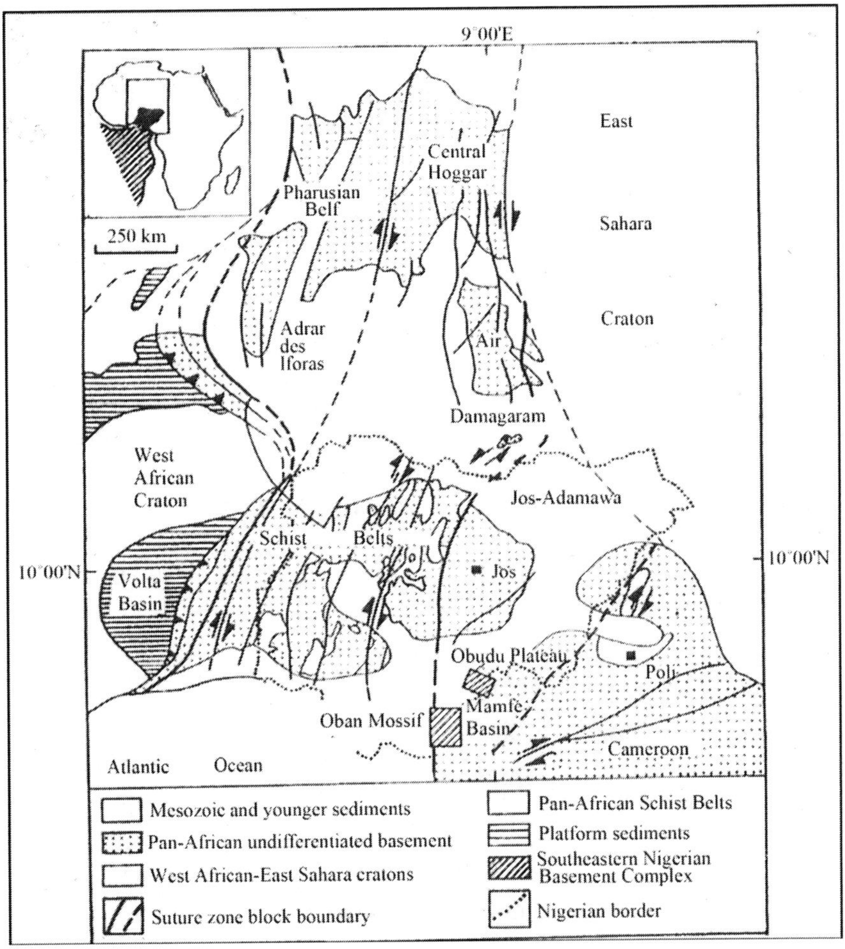

Figure 1. Generalized geological map showing the Pan-African Belts of Nigeria and the location of the southeastern Nigerian Basement Complex [7] .

2. GEOLOGICAL SETTING

The geological map of Nigeria shows that the Basement Complex can distinctively be divided into two zones: a western zone in which the N-S trending elongate schist belts are separated from one another by migmatite gneisses, granite gneisses and granites and an eastern zone in which the schist belts are poorly represented and underlain by migmatites, gneisses and granites [4] . Like the Precambrian of Africa, two models have been proposed: the ensialic deformation and plate tectonic models. The mix-up of rocks lacking a systematic variation and the virtual absence of truly miogeosynclinal sediments until upper Precambrian times do not favor a plate tectonic model of evolution of the Nigerian Basement Complex [5] . Conversely, [6] suggested that the Basement Complex of Nigeria formed as a result of the opening and closing of a small ocean comparable to the Red Sea.

Other researchers inferred that the Pan African belt in the east of the West African Craton is a collision-type orogeny with an eastward dipping subduction zone [8] -[10] . According to them, the closure of the ocean at the craton margin and the intracontinental basins and oceans led to the deformation and metamorphism of the metasediments, partial melting of the mantle and lower crust and the emplacement of the Older Granites. In the Oban Massif, the Basement Complex rocks is dominated by the Uwet granodiorite and charnockite masses which intruded a sequence of schists similar to those described from southwestern and northwestern Nigeria [3] . The scholars who had studied the petrology of Obudu and Oban Massifs, including the works of [3] , [11] agreed that a great similarity exists in characteristics of both massifs.

The rocks of the Oban Massif are comprised of the following suites:

- Migmatitic and sheared gneissic rocks, paraand ortho-schists, phyllites, meta-conglomerates and quartzites, amphibolites and metadolerites, pegmatites, aplites and pyroxenites;

- Synkinematic to late kinematic Older Granite intrusive series comprising rocks of varying compositions;

- Unmetamorphosed dolerite to microdioritic intrusive rocks [12] .

In the Okom-Ita area the dominant intrusive rocks are the pegmatites, quartz veins and dolerites emplaced within the gneiss, schist and phyllite host rocks [3].

3. METHODOLOGY

Twelve (12) representative fresh rock samples were systematically selected from a total of thirty nine (39) samples collected during the field mapping for laboratory analysis. Bulk rock geochemical analysis was carried out using X-Ray fluorescence spectrometry at the geochemistry laboratory of the Technical University of Berlin, Germany. Each of the samples was first crushed and powdered. Fused discs and pressed beads were thereafter produced for major and minor element analyses. Each of the prepared samples was oven-dried overnight at 105˚C to determine their natural moisture contents. The samples were again placed overnight in an oven at 800˚C and the difference in weight calculated to obtain the Loss on Ignition (LOI). 1.5 g of powder from each of the samples for which LOI has been determined was mixed with 6 g of lithium tetraborate ($Li_2B_4O_7$) and lanthanum oxide (La_2O_3) in a platinum crucible and fused at 1200˚C to produce melt beads for the analysis of the major elements.

4. RESULTS

4.1. OCCURRENCE AND PETROGRAPHY OF ROCKS IN OKOM-ITA AREA

In the Okom-Ita area the dominant intrusive rocks are the pegmatites, quartz veins and dolerites emplaced within gneisses, schists and phyllites host rocks (Figure 2). Both the intrusive and their host rocks were petrographically and geochemically analyzed. The results of the petrographic analyses are presented in Table 1, while Table 2 shows the results of the whole rock geochemistry. The quartz veins are leucocratic consisting mainly of quartz minerals. The pegmatites are also leucocratic but consist of feldspar, quartz, mica and tourmaline minerals. The dolerite is dark grey in colour, unmetamorphosed, fine-medium grained, tholeiitic in composition, and mineralogically, dominated by plagioclase, pyroxene and olivine. A metamorphosed basic igneous rock (Amphibolite) was also encountered in the study area. Field and mineralogical properties of the major rock types that underlie Okom-Ita area are presented below:

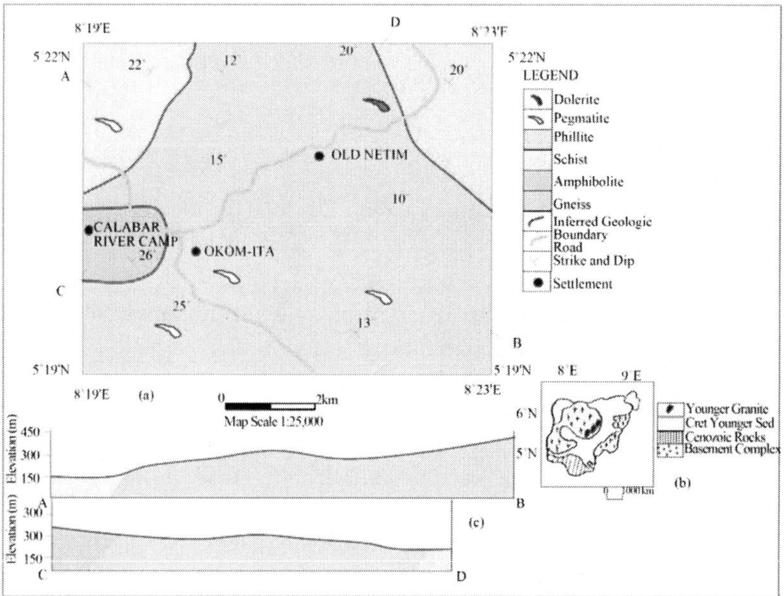

Figure 2. Map showing (a) The geology of Okom-Ita area of the Oban Massif; (b) The inset crystalline Basement Complex of Nigeria; (c) Cross sections drawn from the geologic map (A-B; C-D).

4.1.1. Gneisses

The gneissic rocks observed in the area of study are the quartzo-feldspathic rocks that underlie about 70% of the total mapped area (Figure 1). The rocks are mainly biotite gneiss, feldspathic gneisses and banded gneisses. These rocks are conspicuous around Old Netim and in fact the predominant rock type in the northeastern part of the study area.

In hand specimen, the rocks are medium-coarse grained and in some places dissected with quartzo-feldspathic veins. Modal analysis shows that the biotite gneiss is chiefly comprised of quartz (25%), k-feldspar (25%), biotite (25%), plagioclase (10%) and minor amounts of muscovite and epidote (Table 1). Similarly, the light and dark-coloured minerals representing Quartz and biotite respectively are clearly seen in the photomicrograph of the biotite gneiss sample (Figure 3(a)). The migmatite gneiss is composed of quartz (24%) and K-feldspars (30%) as shown in Table 1. Biotite, muscovite and epidote are also common while plagioclase occur as untwined albite (15%). Banded gneiss shows a well developed foliated texture in which segregation banding predominates. Dark bands typically dominated by ferromagnesian inosillicates (hornblende and pyroxene) and light bands of phyllosilicates typically dominated by mosaic textures of quartz and feldspar were observed in hand specimens of the banded gneiss.

4.1.2. Schists

Schists are strongly foliated rocks formed by regional metamorphism as they readily split into flecks. The conspicuous shistocity of all schists is attributed to mineral orientation. Schist underlies less than 20% of the total mapped area (Figure 2). It is found occurring in areas around Dukwe Camp and the Eng-Haut Rubber plantation. The schists are highly weathered. Schistocity however is strong and trends N-S to NE-SW. Slickensides were also evident indicative of shearing and faulting that occurred as a result of deformation of rocks in the area [3] .

In hand specimens, the schist samples show a foliated texture. Quartz is dominant and occurs together with biotite, chlorite and feldspar.

The preponderance of phyllosilicates, when the schists were observed in thin sections, confirmed the rock type to be pellitic schist [13] . Biotite appears as greenish and elongated flakes. Plagioclase is probably untwined albite indicating low grade metamorphism [14] . The presence of chlorite enhanced the foliation. Opaque minerals occur mainly as scattered grains of iron oxide.

4.1.3. Phyllite

Phyllite is a regionally metamorphosed foliated pellitic rock. They are fine grained micaceous rocks occasionally showing segregation layering. Just like schist, phyllites are restricted to the western part of Oban Massif. Garnetiferous varieties have been reported from the Uwet area [14] . They are characterized by a silky sheen on the cleavage of the surface. In thin section, the main minerals recognized include biotite, quartz, chlorite and plagioclase (Figure 3(b)). Chlorite minerals have impacted a rather peculiar luster to the surface of the schistocity. The mica minerals are mainly biotite and muscovite (20 wt%) (Table 1).

Table 1. Modal composition (Vol %) of rocks in Okom-Ita area, Oban Massive, southeastern Nigeria.

Samples	Quartz	K feldspar	Plagioclase	Hornblende	Biotite	Muscovite	Pyroxene	Olivine	Chlorite	Epidote	Garnet	Opaque	Zircon	Tourmaline
Biotite Gneiss														
BG 1	25	25	10	-	25	10	-	-	-	5	-	-	-	-
BG 2	25	15	12	-	30	8	-	-	-	10	-	-	-	-
BG 3	23	20	10	-	33	7	-	-	-	7	-	-	-	-
Banded Gneiss														
BDG 1	30	30	10	7	10	10	-	-	-	-	1	2	-	-
BDG 2	25	18	15	13	17	-	-	-	-	5	-	1	1	-
BDG 3	25	10	9	16	30	5	-	-	-	3	-	1	1	-
Feldspathic Gneiss														
FG 1	24	30	15	2	15	10	-	-	-	4	-	-	-	-
Mica Schist														
MS 1	32	-	10	-	25	10	-	-	10	-	6	2	-	5
Amphibolite														
AM 1	10	-	15	75	-	-	-	-	-	-	-	-	-	-
Phyllite														
PH 1	25	-	10	20	20	-	25	-	-	-	-	-	-	-
Pegmatite														
PG 1	35	11	20	-	30	12	-	-	-	6	-	-	-	5
Dolerite														
DOL 1	-	-	45	-	15	-	18	19	-	-	-	3	-	-

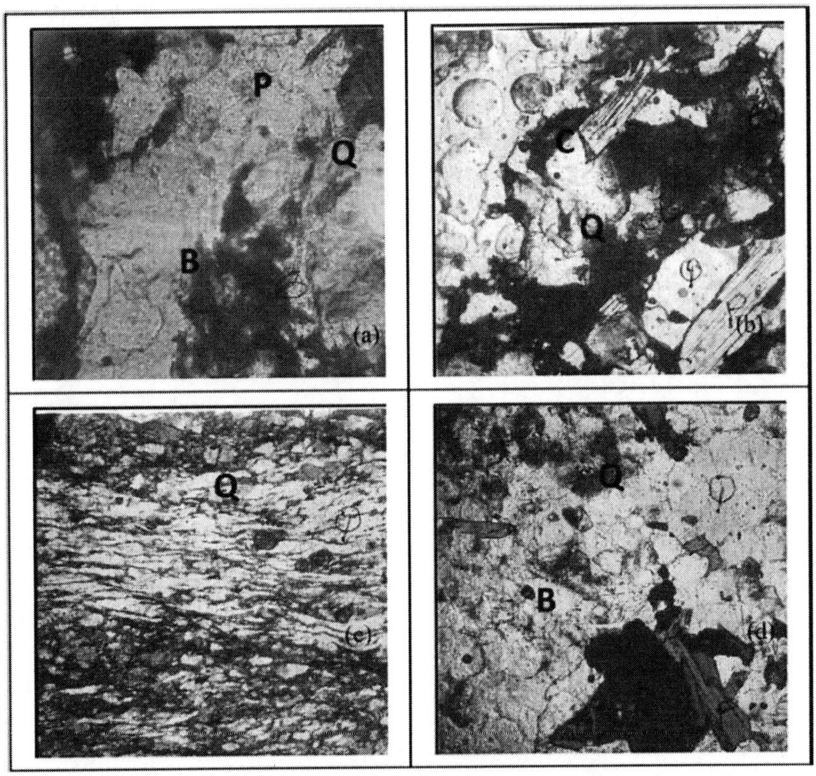

Figure 3. Photomicrographs of rock samples from Okom-Ita area Oban Massive (PPL X30) (a) Biotite Gneiss (BG 1); (b) Phyllite (PH 1); (c) Amphibolite (AM 1); (d) Pegmatite (PG 1), P = Plagioclase; B = Biotite; CL = Chlorite; Q = Quartz.

4.1.4. Amphibolite

Amphibolite occurs as a fine-medium grained metamorphic rock composed mainly of amphibole. It is formed by the regional metamorphism of basic igneous rocks. They may be banded or homogeneous. The banded variety has thin layers of quartzo-feldspathic materials alternating with thick dark bands consisting of mainly hornblende and minor phyllosilicate [15] . In the study area, it constitutes less than 15% of the total area underlain by rocks. Amphibolite is exposed within the Calabar River Camp where it occurred as massive bodies, often along streams and river channels. It has been noted that the Amphibolites are associated with dolerite stocks, which are chemically related, but the amphibolites are older because they have been deformed and metamorphosed [3] .

In thin section, the amphibolites are foliated with hornblende defining the foliation (Figure 3(c)). Hornblende is the dominant mineral present (70%), quartz is minor and plagioclase occurs as anorthite with a composition of An_{45}. Amphibolites consist of about 60% - 70% amphibole; Almandine garnet may be present depending on the rock composition [16] .

4.1.5. Pegmatite

These are very coarse-grained igneous rocks with phenocrysts over 250 mm in length usually of granitic composition and forming at the late stage of crystallization. In the study area, they occur mostly as unmappable veins of varying dimensions within the gneisses and schists. They also occur as boulders (about 3 m × 5 m) around the Calabar River Camp.

Also occurring in varying dimensions in the gneisses and schist are quartz veins. They cut across the whole length and width of the host rock. Their emplacement is controlled by the N-S to NE-SW regional structure of the host rock. In thin sections, quartz occurs as very coarse grained crystal (Figure 3(d)). Other minerals identified in thin sections include biotite, plagioclase, muscovite and epidote. Feldspar occurs mainly as K-feldspar and may grow into zones where they are replaced by sodic plagioclase. Biotite is a minor constituent while muscovite occurs as books or flaky aggregates.

4.1.6. Dolerite

Dolerite is a medium grained mafic igneous rock that is mineralogically and chemically equivalent to basalt and commonly forms minor intrusion. Dolerite is widespread in the basement complex of Nigeria. Their emplacement is believed to have terminated the Pan-African Orogeny in Nigeria [17] . In the mapped area, dolerite was found at Old Netim occurring as dyke bodies of several centimeters thick and cutting the foliation of the gneiss host rock (Figure 4). They displayed sharp contacts with the gneiss host rock showing no evidence of metamorphism. It is possible that dolerites occur at other locations but are buried under the surface because of their hyper-abyssal nature.

In hand specimen, dolerite is a medium grained, dark grey coloured and non-foliated igneous rock. The thin section analysis shows that the dolerite is sub-ophitic in texture with randomly oriented laths of plagioclase and partly enclosed in irregular bounded crystals of pyroxene. Mineralogically, they are composed dominantly of plagioclase (45%). Others minerals occurring in the dolomite include pyroxene, olivine, biotite and opaque minerals (Table 1). Olivine occurs mainly as large equant phenocrysts and irregular cleavage. The pyroxene content is augite and hyperstene. Dominant plagioclase is the labrodorite (An55-An60). The opaque minerals occur as scattered grains of iron oxide along the augite partings.

4.2. Geochemistry of Rocks in Okom-Ita

The results of the geochemical analysis of the rocks in the study area are presented in Table 2, which aided the calculation of the Niggle norm [18] . The Niggle norm was chosen because determination of the norm of metamorphic rocks, highly altered igneous rocks and metasediments can indicate the general nature of the anhydrous igneous equivalent [19] . Again, since a greater proportion of rocks in the study area were metamorphic rocks, Niggle norm became useful in the interpretation [16] .

The geochemical data (expressed in weight percent and molecular proportion) of the oxides were further plotted in variation/discriminatory diagram such as:

- SiO_2 (wt%) versus $Na_2O + K_2O$ (wt%) diagram [20] ;
- ACF plot of [21] ;
- AFM plot ($Na_2O + K_2O$-FeO-MgO);
- Na_2O/Al_2O_3 versus K_2O/Al_2O_3 diagram of [22] ;
- al-alk versus Niggle diagram of [23] ;
- TiO_2-K_2O-P_2O_5 plot of [24] .

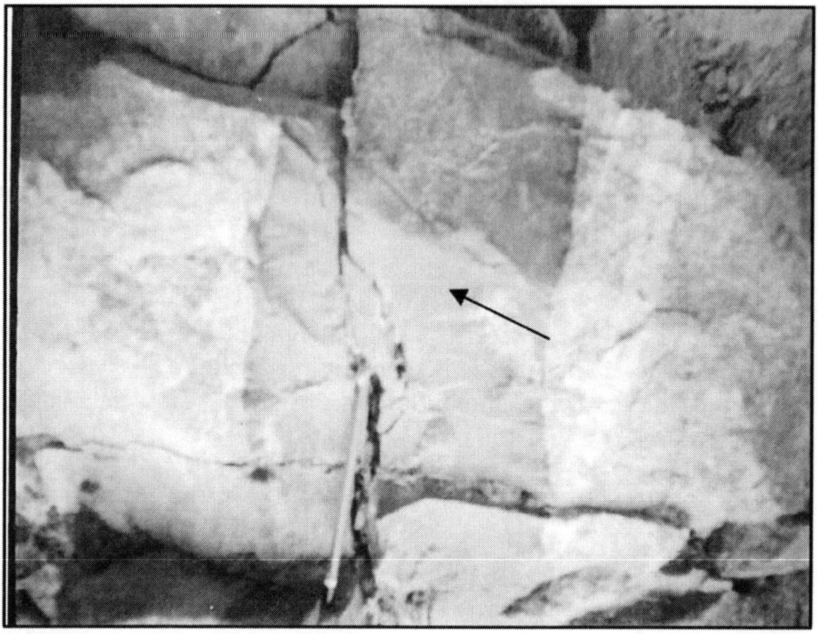

Figure 4. Dolerite intrusion (dark portion) truncating the foliation of the host gneiss in a N-S trend, Old Netim Oban Massive.

The chemical composition of rocks is a direct reflection of the modal composition of the rocks [3] . Phyllite and schist are characterized by SiO_2 in the range of (60% - 66%), excess Na_2O over K_2O, high Al_2O_3 to alkali ratio and low content of CaO and MgO. All the rocks exhibited high molecular trend of the order: Al_2O_3 > (CaO + Na_2O + K_2O), reflecting their preliminary characteristics. The gneisses of Okom-Ita contain a high percentage of SiO_2 in the range of (59 wt% - 73 wt%). $Al_2O_3$is also high (13% - 17%). The gneisses contain higher silica and alkali concentrations than the associated metasediments. They are also enriched in K_2O compared to other rocks in the area. The biotite gneisses differ in composition from the other gneisses in the area, in that they are impoverished in SiO_2 but enriched in Al_2O_3, Fe_2O_3, and MgO when compared with the migmatitic and banded gneisses. Generally, the TiO_2 is low, which suggests a compensation for the high silica and alumina values.

The amphibolite is impoverished in silica (50.05 wt%), but enriched in MgO, Fe_2O_3, TiO_2, and CaO, which is consistent with mafic character. Their low silica value is compensated by its enrichment in TiO_2 (1.2 wt%).

Table 2. Major element composition of rocks in the Okom-Ita area Oban Massive.

Samples	SiO_2	TiO2	Al_2O_3	Fe_2O_3	Foe	Mino	MgO	CaO	Na_2O	K_2O	P_2O_5	LOI
Biotite Gneiss												
BG 1	63.55	0.84	16.1	5.85	ND	0.67	2.09	3.64	4.36	2.29	0.34	0.29
BG 2	63.00	0.20	16.06	0.20	2.77	0.25	0.91	5.70	2.80	1.41	0.41	0.47
BG 3	59.04	1.06	16.28	5.90	ND	0.06	3.54	4.88	3.65	3.60	0.43	1.35
Banded Gneiss												
BDG 1	64.88	0.20	14.80	4.42	3.92	0.27	0.72	5.90	2.90	1.41	0.31	0.29
BDG 2	64.60	1.14	15.00	4.80	ND	0.05	1.55	3.20	3.86	3.53	0.32	0.29
BDG 3	70.00	0.60	13.10	4.71	ND	0.08	0.80	2.20	2.72	4.48	0.15	0.35
Feldspathic Gneiss												
FG 1	73.70	0.33	13.26	2.81	ND	0.05	1.00	1.98	3.70	3.11	0.16	0.18
Mica Schist												
MS 1	60.10	0.86	7.75	8.50	0.20	4.30	0.70	1.30	4.10	0.15	0.15	1.98
Amphibolite												
AM 1	50.05	1.26	12.20	12.21	ND	0.81	8.85	8.30	1.84	2.35	0.15	1.35
Phyllite												
PH 1	65.60	0.80	16.20	3.70	0.07	2.00	3.50	3.45	3.70	0.18	0.12	0.64
Pegmatite												
PG 1	75.30	0.10	13.40	0.67	ND	0.03	0.18	3.30	3.33	3.28	0.17	0.74
Dolerite												
DOL 1	50.00	2.33	13.10	2.12	9.58	0.15	7.42	8.60	3.34	0.27	0.77	1.72

The major intrusive rocks in the area include pegmatite and dolerite. The pegmatite contains high percentage of SiO_2 (>75 wt%) and alkali elements (of about 6.61 wt%; Table 2). Dolerite on the other hand is impoverished in SiO_2 (50%), but high CaO (8.60 wt%) MgO (7.42 wt%) contents and relatively rich in Na_2O (63.34 wt%) but highly depleted in K_2O (0.27 wt%), which is consistent with its mafic characteristics.

5. DISCUSSION

5.1. Chemical Affinity

The high SiO_2 content of mica schist in the study area is typical of pellitic to semi-pellitic schists [2] [3] [25] . Also the excess of Na_2O over K_2O may be attributed to possible metasomatic reaction in which Na_2O was preferably enriched and K_2O depleted [3] [26] [27] . The higher CaO (3.45%) in the phyllite at Old Netim could be due to the contamination of the rock by secondary calcite [3] . The enrichment in TiO_2, Al_2O_3, Fe_2O_3, MgO and CaO of biotite gneiss compared to the other varieties of gneiss reflects a high modal proportion of Fe-Mg bearing silicates such as biotite and hornblende [3] . The same applies to amphibolites. The relatively high alkali contents of intrusive in the area might be attributed to widespread alkali metasomatism in the Nigerian basement due to its rejuvenation during the Pan African Orogeny [28] [29] . Finally, amphibolites and dolerite are characterized by a negative Niggli (−q) value compared to the positive (+q) values of other rock types. This negative value shows that the rock is none quartz normative, and therefore non-siliceous rock (Table 3).

5.2. Tectonic Setting and Petrogenesis

The tectonic setting and petrogenesis of rocks in the study area were constrained using the AFM triangular discriminant and the TiO_2-K_2O-P_2O_5 diagrams. The AFM plot is particularly useful in discriminating between tholeiitic and Calc-alkaline suites considering the Fe_2O_3 and MgO components [30] and hence the tectonic setting, since no trace element composition was determined for the rocks. From the plot, the pegmatite plotted in the calc-alkaline magmatic field, while the dolerite and amphibolite plotted in the tholeiitic field (Figure 5). On the alkali versus silica diagram of [20] , dolerite and amphibolites also plotted in the tholeiitic field, indicating generation from tholeiitic magma source (Figure 6). On the TiO_2-K_2O-P_2O_5 diagram of [24] , the amphibolite and dolerite plotted in the field of continental basalt (Figure 7). Representatives of the analyzed samples were plotted in other variation diagrams to determine their protolith. On the Na_2O/Al_2O_3-K_2O/Al_2O_3 diagram of [22] , the mica schist and phyllite plotted in the metasedimentary field, while all the gneisses and pegmatites plotted in the igneous field (Figure 8). The Niggli al-alk-Niggli C diagram of [23] shows the same character by having schist and phyllite plot on the outside of the igneous field, while gneiss and pegmatite plotted within the igneous field (Figure 9). The chemical composition of metaigneous and metasediments in the area were also plotted in the ACF diagram of [21] to know their progenitors. From the plot (Figure 10) biotite and feldspathic gneiss plotted in the field of clay-shale either free from carbonates or containing up to 30% carbonates or in the andesitic-basaltic field. The banded gneiss had one sample each plotting on the field of andesitic-basaltic and field of clayshales containing up to 30% carbonates or carbonate free mud rocks. These rocks were believed to originate from the metamorphism of intermediate igneous rock such as granodiorite. Mica schist plotted in the field of clay-shales containing up to 30% carbonate while phyllite plotted in the greywacke field. These rocks are thought to originate from greywackes [31] . Amphibolites on the other hand plotted in the andesitic-basaltic field indicative of metamorphism from magmas derived from basaltic source.

Mica schists are characterized by the preponderance of phyllosilicate quartz, and feldspars [16] . The phyllosillicates are indication of metamorphism from shale or pellitic rocks. Gneisses are formed by deep seated high grade regional metamorphism. They are generally confined to eroded folded mountain belts and Precambrian terrain. Generally, they are made up of mineral constituents suggestive of metamorphic processes either of low grade or high grade. Phyllites are formed by low grade metamorphism of pellitic material and occasionally tuff or tuffaceous sediments. Also with higher grade metamorphism caused by increasing temperature or greater activity of permeating fluid, slates pass into phyllites [16] . Amphibolites are formed by regional metamorphism of basic rocks such as dolerites and basalts. The metamorphism is typical of the amphibolite facies which according to [32] [33] is characterized by plagioclase containing at least 15% anorthite. The mineralogy of pegmatite is indicative of the residual nature of magmatic melt. It is likely that most pegmatites owe their character to crystallization of either an anatectic melt or differential magma. It has been envisioned the pegmatites crystallize from hydrous granite magma [34] , which were possibly emplaced during the main phase of Pan African Orogeny [3] .

Table 3. Calculated Niggli Norm for the rocks of Okom-Ita area.

Samples	Al	Fm	C	All	Si	K	Mg	Q
Biotite Gneiss								
BG 1	30	40	12	18	201	0.25	0.39	29
BG 2	41	17	26	16	271	0.25	0.35	107
BG 3	27	41	15	17	168	0.39	0.78	0.1
Banded Gneiss								
BDG 1	32	30	24	14	243	0.24	0.24	87
BDG 2	32	34	12	22	231	0.37	0.38	43
BDG 3	32	35	10	23	292	0.52	0.25	100
Feldspathic Gneiss								
FG 1	37	26	10	26	350	0.35	0.41	146
Mica Schist								
MS 1	38	41	5	15	221	0.03	0.21	61
MS 2								
Amphibolite								
AM 1	14	62	17	7	98	0.45	0.59	−30
Phyllite								
PH 1	36	36	14	14	245	0.03	0.75	89
Pegmatite								
PG 1	44	7	20	20	42	0.39	0.32	202
Dolerite								
DOL 1	19	50	22	22	121	0.05	0.58	−11

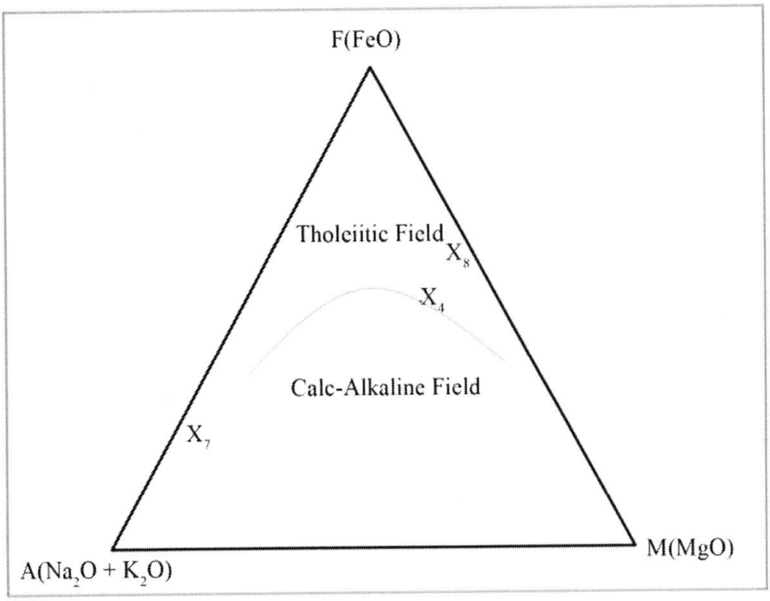

Figure 5. AFM diagram for rocks in the study area discriminating pegmatite of calc-alkaline field from mafic rocks of tholeiitic magma field (X_4—amphibolite, X_7—pegmatite, X_8—dolerite) adapted from [35] .

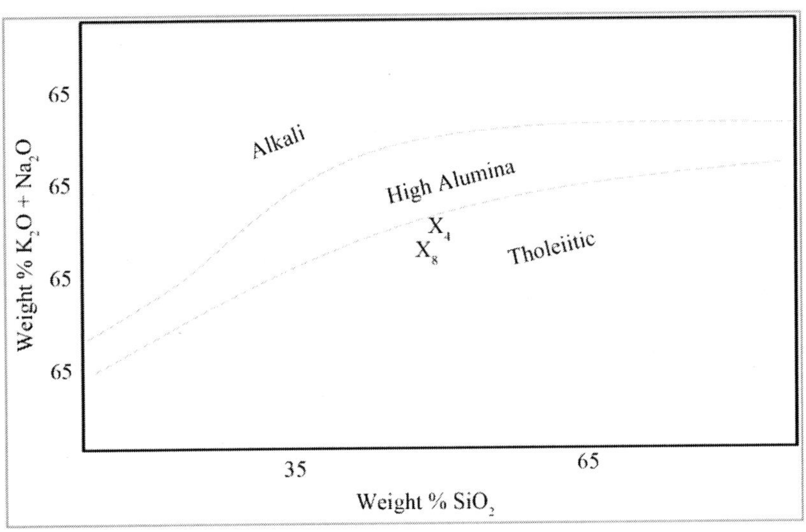

Figure 6. Alkali-Silica diagram of mafic rocks in the study area, adapted from [20] .

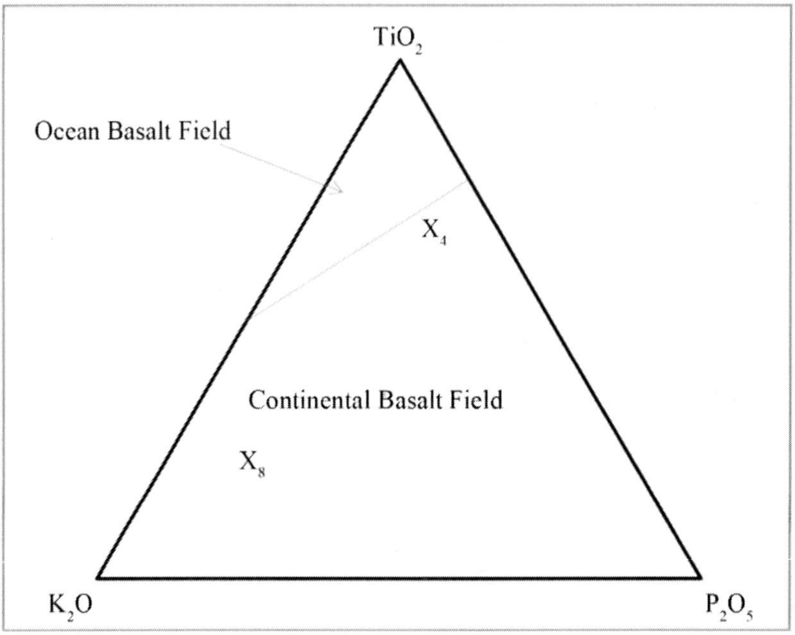

Figure 7. TiO_2-K_2O-P_2O_5 diagram of mafic rocks in the study area, adapted from [24] .

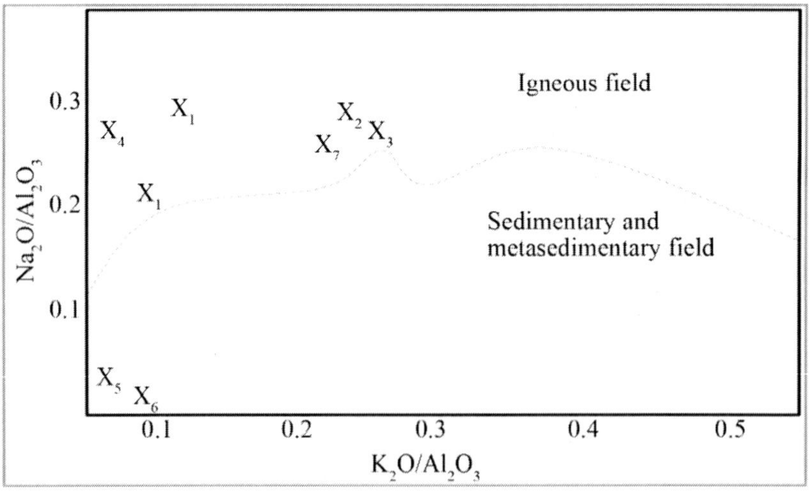

Figure 8. Alkali-Silica diagram of mafic rocks in the study area, modified after [20] , X_1—banded gneiss, X_2—feldspathic gneiss, X_3—banded gneiss, X_5—phyllite, X_6—mica schist.

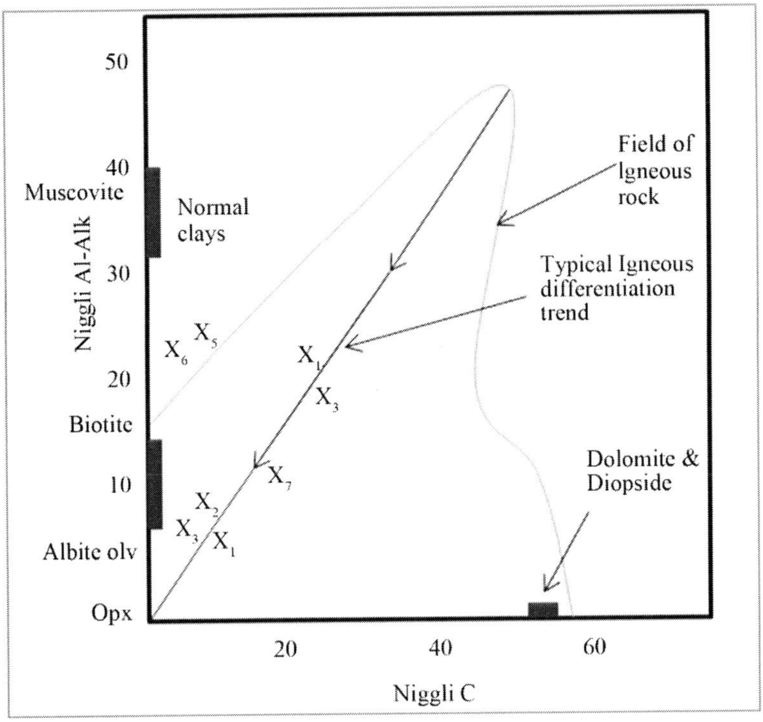

Figure 9. Niggli Al-Alk versus Niggli C diagram for rocks of the study area, modified after [23] .

Dolerite must have been derived from basaltic magma, thus they are tholeiitic in composition. This primary basic magma may have been derived from the upper mantle at considerable depths. Their fine texture is a reflection of the relatively shallow level of emplacement which favored fast cooling. They are unaffected by metamorphism and deformation. Their emplacement is believed to have terminated the Pan-African Orogeny in Nigeria [17] . Hence, they are often referred to be the youngest rock in the Oban Massif [35] .

6. SUMMARY AND CONCLUSION

Detailed study of the petrology and geochemistry Basement Complex rocks in Okom-Ita area, Oban Massif, southeastern Nigeria, revealed that the dominant intrusive rocks in the area comprised of pegmatites, quartz veins and dolerites emplaced within gneiss, schist and phyllite host rocks. Amphibolite also occurs in the area. Bulk-geochemical rock analysis carried out using X-Ray fluorescence spectrometry, revealed that the geochemical properties of the rocks are a direct reflection of the characteristics observed in thin sections.

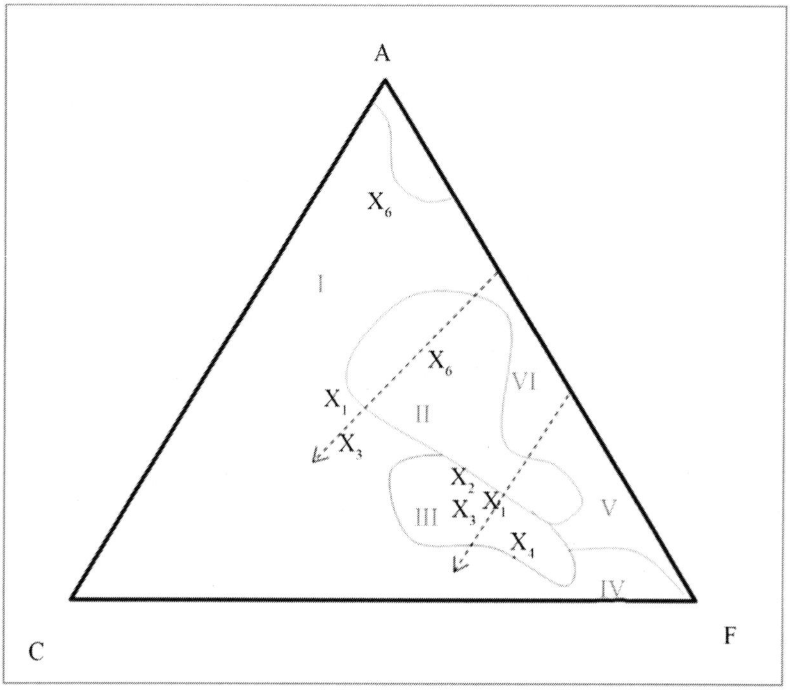

Figure 10. ACF diagram for determination of the chemical affinity of metamorphosed rocks of Okom-Ita area, modified after [21] , I = Clay-shales free of carbonate or containing up to 30% carbonate, II = greywacke, III = andesitic-basaltic field, IV = ultra basic field, V = clay or shale rich in carbonates, VI = Al-rich clays and shales.

The gneisses, metasediments and pegmatites are dominated by quartz and feldspar, with subordinate biotite and plagioclase. The dolerites on the other hand are composed dominantly of plagioclase (45%), olivine and pyroxene, with no quartz in their norm, reflecting their mafic character.

The chemical data (expressed in weight percent and molecular proportion) of the oxides were further plotted in chemical variation/discriminatory diagrams to determine their protoliths. The discriminant shows that the pegmatite was derived from late stage partial melting and differentiation of calc-alkaline magma, while the dolerite was generated from tholeiitic magma sources. On the alkali versus silica diagram, the dolerite plotted in the tholeiitic field corroborating origin from tholeiitic magma source. However, on the TiO_2-K_2O-P_2O_5 diagram, the dolerite plotted in the field of continental basalt.

REFERENCES

1. Kroner, A., Ekwueme, B.N. and Pidgeon, R.T. (2001) The Oldest Rock in West Africa: SHRIMP Zircon Age for Early Archaean Migmatitic Orthogneiss at Kaduna, Northern Nigeria. The Journal of Geology, 109, 399-406.

2. Ekwueme, B.N. (1990) Rb-Sr Ages and Petrologic Features of Precambrian Rocks from the Oban Massif, Southeastern Nigeria. Precambrian Research, 47, 271-286.

3. Ekwueme, B.N. (2003) The Precambrian Geology and Evolution of the Southeastern Nigerian Basement Complex. University of Calabar Press, Calabar, 1-57.

4. Okezie, C.N. (1974) Geological Map of Nigeria, Scale 1:2,000,000. Geological Survey of Nigeria, Abuja.

5. Odeyemi, I. (1982) A Review of the Orogenic Events in the Precambrian Basement of Nigeria, West Africa. Geologische Rundschau, 70, 897-909.

6. McCurry, P. (1976) The Geology of the Precambrian to Lower Paleozoic Rocks of Northern Nigeria—A Review. In: Kogbe, C.A., Ed., Geology of Nigeria, Elizabethan Press, Lagos, 15-39.

7. Ferre, E., Deleris, J.I., Lar, A.U. and Peuct, J.J. (1996) The Pan-African Reactivation of Eburnean and Archaen Provinces in Nigeria: Structural and Isotopic Data. Journal of The Geological Society (London), 153, 719-728.

8. Burke, K.C. and Dewey, F.J. (1972) Orogeny in Africa. In: Dessauvagie, T.F.J. and Whiteman, A.J., Ed., African Geology, Ibadan University Press, Ibadan, 583-608.

9. Black, R., Caby, R., Pouchkine, A., Bayer, B., Bertrand, J.M., Boullier, A.M., Fabre, J. and Lesquer, A. (1979) Evidence for the Late Precambrian Plate Tectonics in West Africa. Nature, 278, 223-227.

10. Caby, R., Bertrand, J.M.I. and Black, R. (1981) Pan African Ocean Closure and Continental Collision in the HoggarIforas Segment, Central Sahara. In: Kroner, A., Ed., Precambrian Plate Tectonics, Elsevier, Amsterdam, 407-434.

11. Ukaegbu, V.U. and Beka, F.T. (2007) Petrochemistry and Geotectonic Significance of Enderbite-Charnockite Association in the Pan-African Obudu-Plateau, Southeastern Nigeria. Journal of Mining and Geology, 43, 1-14.

12. Rahaman, M.A., Ukpong, E.E.S. and Azmatullah, M. (1981) Geology of Parts of the Oban Massif Southeastern Nigeria. Journal of Mining and Geology, 18, 60-65.

13. Ekwueme, B.N. (1983) Regional Metamorphism of Pellitic Rocks Southeast of Lokoja, Nigeria. Geology, 20, 71-77.

14. Ekwueme, B.N. and Onyeagocha, A.C. (1985) Metamorphic Isograds of Uwet Area, Oban Massif Southeastern Nigeria. Journal of African Earth Sciences, 3, 443-454.

15. Ekwueme, B.N. and Ekwere, S.J. (1989) The Geology of Eastern Section of Oban Massif, SE Nigeria. Journal of Mining and Geology, 25, 317-329.

16. Ekwueme, B.N. (1993) An Easy Approach to Metamorphic Petrology. University of Calabar Press, Calabar, 169 p.

17. Ekwueme, B.N. and Shilling, H. (1995) Occurrence, Geochemistry and Geochronology of Mafic-Ultramafic Rocks in the Obudu Plateau, SE Nigeria. In: Srivastava, R.K. and Chandra, R., Eds., Magmatism in Relation to Diverse Tectonic Setting, Oxford & IBH publishing Co. Pvt Ltd., New Delhi, 291-307.

18. Niggli, P. (1948) Gesteine und Minerallgerstatten. Birkhauser, Basel, 540 p.

19. Ehlers, E.G. and Blatt, H. (1982) Petrology: Igneous, Sedimentary and Metamorphic. W. H. Freeman, New York, 732.

20. MacDonald, G.A. and Katsura, I. (1964) Chemical Composition of Hawaiian Lavas. Journal of Petrology, 5, 82-133.

21. Winckler, H.G.P. (1967) Petrogenesis of Metamorphic Rocks. Springer-Verlag, New York, Inc., 237 p.

22. Garrels, R.M. and Mackenzie, F.T. (1971) Evolution of Metasedimentary Rocks. Northern and Co., New York, 394 p.

23. Leake, B.E. and Singh, D. (1986) The Delaney Dome Formation, Connemara, W. Ireland and the Geochemical Distinction of Orthoand Para-Quartzofeldspathic Rocks. Mineralogical Magazine, 50, 205-215.

24. Pearce, T.H., Gorman, B.E. and Birkett, T.C. (1975) The TiO_2-K_2O-P_2O_5 Diagram: A Method of Discriminating between Oceanic and Non-Oceanic Basalts. Earth and Planetary Science Letters, 24, 419-426.

25. Obioha, Y.E. and Ekwueme, B.N. (2012) Geochemistry of Schists of Northwest Obudu Plateau, Southeastern Nigeria. Global Journal of Geological Sciences, 10, 157-165.

26. Rajamani, V., Krogstad, E.J., Hanson, G.N. and Balakrishnan, S. (1987) Are Patna and Bisansttam Granites (Adjoining the Kolar Schist Belt) Cogenetic. Geological Society of India, 30, 98-105.

27. Obioha, Y.E. and Ekwueme, B.N. (2011) Petrology and Chemical Composition of Gneisses of Northwest Obudu Plateau, Southeastern Nigeria. Global Journal of Pure and Applied Sciences, 17, 215-226.

28. OYAWOYE, M.O. (1972) THE BASEMENT COMPLEX OF NIGERIA. IN: DESSAUVAGIE, T.F.J. AND WHITEMAN, A.J., EDS., AFRICAN GEOLOGY, UNIVERSITY OF IBADAN PRESS, IBADAN, 67-99.

29. Ekwueme, B.N. and Onyeagocha, A.C. (1985) Petrology of Intrusive Rocks in Uwet Area, Oban Massif, SE Nigeria. Journal of Mining and Geology,, 22, 71-78.

30. Cox, K.G., Bell, J.D. and Pankhurst, R.J. (1979) The Interpretation of Igneous Rock. Allen & Unwin, London, 450 p.

Oyawoye, M.O. (1972) The Basement Complex of Nigeria. In: Dessauvagie, T.F.J. and Whiteman, A.J., Eds., African Geology, University of Ibadan Press, Ibadan, 67-99.

31. Pettijohn, F.J. (1957) Sedimentary Rocks. 3rd Edition, Harper & Row, New York, 628 p.

32. Eskola, P. (1920) The Mineral Facies of Rocks. Norsk Geologisk Tidsskrift, 6, 143-194.

33. Turner, F.J. and Verhoogen, J. (1960) Igneous and Metamorphic Petrology. 2nd Edition, McGraw-Hill, New York, 694 p.

34. Jahns, R.H. and Burnham, C.W. (1969) Experimental Studies of Pegmatites Gneiss: A Model for the Derivation and Crystallization of Granitic Pegmatities. Economic Geology, 64, 843-864.

Chapter 10

Geochemistry and Petrology of Basic Volcanic Rocks of Jabal Al Haruj Al-Aswad, Libya

N. M. Al-Hafdh[1*], Abd El-Salam S. El-Shaafi[2]

[1]Department of Earth Science, Faculty of Sciences, Mosul University, Mosul, Iraq

[2]Department of Earth Science, Faculty of Sciences, Benghazi University, Benghazi, Libya

ABSTRACT

The Al Haruj Intra-continental Volcanic Province is the largest part of the extensive volcanic activity in Libya which is considered to be a typical within plate basalts. The volcano-tectonics evolution of this province, as well as its origin, are still widely disputed. According to K-Ar dating previously studied, the volcanic activity started in the Late Miocene and lasted until at least the Late Pleistocene. The field may still be volcanically active. The mafic rocks of Jabal Al Haruj have been classified into six major phases or groups. These phases have been differentiated using Landsat images together with aerial photographs of different scales as well as field observations. The topographic forms of the earliest phase are highly eroded while the forms of the latest phase are usually fresh and very well preserved as regards primary features. Mafic lavas of this field consist of alkali basalts to olivine tholeiites (transitional basalt) which contain olivine as essential constituent together with clinopyroxene, plagioclase and glass. The basalt exhibits intergranular, intersertal, ophitic and subophitic relations. Amygdaloidal and glomerporphyritic textures are also observed. The basaltic rocks of different ages and from different localities are petrographically rather similar. Phenocrysts of olivine probably the result from slow cooling in crustal magma chambers prior to eruptions, suggesting that magmas ascended slowly through the crust. 109 sam- ples have been carefully collected from various phases, some of these samples have been chosen for major and trace elements analyses, using XRF in order to determine the characteristics of the mantle source and investigate crustal interaction. The major and trace elements revealed a slightly significant chemical diversity among the phases and within each phase. The normative classification of most of these rocks shows close agreement with their modal classification. A vague correlation between MgO and most major oxides in the studied samples suggests different degrees of partial melting rather than fractional crystallization. A characteristic feature of the studied volcanic rocks is the relatively constant ratios of certain incompatible trace elements (Nb/Zr, Rb/Zr), which provides strong evidence of a common source. In addition, the rocks display similar patterns of the peaks and troughs; this strongly suggests that they have a common parent and common subsequent processes. The compatible transitional metals Ni (81 - 193 ppm) and Cr contents (238 - 361 ppm) and relatively low Mg# (Mg/(Mg + Fet)) (52 - 62) give an indication that the studied basaltic rocks have slightly to moderately fractionated olivine and/or spinel. The magmatism of this volcanic field seems to be related to reactivation of pre-existing structures during the passive rifting of the Sirt Basin that most likely produced in response to convergence between European and African plates since Jurassic until Holocene times.

KEYWORDS

Jabal Al Haruj Al Aswad Area, Basaltic Rocks, Geochemistry, Petrology, Petrogenesis, Libya

1. INTRODUCTION

The Al Haruj (Jabal Al Haruj al Aswad and al Abiyad) area is located in central part of Libya, and is considered one of the main occurrences of young extrusive rocks. It is the largest of four volcanic provinces in Libya; the other three occurrences are in Gharyan province in north-western Tripoli, Jabal as Sawda between Gharyan and Jabal Al Haruj, and Jabal Nuqy farther south, close to the Tibesti massif at the Chadian-Libyan border (Figure 1).

All these occurrences represent the main Tertiary-Recent volcanics in Libya, and are arranged with older volcanics in the NNW to the youngest in the SSE. Moreover, the smallest sizes of these volcanics are encountered in the extreme south-east of Libya, in Jabal Awayant, Busrewil and Wadsworth [1] . In general, these volcanics are composed essentially of voluminous alkaline basalts and less extensive transitional and theoliitic basalts; more differentiated rocks are less common, Busrewil and Suwesi [2] .

Al Haruj al Abiyad falls within the Al Haruj basalt field; it is located south-west of the main body and is the youngest region of the field, which is called Al Haruj Aswad. The extensive lava plateaus cover an area of at least 16,000 km^2 in Al Haruj al Aswad, preserved as variably eroded lava fields. The entire Al Haruj has a volcanic area of 45,000 km^2; thus, it is significantly larger than the Tibesti volcanic region, Nemeth et al. [3] . Most of the volume of the lava involved in the building of the lava plateau seems to have been extruded in an extremely fluid condition through thousands of fissures, which are not visible in the field, Busrewil and Wads- worth [4] .

The study area represents the north-eastern part of Jabal Al Haruj al Aswad and is limited by latitudes 27°00'00"N to 27°50'00"N and longitudes 18°00'00"E to 18°39'58.5"E (Figure 2). The landscape of the study area is characterized by a hilly morphology with frequent steep lava fronts, pressure ridges and blocky lava fields, with transverse intermittent desert sands dotted by balta (kipukas), but in other parts the relief is more or less flat.

Field geological mapping was carried out using Landsat TM multispectral satellite imagery with a scale of 1: 100,000, which can be fitted with any kind of coordinate system with an accuracy of at least 20 m, using a few calibration points that were measured during the fieldwork by GPS (Figure 2). This image formed the base of the geological map completed with altitude contours, locality names and other relevant data, Peregi et al. [5] .

As the study area is remote, rather inaccessible and extremely difficult terrain, the fieldwork was carried out with the Industrial Research Centre (IRC), Tripoli as part of the geological mapping project of SE Libya in March

2006. During fieldwork, systematic field traverses were carried out by researchers equipped with GPS units, digital cameras and so on.

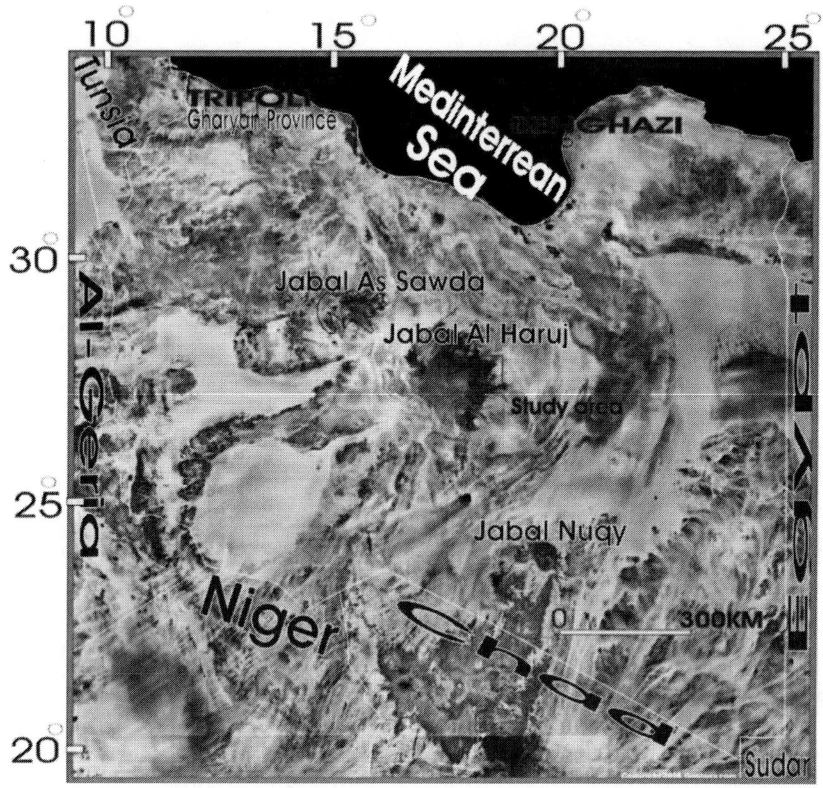

Figure 1. Landsat image showing the main occurrences of Tertiary volcanic provinces in Libya, in a trend NNW to SSE.

During the field trip, all basaltic exposures were investigated. About 109 samples were carefully collected from the flows for each phase (Figure 3). Of these, 42 samples were prepared for thin sections, and 29 samples were chosen for the major and trace element analysis using the XRF (Philips 2440) technique carried out by the Central Laboratories of the Egyptian Geological Survey, Cairo.

Figure 2. Landsat image of the north-eastern part of Al Haruj al Aswad, mosaiced and overlain by topographic contour lines to depict the spatial distribution of major volcanic phases.

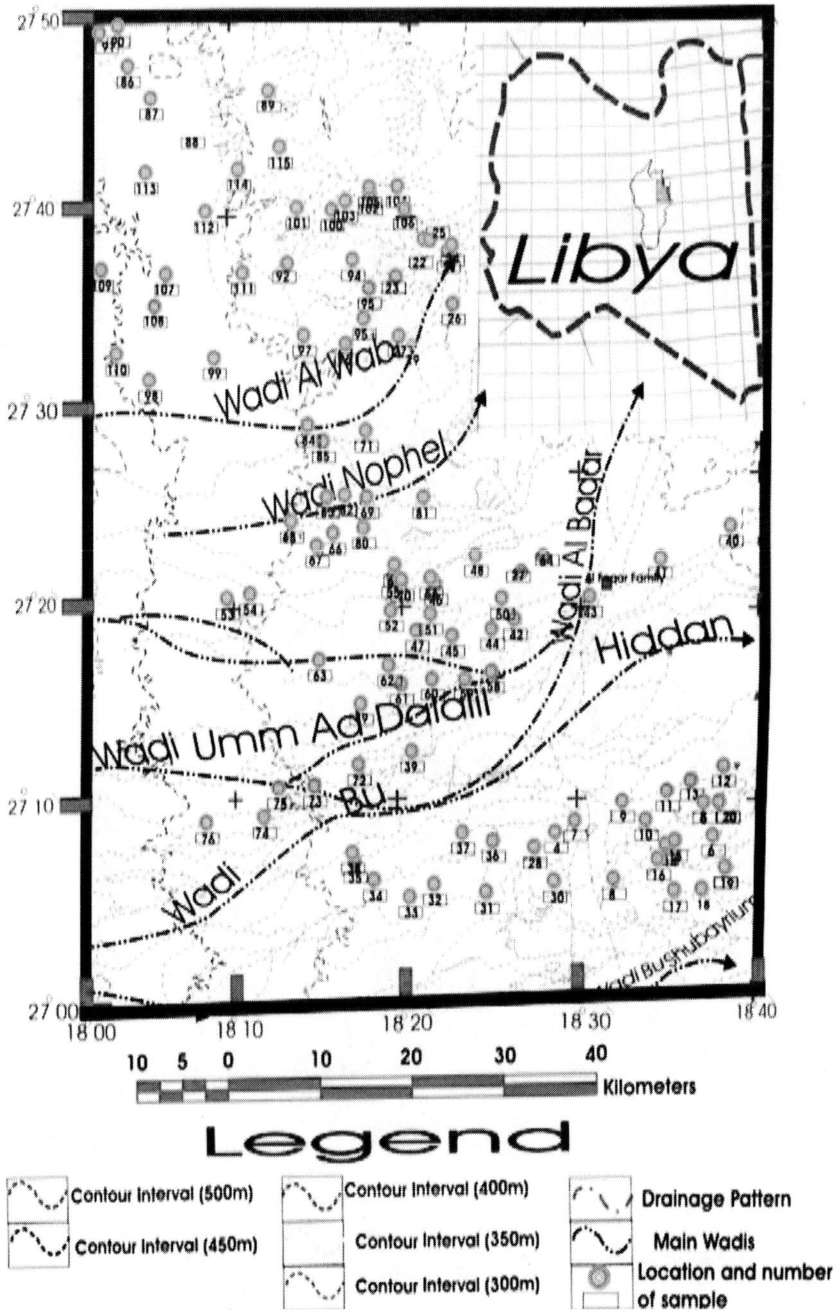

Figure 3. Topographic contour map showing the main wadis and their drainage patterns, as well as the localities of the samples.

The drainage patterns developed on the surfaces of the basaltic lava flows show systematic changes with the increasing age of basaltic lava flows, Busrewil and Suwesi [2] . The Landsat TM image shows the drainage pattern frequently increases in density as the basaltic rock become progressively older or decreases as the flow become younger, as in Phase 5, El-Shaafi [6] . The area investigated is cut by many wadis flowing from west and south-west towards the east and south-east, most of them starting from the central part of Al Haruj Al Aswad, outside the study area.

In the study area, the most common type of drainage pattern is dendritic, which is a well-integrated pattern formed by the main wadis (Wadi Bu al Hiddan, Wadi Al Baqar and Wadi Umm daladil), with tributaries branching and rebranching in several directions without systematic arrangement, Busrewil and Suwesi [2] .

The core of southern Libya was formed by Precambrian basement massifs represented by Jabal Awayant, the Tibesti massif and the Tihemboka Arch, which are parts of the Nile craton, East Saharan craton and Pan-African remobilized belt, Hallett [7] .

During the Pan-African orogeny, the rocks within these terrains were uplifted and accompanied by intense deformation, shearing and metamorphism. The main phase of the pan-African orogeny was completed about 540 Ma, followed by post-orogenic magmatism until about 440 Ma, interpreted by Klitzsch [8] as the early manifestation of the Caledonian orogeny.

This event was accompanied by the formation of NNW-SSE trending extensional structures more than a thousand km long such as the Kalanshio trough, south Haruj uplift, Dor el Qussah trough and the Tripoli-Tibesti uplift, among others. These directly affected the geological history of Al Haruj al Aswad and al Abiyad, Peregi et al. [5] .

Toward, the Haruj uplift, the Palaeozoic geological development is much less clear than in the Dor el Qussah trough, as this area is covered by the volcanic pile of Al Haruj al Aswad. Following the Pan-African orogeny, Lower Paleozoic sediments were deposited and probably overlain by the Middle-Upper Devonian and Carboniferous sedimentary formations, Hallett [7] .

After the Hercynian orogeny (Figure 4) almost the entire Kalanshio trough became part of Sirt Arch and most of the Palaeozoic sequence was destroyed by erosion, Peregi et al. [5] .

The first data on the radiometric age lava of Jabal Al Haruj was given by Pesce [10] , who considered it to be Late Pliocene based on K/Ar determination.

Ade Hall et al. [11] published the first results of a detailed K/Ar whole rock radiometric age determination survey combined with a palaeomagnetic measurement of the Al Haruj volcanic area. The measurements give an interval from 6 to 0.4 Ma for the formation of the volcanic rocks, Less et al. [12], with the majority of the flows being younger than 2.2 Ma. The modern study of age dating by Peregi et al. [5] was based on K/Ar analyses of 20 basalt samples carried out at Atomki in Hungary.

It shows that the volcanic activities continued from 5.27 Ma to about 0.1 Ma, which means that they can be assigned to the Pliocene and Pleistocene epochs, taking the limit of the Miocene at 5.3 Ma according to the Global Standard Stratigraphic Age (Table 1).

From this review it is clear that the study area has not been thoroughly studied so far, El-Shaafi [6] , except for the work of Klitzsch [13] , which deals with the regional geological mapping of Jabal Al Haruj (Al Abiyad and Al Aswad), showing the distribution of six major phases and one minor phase.

On the other hand, Farahat et al. [14] mentioned that the Libyan low volcanicity rift is probably related to the large Afro-Arabian rift system that extends from Turkey to Mozambique. This rift system is thought to have developed episodically rather than continuously. Age dating has suggested that the eastern rift in Kenya has evolved in three major pulses as follows: Eocene (44 - 38 Ma), Middle Miocene (16 - 11 Ma) and Pliocene to Pleistocene (5 - 0 Ma). These ages are generally contemporaneous with the ages of the Libyan volcanics; for example, Jabal Al Haruj was contemporaneous with the third pulse and the Gharyan volcanic province with the three pulses.

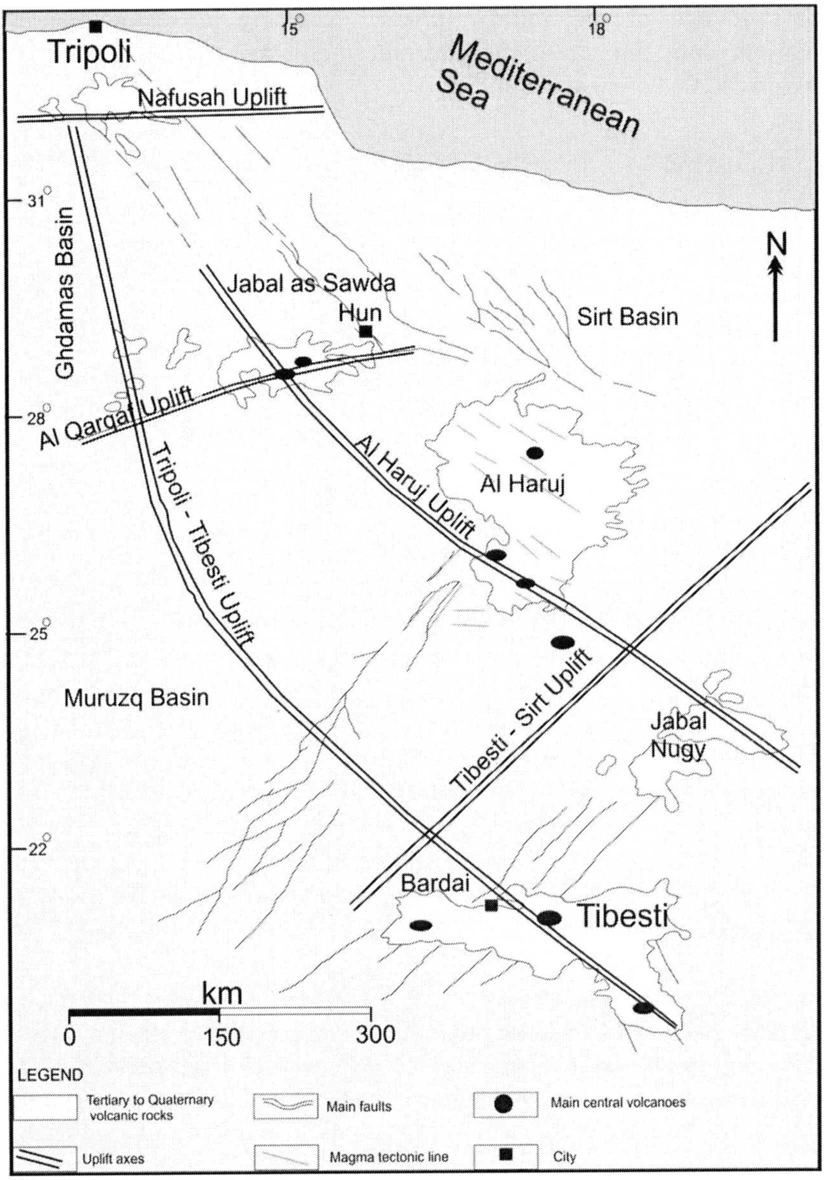

Figure 4. The volcanic distribution and relation between NW trending (Caledonian) and NE trending (Hercynian) structural elements and the volcanic eruptions, after Woller and Feduick [9] .

Table 1. Age of the volcanic phases according to radiometric age determinations and paleomagnetic measurements, and their geological interpretation, Peregi et al. [5] .

Volcanic Phases	Age in Ma	Presumed geological age
6	0.10 - 0.30	Late to Middle Pleistocene
5	0.30 - 0.50	Middle Pleistocene
4	0.50 - 1.22	Early to Middle Pleistocene
3	1.22 - 1.77	Early Pleistocene
2	1.77 - 2.24	Late Pliocene
1	2.73 - 5.27	Early to late Pliocene

The linear trend of the Libyan volcanics and their decrease in age towards the SSE is probably caused by intraplate extensional stresses. Continental rifting is normally associated with asthenospheric upwelling and high heat flow, but there is some controversy as to the mechanism involved, Farahat et al. [14] . It is difficult to decide whether the volcanic activity in Jabal Al Haruj was related directly to an uprising mantle plume (active rifting) or to the extension of the lithosphere (passive rifting), Peregi et al. [5] . The present work deals mainly with the different exposures in the study area. The exposed strata range in age from the Lower Miocene to Upper Pleistocene. Stratigraphically, the basalt flows in the study area disconformably overlie the older sequence of sedimentary rocks. The investigated volcanic rocks are overlay Maradah Formation and have a varied morphology, implying that they are produced episodically in consistency with available age data of 5.3 to 0.10 Ma from the Upper Miocene to Upper Pleistocene, Peregi et al. [5] .

The study area is not mountain range but rather a vast plateau. The commonest type of volcanic occurrence is represented by basalt lava flows, which shaped the present morphology. Generally, the central eruption of these lava flows are most markedly shown in the central part of Al Haruj, while elsewhere it is on its margins. In contrast, and despite the study area being located at the margin of Al Haruj, it is difficult to locate these features.

The Jabal Al Haruj Al Aswad volcanic province is characterized by the prevalence of lava of an effusive form rather than an intrusive form. Although some of the lava may have come from central volcanoes, the great volume of lava and the remarkable uniformity in thickness of individual lava sheets over a wide area indicate that the lava must have extruded from numerous fissures, Farahat et al. [14] . In the area investigated, plateau basalts are the most widespread; they are composed of several lava flows, whose thickness does not exceed several metres. Five major volcanic phases of different field appearances (including weathering surface, morphology, and primary structures) have been documented in the investigated area in the north-eastern part of Al Haruj Al Aswad province (Figure 5).

2. PETROGRAPHY

The petrographic characteristics of the Tertiary volcanic rocks of Jabal Al Haruj (al Aswad and al Abiyad) have been studied by many authors, among them Busrewil and Suwesi [2] , Less et al. [12] , Peregi et al. [5] , Vesely [15] and Woller [16] .

In the present study, 109 samples from the investigated area were described, and on the basis of preliminary petrographic descriptions 42 samples were selected for detailed petrographic study. Modal analyses of the various phases of basaltic rocks were carried out using an electronic point counter (Model E). The results of the modal analyses are presented in Table 2.

Microscopically, the rock samples are variably vesicular, amygdaloidal and almost medium grained and subordinately fine grained. Some of the samples are characterized by the presence of variable amounts of glass in the groundmass, but most of them are holocrystalline with an intergranular to intersertal texture.

The phenocrysts consist predominately of olivine and a subordinate amount of clinopyroxene, while plagioclase is not found. Olivine phenocrysts form about 7% to 21% by volume of the rock and occur as subhedral and sometimes as euhedral crystals without zoning; therefore, they indicate the presence of enough time to equilibrate with the melting at mantle depth.

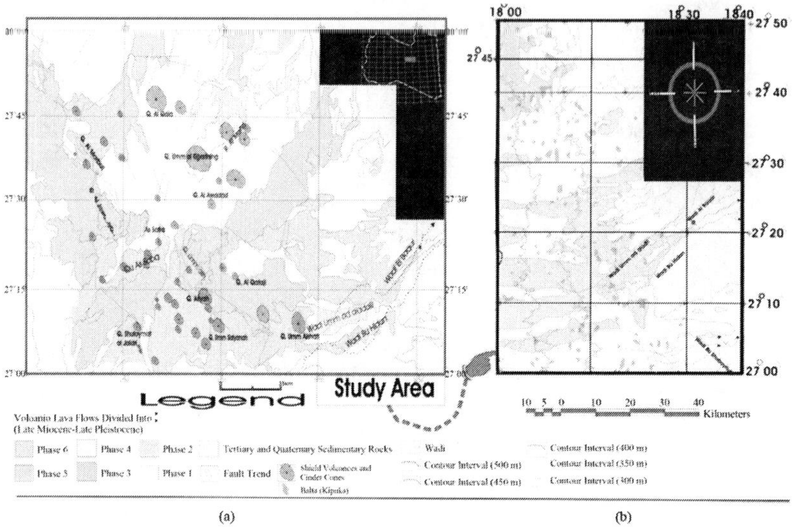

Figure 5. Geological map showing (a) The central part of al Haruj al Aswad, studied by Busrewill and Suwesi [2] and the north-eastern part of al Haruj al Aswad (study area) investigated in the present work; (b) Magnified geological map of the study area.

Olivine phenocrysts sometimes occur as clots displaying a glomerporphyritic texture. Pyroxene phenocrysts were found in some samples as elongated and prismatic crystals. The matrix consists mostly of medium grained plagioclase laths, olivine, clinopyroxene (augite) and opaque minerals (probably ilmenite).

Plagioclase laths are the most abundant, representing 32% to 48% of the matrix in most of the studied samples. They are both pilotaxitically and randomly arranged, and sometimes they exhibit a trachytic arrangement around olivine phenocrysts. Cryptocrystals fill the interstices between plagioclase laths. Some samples showed vesicles filled with carbonate secondary minerals.

The different textures and mineralogical composition of basaltic rocks of each phase are described in more detail in El-Shaafi [6] . A detailed petrographic description of Phase 1 follows.

Phase 1

This phase represents the earliest volcanic activity in the area. The lava flow of this phase is highly weathered, but although weathered in the field, the plagioclase and pyroxene (augite) are nearly fresh under the microscope.

Olivine phenocrysts are partially altered to iddingsite, which has stained the rocks a red colour. The most common texture of the rocks of this phase is an intersertal to intergranular texture (Plate 1). The intergranular texture is one in which the dimensions of the plagioclase and pyroxene are similar and the glass is relatively minor, which grades into intersertal when interstitial glass is a significant component.

The rocks are composed mainly of plagioclase, clinopyroxene and olivine. Opaque minerals that might be ilmenite are present as accessory minerals. Reddish brown iddingsite and carbonate minerals are secondary products. Volcanic glass is also observed. The detailed petrography of Phase 1 as follows:

A?Plagioclase is the most abundant, occurring in the form of laths of lamellar twinned crystals, ranging in size from 0.40 to 0.95 mm in length and from 0.10 to 0.20 mm in width, and occurring in the form of irregular grains. The interference colour of plagioclase laths of the first order is grey to straw-yellow with broad twin lamellae. Generally, the relief of plagioclase is quite high, so this is a calcium-rich plagioclase, Mackenzie and Adams [17] .

Plagioclase crystals contain opaque crystals as inclusions, El-Shaafi [6] . There is more than one generation of plagioclase: the first generation produced small plagioclase crystals which occur as inclusions within clinopyroxene; the second produced lath-shaped crystals with lamellar twinning. Optical measurements indicate that the plagioclase compositions are generally in the range of An 61% to An 72% (labradorite-bytownite), which is characterized by broad twin lamellae under cross-polarized light. Laths of plagioclase are commonly seen to flow around phenocrysts and form a trachytic texture, which may have resulted from the flow of the magma before solidification, El-Shaafi [6] .

B-Olivine is the third of the abundant minerals which occur in the euhedral to subhedral form, or as rounded crystals. The olivine crystals range in size from 0.40 to 1.2 mm in length and from 0.40 to 0.70 mm in width, although rounded crystals averaging 0.38 mm diameter are recorded. It is recognized in the examined samples by its colourless, high relief and high birefringence, and it is commonly traversed by randomly oriented cracks.

Table 2. Model analyses of representative basaltic rocks from north-eastern part of Al Haruj al Aswad.

Phase Number	Sample Number	Olivine %	Clino-Pyroxene %	Plagioclase %	Opaque %	Glass %	Carbonate %	Texture Type
1	17	18.57	21.76	37.28	12.96	5.69	3.74	Porphyritic Intergranular
1	54	21.55	24.84	39.14	9.09	2.21	3.17	Porphyritic Intergranular
2	23	13.77	19.55	41.40	8.11	8.20	8.97	Porphyritic Intergranular
2	95	12.29	15.01	37.41	7.28	18.30	9.71	Intersertal, Amygdaloidal
3	35	15.87	24.11	45.31	8.21	3.99	2.51	Porphyritic Intergranular
3	51	16.21	20.01	43.72	6.25	10.41	3.41	Intergranular to Intersertal
3	68	16.45	19.92	48.46	4.29	7.48	3.40	Intergranular to Intersertal
4	30	18.01	19.56	46.09	5.02	10.71	0.61	Intergranular to Intersertal
4	69	19.10	21.14	47.16	4.77	3.68	4.15	Intergranular to Intersertal
4	108	6.71	26.88	46.43	2.56	13.86	3.56	Intergranular to Intersertal
5	86	7.83	12.99	32.92	0.78	42.76	2.72	Intersertal to Hyalo Ophitic

Olivine phenocrysts are partially or completely converted into a strongly coloured reddish brown along their cracks and boundaries, known as iddingsite, El-Shaafi [6] . Multiple grains as clusters of adhering phenocrysts of olivine were observed and formed a glomerporphyritic texture, El-Shaafi [6] . Skeletal crystals of olivine were observed, where the edges and corners grow more rapidly than the faces, El-Shaafi [6] . More than one generation of olivine were observed: the first generation were the phenocrysts, which appear as highly cracked crystals, whereas the second generation produced small scattered crystals associated with clinopyroxene (augite) occupying the spaces between the plagioclase.

C-Clinopyroxene occurs in a subhedral to anhedral form, and as prismatic crystals varying from 0.35 mm to 0.85 mm in length and from 0.10 to 0.60 mm in width, which have a maximum extinction angle of 42° to 51°. They are identified as clinopyroxene (augite). Augite crystals are colourless to pale brown, and brown in some specimens. Rarely, the crystals show simple twining. The crystals are generally fresh, but may be slightly altered along the cracks and cleavages. Some of them contain inclusions of plagioclase and opaque minerals (most probably ilmenite).

A subophitic texture is well developed in the rock, where plagioclase laths are partly enclosed with clinopyroxene, whereas the carbonate (most probably calcite) iddingsite occurs and secondary constituents.

D-Opaque minerals are present in considerable amounts, being in the form of discrete crystals that exhibit various forms. They are generally of a smaller size than associated silicate minerals, where they range in length from 0.05 mm to 0.20 mm and from 0.035 mm 0.10 mm in breadth, El-Shaafi [6] . They are the primary constituents, being crystallized straight from the melt.

E-Volcanic glass is present in some of the examined samples, but in small amounts between the plagioclase laths, manifesting an intersertal texture.

F-Carbonate occurs plentifully in some samples in the interstitial spaces and it fills irregular vesicles and forms an amygdaloidal texture, El-Shaafi [6]. Carbonate minerals have higher birefringence, so that when viewed in cross-polarized light (XPL), they do not show interference colours; instead, they show delicate pastel shades of colours because of the extreme birefringence, which reaches up to the sixth order of interference colour.

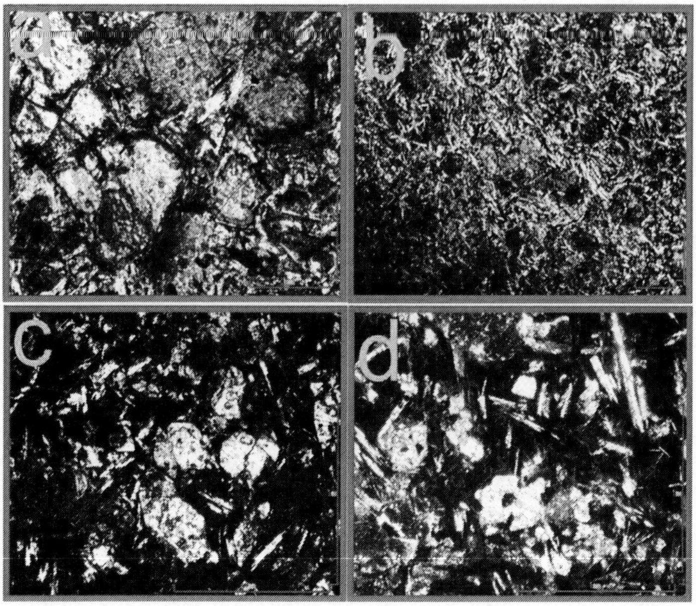

Plate 1. Microphotographs of olivine basalt of phase one, showing fine-grained rock with microphenocrysts of olivine (highest interference colour) and clinopyroxene (augite). The groundmass is composed of olivine, laths shaped of plagioclase and clinopyroxene (augite), with roughly rectangular crystals of opaque minerals. A higher magnification view reveals the olivine crystals and laths of plagioclase in more detail. Sometimes clinopyroxene crystals occur as a large poikilitic plate, enclosing several plagioclase and olivine crystals and forming an ophitic texture.

G-Iddingsite and bowlingite. In certain samples of the olivine basalts of Phase 1, the olivine phenocrysts are partially or completely altered into a strongly coloured yellowish or reddish brown mineral called iddingsite, El-Shaafi [6] . Bowlingite may be similar to iddingsite, but bowlingite is more strongly coloured than iddingsite and is rich in red colour, Hatch et al. [18] .

3. GEOCHEMISTRY

The geochemistry of Late Miocene-Pleistocene basaltic rocks of Al Haruj area has previously been treated by investigators. Busrewil and Wadsworth [1] , made the first attempt to undertake a preliminary study of chemical data on Al Haruj volcanics. These authors indicated that the rocks of Al Haruj belong to the alkali olivine basalt association, ranging in composition from basalts to hawaiites. Vesely [15] revealed that the rocks there are com- posed of alkaline to tholeiitic basalts. Busrewil and Suwesi [2] suggested the clustering of Al Haruj volcanics into four trends of differentiation, which may be described as follows: alkali basalts of the block lava series, alkali basalts of the main lava series, subalkali basalts and dolerites.

Peregi, et al. [5] studied the geochemistry of basaltic rocks from Al Haruj al Abiyad; they concluded that they were formed in an intracontinental (within plate) tectonic setting. They classified them into three main mafic groups based on their chemical composition: basanites, alkali basalts and olivine tholeiites. Moreover, they originated from different degrees of partial melting of a similar mantle source. The basanite magmas could have been formed by 0.5% to 1% partial melting, whereas the primary magmas of the alkaline basalts could have been generated by 2% to 6% partial melting, Peregi, et al. [5] .

The present study is an attempt to identify the geochemical characteristics of the Late Miocene?Pleistocene basaltic rocks of the NE part of Al Haruj al Aswad, evaluate the chemistry of these rocks within the framework of established magma types and determine the tectonic environment of their eruption. A major step in the development of evaluating the studied basaltic rocks involves choosing a method that enables rapid and accurate analysis to be conducted of major and trace elements present at low concentrations in silicate materials.

Twenty-nine samples were chemically analysed using X-ray fluorescence (Philips 2404) at the Central Laboratories of the Geological Survey of Egypt to determine the major and trace elements of the chemically analysed samples (Figure 6). Details of the analytical procedure and precision can be found in Appendix 1 of El-Shaafi [6] . The geochemical data of the 29 samples and the CIPW norms are presented in Table 3.

The chemical composition ranges of the investigated samples as a whole are relatively narrow within each individual basaltic phase and also within the different phases that existed there. The chemical ranges of the studied samples are somewhat restricted.

According to the CIPW norm, the analysed rocks are characterized by a high normative olivine content, with at most a low content of normative nepheline. Hence, these rocks can be classified as olivine basalt. The occurrence of normative hypersthene leads to rock being classified as normal olivine transitional basalt. SiO_2 ranges from 49.5% to 52.7% together with Mg# (Mg/(Mg + Fet)) values of 52 to 62 indicate that the studied basaltic rocks are slightly to moderately evolved (less primitive). Thus, they are not considered as the primary melt of the primitive mantle. The studied basaltic rocks are actually basalt on the basis of their silica content.

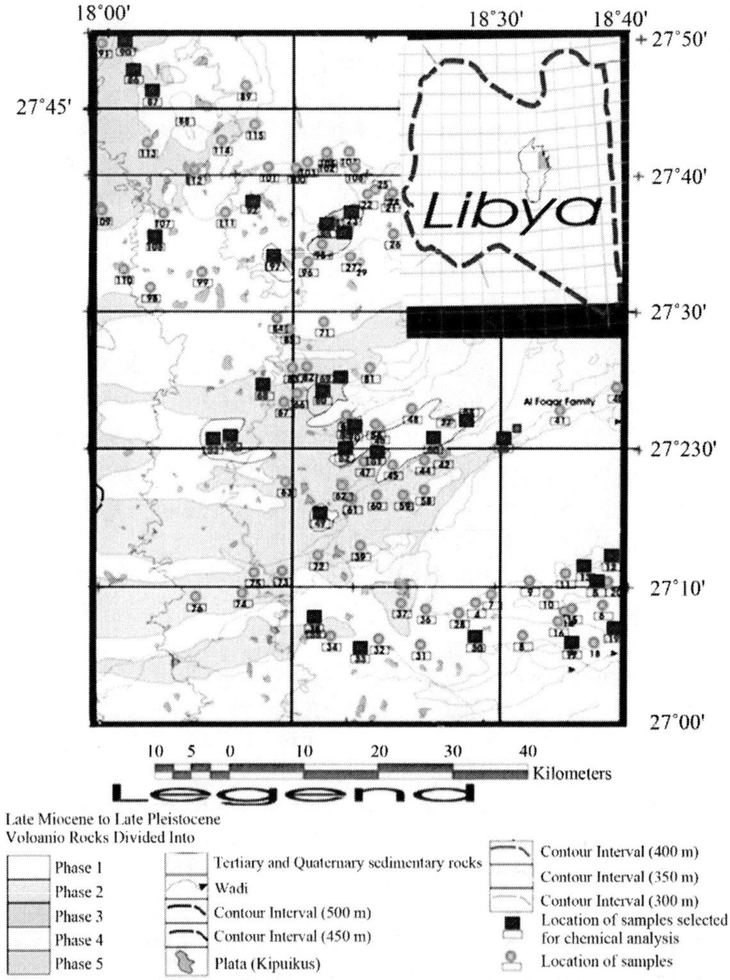

Figure 6. Geological map of the investigated area showing the sample localities of all collected samples during field inspection and location of samples selected for chemical analysis.

Table 3. Representative analyses of the major elements (wt%), trace elements (in ppm) and (CIPW) norms of the studied basaltic rocks. CIPW normative components are calculated for the volatile free data normalized to 100%. Fe_2O_3 and FeO were estimated based on ratio of $Fe_2O_3/FeO = 0.2$.

(a)

Representative analyses of the major elements (wt%) of Jabal al Haruj al Aswad																	
	Intrusion	SiO_2	TiO_2	Al_2O_3	Fe_2O_3	FeO	MnO	MgO	CaO	Na_2O	K_2O	P_2O_5	LOI	SO_3	Cl	Sum	Total Fe
DN19	Phase 1.1	49.85	1.79	15.81	1.65332	8.2666	0.15	6.49	9.26	3.65	0.98	0.25	0.01	0.14	0.13	98.43	10.84
DN49	Phase 1.1	48.23	1.55	15.54	1.68688	8.4344	0.16	6.38	9.92	3.38	0.74	0.22	1.17	0.85	0.16	98.421	11.06
DN50	Phase 1.1	48.13	1.62	15.77	1.62282	8.1141	0.16	5.54	10.02	3.6	0.73	0.19	1.39	1.57	0.24	98.697	10.64
DN52	Phase 1.1	47.9	1.87	15.63	1.53	8.35	0.15	5.2	12.33	3.45	0.96	0.32	1.21	0.31	0.11	99.32	10.13
DN53	Phase 1.1	47.63	1.73	15.11	1.67467	8.3734	0.16	6.56	10.17	3.39	0.97	0.31	0.77	1.69	0.18	98.718	10.98
DN54	Phase 1.1	50.06	1.64	16.33	1.57554	7.8777	0.16	6.02	9.37	3.7	0.91	0.23	0.55	0.1	0.2	98.723	10.33
DN97	Phase 1.1	50.18	1.64	16.41	1.67162	8.3581	0.16	5.72	9.36	3.96	0.68	0.19	0.01	0.18	0.16	98.68	10.96
DN64	Phase 1.1	46.93	1.56	15.27	1.67162	8.3581	0.16	5.7	11.82	3.24	0.69	0.28	1.03	1.81	0.1	98.622	10.96
DN5	Phase 1.1	48.45	1.64	15.08	1.68993	8.4496	0.16	6.7	10.4	3.22	0.77	0.23	0.94	0.69	0.26	98.68	11.08
DN12	Phase 1.1	50.85	1.83	15.67	1.62892	8.1446	0.15	5.59	9.21	3.65	0.79	0.14	0.4	0.41	0.23	98.694	10.68
DN13	Phase 1.1	50.76	1.7	16.35	1.58164	7.9082	0.15	5.56	9.46	3.82	0.95	0.2	0.01	0.12	0.16	98.73	10.37
DN17	Phase 1.1	47.66	1.61	14.44	1.62282	8.1141	0.15	6.65	10.9	3	0.54	0.15	1.59	1.9	0.2	98.527	10.64
DN23	Phase n.2	48.25	1.88	15.81	1.61062	8.0531	0.16	5.26	10.46	3.55	0.81	0.31	1.78	0.7	0.17	98.804	10.56
DN80	Phase n.2	47.89	1.68	14.63	1.63197	8.1598	0.15	6.95	10.44	3.29	0.78	0.22	1.83	0.91	0.15	98.712	10.7
DN94	Phase n.2	49.95	1.71	17.5	1.52063	7.6031	0.15	4.75	10.56	4	0.54	0.27	0.19	0.09	0.05	98.884	9.97
DN95	Phase n.2	50.06	1.87	15.74	1.60451	8.0226	0.16	6.05	9.54	3.6	0.86	0.23	0.39	0.53	0.12	98.777	10.52

DN35	Phase n.3	48.75	2.07	15.22	1.7189	8.5945	0.16	6.32	9.9	4	0.96	0.37	0.19	0.4	0.1	98.753	11.27
DN43	Phase n.3	48.3	1.57	15.81	1.54961	7.748	0.15	5.66	10.38	3.53	0.77	0.19	1.26	1.67	0.18	98.768	10.16
DN51	Phase n.3	48.48	1.8	14.88	1.61977	8.0988	0.15	7.51	9.89	3	0.94	0.24	0.64	1.26	0.21	98.719	10.62
DN55	Phase n.3	50.12	1.83	16.62	1.61062	8.0531	0.16	4.71	10.34	3.85	0.74	0.29	0.15	0.14	0.09	98.704	10.56
DN68	Phase n.3	49.51	1.87	15.16	1.63502	8.1751	0.17	7.66	9.23	3.45	0.89	0.34	0.12	0.39	0.07	98.67	10.72
DN30	Phase n.4	51.63	1.82	16.54	1.45047	7.2523	0.15	5.36	8.97	4.03	1.06	0.3	0.01	0.05	0.13	98.753	9.51
DN33	Phase n.4	49.45	1.62	14.80	1.63654	8.1827	0.15	7.00	10.23	2.96	0.56	0.14	1.01	0.79	0.18	98.709	10.73
DN69	Phase n.4	49.65	1.68	16.25	1.59689	7.9844	0.16	5.82	10.24	3.75	0.83	0.26	0.13	0.16	0.18	98.691	10.47
DN92	Phase n.4	51.60	1.81	15.65	1.61977	8.0988	0.15	6.17	8.59	3.34	0.76	0.14	0.00	0.18	0.18	98.289	10.62
DN108	Phase n.4	48.44	1.87	14.75	1.7189	8.5945	0.17	7.15	10.42	3.29	0.80	0.27	0.84	0.12	0.16	98.593	11.27
DN86	Phase n.5	48.87	1.76	14.91	1.81499	9.075	0.17	7.35	10.09	2.99	0.68	0.15	0.11	0.49	0.20	98.66	11.90
DN87	Phase n.5	48.77	2.06	14.07	1.91261	9.563	0.17	6.96	10.01	3.22	0.86	0.21	0.14	0.46	0.13	97.946	12.54
DN90	Phase n.5	48.66	1.67	15.32	1.80737	9.0368	0.18	7.14	10.05	3.11	0.61	0.19	0.18	0.59	0.16	98.704	11.85

(b)

CIPW norms of the studied Jabal al Haruj al Aswad basaltic rocks. CIPW normative components are calculated for the volatile free data normalized to 100% Fe_2O_3 and

	Intrusion	Q	Or	Ab	An	Ne	Di	Hy	Ol	Mt	Il	Ap
DN19	Phase 1.1		5.91	30.89	23.86	0	17.1	2.03	12.72	2.44	3.46	0.58
DN49	Phase 1.1		4.55	28.6	25.045	0	19.3	0.45	13.8	2.54	3.06	0.53
DN50	Phase 1.1		4.49	29.04	24.714	0	20.4	0	11.8	2.46	3.23	0.46
DN52	Phase 1.1		5.79	20.94	24.326	4.87	29.3	0	7.6	2.28	3.63	0.74
DN53	Phase 1.1		5.79	24.45	23.147	1.85	21.6	0	13.41	2.52	3.42	0.74
DN54	Phase 1.1		5.5	31.31	25.262	0	16.4	3.2	10.98	2.32	3.19	0.56
DN97	Phase 1.1		4.08	33.51	24.992	0	16.9	1.155	12.28	2.46	3.17	0.44
DN64	Phase 1.1		4.25	21.79	25.084	3.22	27.3	0	10.01	2.51	3.1	0.67
DN5	Phase 1.1		4.73	27.25	24.419	0	21.7	0.84	13.04	2.54	3.21	0.56
DN12	Phase 1.1		4.79	30.89	24.04	0	17.4	10.84	4.4	2.41	3.55	0.32
DN13	Phase 1.1		5.73	32.32	24.659	0	17.5	3.3	9.53	2.33	3.29	0.46
DN17	Phase 1.1		3.37	25.39	24.34	0	24.7	3.29	10.11	2.48	3.23	0.37
DN23	Phase n.2		4.96	28.56	24.811	0.81	21.4	0	10.39	2.41	3.72	0.74
DN80	Phase n.2		4.85	26.25	22.847	0.88	23.5	0	13.17	2.45	3.34	0.53
DN94	Phase n.2		3.19	32.36	28.2	1.03	18.6	0	10.23	2.36	3.27	0.63
DN95	Phase n.2		5.2	30.46	24.248	0	18	0	8.68	2.36	3.63	0.56
DN35	Phase n.3		5.79	27.68	20.739	3.42	21.7	0	12.37	2.52	4.01	0.88

DN43	Phase n.3		4.79	28.12	25.019	1	21.6	0	11.17	2.33	3.11	0.46
DN51	Phase n.3		5.73	25.39	24.359	0	19.5	0	11.54	2.39	3.35	0.58
DN55	Phase n.3		4.43	32.58	25.882	0	19.8	0	9.31	2.35	3.53	0.67
DN68	Phase n.3		6.26	29.19	23.25	0	17	4.29	13.05	2.42	3.63	0.72
DN30	Phase n.4		6.38	34.1	23.91	0	15.5	6.03	6.91	2.13	3.51	0.7
DN33	Phase n.4		3.43	25.05	25.442	0	20.6	0	4.89	2.45	3.17	0.32
DN69	Phase n.4		5.02	30.06	25.055	0	20.1	0	11.68	2.35	3.25	0.6
DN92	Phase n.4	0.65	4.61	28.26	25.465	0	13.6	20.07	0	2.39	3.51	0.32
DN108	Phase n.4		4.85	26.97	23.116	0.5	22.5	0	14	2.54	3.65	0.65
DN86	Phase n.5		4.08	25.3	25.253	0	19.9	7.18	10.67	2.68	3.42	0.35
DN87	Phase n.5		5.2	27.25	21.397	0	22.6	2.53	12.67	2.83	4.01	0.49
DN90	Phase n.5		5.2	26.32	26.04	0	22.6	2.53	12.67	2.83	4.01	0.49

(c)

Representative analyses of trace elements (ppm) of Jabal al Haruj al Aswad basaltic rocks

	Intrusion	Sc	V	Cr	Co	Ni	Cu	Zn	Rb	Sr	Y	Zr	Nb	Mo
DN19	Phase 1.1		195.7	337.5	56.8	190.5	69.2	70.6	4.9	332.8	<1	86.8	10.1	1.6
DN49	Phase 1.1		227.4	319.9	52.9	120.4	48.0	70.2	5.2	179.2	<1	72.9	10.0	2.4
DN50	Phase 1.1		203.6	295.6	51.7	134.8	47.1	63.5	6.7	179.9	<1	70.4	9.7	2.1
DN52	Phase 1.1		207.5	281.4	50.7	116.9	54.4	59.6	4.2	184.0	<1	66.1	9.2	2.0
DN53	Phase 1.1		199.1	277.3	58.6	117.6	34.4	61.6	4.4	174.5	<1	64.0	7.8	2.3
DN54	Phase 1.1		201.6	281.1	53.9	160.5	73.0	63.8	6.3	197.3	<1	74.4	8.7	1.7
DN97	Phase 1.1		205.7	292.1	47.2	186.6	68.1	61.0	6.4	220.1	<1	81.0	11.0	2.7
DN64	Phase 1.1		206.9	322.1	51.1	184.9	53.4	62.6	5.5	306.2	<1	88.1	10.7	2.3
DN5	Phase 1.1		211	289.0	59.6	150.4	47.2	66.3	5.8	198.9	<1	76.5	11.1	2.5
DN12	Phase 1.1		196.5	273.4	54.6	141.5	46.0	66.0	5.1	238.5	<1	79.8	10.8	2.4
DN13	Phase 1.1		196.8	287.4	51.2	134.1	41.3	62.3	5.3	221.6	<1	71.0	9.5	1.7
DN17	Phase 1.1		204.7	325.3	58.0	124.6	47.0	63.8	5.4	318.8	<1	84.8	9.1	2.8
DN23	Phase n.2		193.2	267.8	53.5	118.6	41.5	52.5	5.2	240.5	<1	80.5	11.2	2.0
DN80	Phase n.2		210.3	305.3	54.2	136.9	68.5	70.0	5.6	206.1	<1	83.2	11.9	1.5
DN94	Phase n.2		219.2	292.7	51.7	141.6	45.5	59.7	5.4	266.2	<1	82.8	10.7	1.8
DN95	Phase n.2		196.7	238.7	42.6	81.9	52.9	58.7	3.4	176.9	<1	59.5	7.2	1.8

DN35	Phase n.3	197.3	339.1	53.5	193.1	55.0	66.8	5.6	306.4	<1	84.2	8.6	1.6
DN43	Phase n.3	208.3	329.4	47.1	114.0	60.2	62.2	4.9	405.6	<1	99.3	12.4	1.5
DN51	Phase n.3	186.1	265.0	45.2	124.2	48.1	56.9	4.6	247.0	<1	71.0	9.0	1.9
DN55	Phase n.3	201.2	282.1	43.8	137.3	50.5	60.0	5.6	235.9	<1	80.4	10.4	2.0
DN68	Phase n.3	228.1	316.0	61.2	153.8	47.7	65.0	5.2	388.3	<1	101.4	12.4	1.9
DN30	Phase n.4	217.1	268.5	48.8	108.3	32.6	62.5	7.1	186.6	<1	76.6	9.8	1.9
DN33	Phase n.4	211.9	304.1	53.4	114.7	44.5	74.7	6.0	183.1	<1	68.9	8.3	1.5
DN69	Phase n.4	211.2	283.3	46.5	103.0	35.2	58.9	5.8	187.9	<1	71.5	10.4	2.1
DN92	Phase n.4	237.1	358.4	56.9	133.6	47.9	70.7	6.3	176.9	<1	75.4	9.8	2.6
DN108	Phase n.4	215.0	345.2	56.7	172.8	49.4	71.7	4.7	202.9	<1	75.1	10.7	1.6
DN86	Phase n.5	243.2	327.6	57.3	139.1	52.4	68.8	6.1	200.1	<1	75.1	9.6	1.0
DN87	Phase n.5	260.0	361.3	60.1	165.3	59.1	70.8	7.0	236.6	<1	91.6	13.9	1.7
DN90	Phase n.5	229.1	300.1	57.1	130.6	54.3	76.8	6.1	178.5	<1	71.0	9.2	2.4

(d)

Representative analyses of trace elements (ppm) continued									
	Intrusion	Sn	Ba	La	Yb	Hf	Ta	Pb	Th
DN19	Phase 1.1	3.2	333.4	16.5	4.4	3.7	3.7	3.9	<1
DN49	Phase 1.1	5.9	176.6	6.0	4.1	2.8	4.2	3.9	<1
DN50	Phase 1.1	2.2	177.5	17.4	4.2	3.7	3.7	4.7	2.3
DN52	Phase 1.1	2	191.6	13.5	3.9	1.8	4.0	4.8	4.4
DN53	Phase 1.1	2.4	166.4	15.0	4.1	3.4	4.0	4.1	<1
DN54	Phase 1.1	2.4	241.3	11.4	4.2	3.0	3.5	7.7	<1
DN97	Phase 1.1	3.2	194.7	13.7	4.3	3.4	4.0	<1	<1
DN64	Phase 1.1	2.4	616.0	7.1	4.1	2.2	3.6	4.8	3.0
DN5	Phase 1.1	0.6	203.6	19.0	4.2	3.4	5	6.4	4.0
DN12	Phase 1.1	3.7	930.6	15.3	4.2	4.6	3.6	4.3	2.0
DN13	Phase 1.1	2.9	157.7	17.9	4.2	3.5	3.8	4.4	<1
DN17	Phase 1.1	2.5	214.2	15.5	4.1	2.2	3.4	4.5	1.0
DN23	Phase n.2	1.9	194.3	10.2	4	2.8	3.6	2.8	2.2
DN80	Phase n.2	3.4	189.8	12.3	4	3.1	3.6	5.5	0.8
DN94	Phase n.2	2.0	1070	13.4	4	4.9	2.8	2.2	2.1
DN95	Phase n.2	4.3	164.1	7.2	3.9	1.4	3.2	4.9	2.8
DN35	Phase n.3	2.4	1062	15.7	4.1	4.4	3.3	6.3	1.0
DN43	Phase n.3	1.5	764.0	15.4	4	3.4	4.1	3.8	2.3
DN51	Phase n.3	2.5	310.3	10.5	3.9	3.5	3.2	6.4	<1
DN55	Phase n.3	2.2	321.3	12.3	4	2.3	3.1	5.0	2.1
DN68	Phase n.3	3.8	663.0	17.0	4.1	3.8	3.3	3.7	<1

DN30	Phase n.4	3.7	219.8	12.6	3.9	3.5	3.6	5.5	3.0
DN33	Phase n.4	4.1	352.8	16.4	4.1	1.5	4.6	4.9	<1
DN69	Phase n.4	3.5	199.4	15.6	4.1	3.3	4	6.7	1.4
DN92	Phase n.4	2.1	176.8	19.5	4.1	4	4	4.6	<1
DN108	Phase n.4	2.5	292.0	11.3	4.2	4.1	3.9	5.9	1.0
DN86	Phase n.5	1.1	180.9	14.7	4.2	3.4	4	4.3	2.0
DN87	Phase n.5	2.4	182.0	29.9	4.2	3.3	3.4	6.4	2.5
DN90	Phase n.5	2.8	173.0	22.9	4.1	2.5	3.5	6.1	2.0

The petrographical criteria obtained from the studied basaltic rocks are not strictly conclusive in their classification. However, the modal contents cannot be accurately determined in most cases in volcanic rocks because of the fine grain size and presence of glassy materials. Therefore, some chemical criteria should be used to determine the basalt type of the studied rocks. The chemical classifications use parameters such as the existence of oxides, functions of oxides and ratios. Various suggestions have been proposed in this domain by different authors.

The volcanic rocks may be subdivided into two members of the major magma series, alkalic and subalkalic; however, although the boundary between them is marked as a solid line, it is actually transitional.

From the various literature dealing with petrochemical classification, it is clear that four types of basaltic rocks can be distinguished: tholeiitic basalts, transitional basalts, alkali basalts and high alumina basalts or calc-alkaline basalts, Cox et al. [19] . The concept of transitional basalts between the tholeiitic and alkali basalt type has been clarified by many authors, e.g. Middlemost, [20] . For the chemical classification of the basaltic rocks under consideration variation diagrams are employed using major elements, which are introduced in the following.

According to Middlemost [21] the investigated samples are transitional basalts of either a soda or potash series. This can be accomplished by an empirical boundary line on a soda vs potash diagram. The studied transitional basalts were found to belong to the soda series (Figure 7).

In addition to the classifications based on major elements, certain variation diagrams have been suggested by Winchester and Floyd [23] to classify the basaltic rocks using five minor and trace elements (Ti, P, Zr, YNb) to produce a set of binary diagrams that discriminate between tholeiitic and alkali basalts (Figure 8).

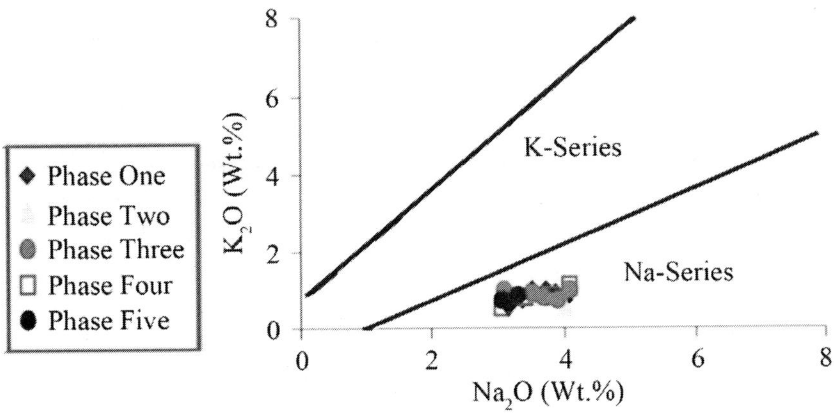

Figure 7. K_2O versus Na_2O (wt%) diagram, showing the subdivision of the transitional magma series into the Na sub-series of the samples in question, after Winter [22] .

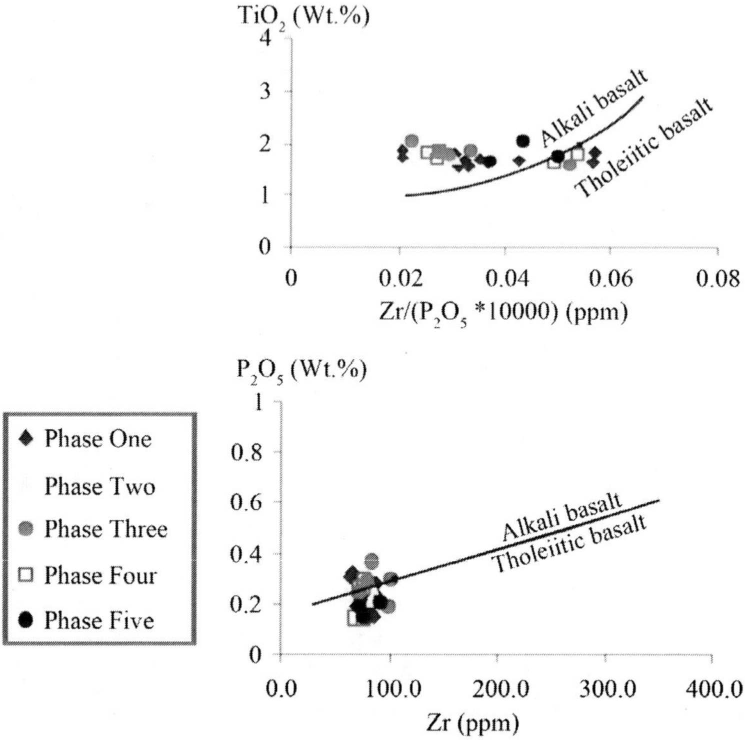

Figure 8. Minor (TiO_2, P_2O_5) and trace (Zr) element variation plots for the classification of the studied rocks as alkalic and tholeiitic basalts, after Winchester and Floyd [23] .

The P_2O_5-Zr diagram shows that the investigated samples are plotted close to the boundary of the alkali between the alkali basalt and tholeiitic basalt, whereas the TiO_2-Zr/P_2O_5 diagram shows that most of samples are situated in alkali field (Figure 8). This is expected since these diagrams use phosphorous as a principal component, which shows varying degrees of mobility, as stated by Rollinson [24] . Moreover, the Zr values of the studied samples are generally low.

The major element data contains only Fe_2O_3 as total iron. To obtain appropriate CIPW norms these values had to be recalculated into Fe_2O_3 and FeO. In the current study the ratio of $Fe_2O_3/FeO = 0.2$ was used (Less et al. [12] ; Peregi et al., [5]). Middlemost [20] has proposed a range of oxidation ratios (Fe_2O_3/FeO) for use with volcanic rocks drawn from geological literature; that of basalts is $Fe_2O_3/FeO = 0.2$, Rollinson, [24] .

There are various published programs for the calculation of the CIPW. The CIPW normative calculation in the current study was carried out by Norm4 software, written by Kurt Hollocher of the Geology Department, Union College, New York State. Several methods for the normative classification of basaltic rocks have been proposed. Yoder and Tilley [25] devised a scheme for classifying basaltic rocks by their CIPW norms using a basalt tetrahedron. Several modifications have been suggested to this classification and various schemes have been proposed by later investigators. Among these, Thompson [26] proposed a classification scheme for basalts based upon CIPW normative proportions of Ne, Ol, Di, Hy and Q. The three equilateral triangles of Thompson's diagram [26] represent basaltic and related rocks, which are respectively undersaturated, saturated and oversaturated: silica-undersaturated basalts (alkali basalts) are characterized by normative Ol + Ne; silica-saturated basalts (olivine tholeiite) are characterized by Hy + Ol; and, silica-oversaturated basalts (quartz tholeiites) are characterized by normative Q + Hy. With regard to normative minerals, the most studied samples are classified into hypersthene-normative; Wilkinson [27] considered hypersthene-normative rocks to be transitional or mild alkaline.

Busrewil and Oun [28] suggested that the transitional basalt is applied to basaltic rock that exhibits a nepheline-free norm. Three samples belonging to Phase 1, two samples from Phase 2 and two samples from Phase 3 were classified into nepheline-normative alkaline basalts (ne = 0.10 - 4.8; alkali-ol basalts), and one sample from Phase 4 was located at the boundary between ol-tholeiite and quartz tholeiite; this latter sample is near to oversaturation in SiO_2, which is a common feature of tholeiitic basalts (Figure 9).

The two main types of variation diagrams currently used are bivariate plots and triangular diagrams. A bivariate plot shows the variation between samples and identifies trends. Elements plotted along the x-axis of the diagram should be selected either to show maximum variability between samples or illustrate a particular geochemical process. Normally, in basic igneous rock MgO might be selected, Rollinson, [24] .

One of the most commonly used alternatives to the Harker diagram is MgO plot. This is most appropriate for a rock series which includes abundant mafic members, as in this case the range of SiO_2 concentrations may be small. The changes in selected major and minor element oxides are plotted against MgO for representative samples from the study area. The analyses exhibit an evolutionary trend representing slight increases in SiO_2, K_2O and

Na_2O, along with decrease CaO, FeOt and MgO. Clearly, MgO increases with increased FeOt. However, in spite of this overall trend there is no consistent chemical pattern related to each phase.

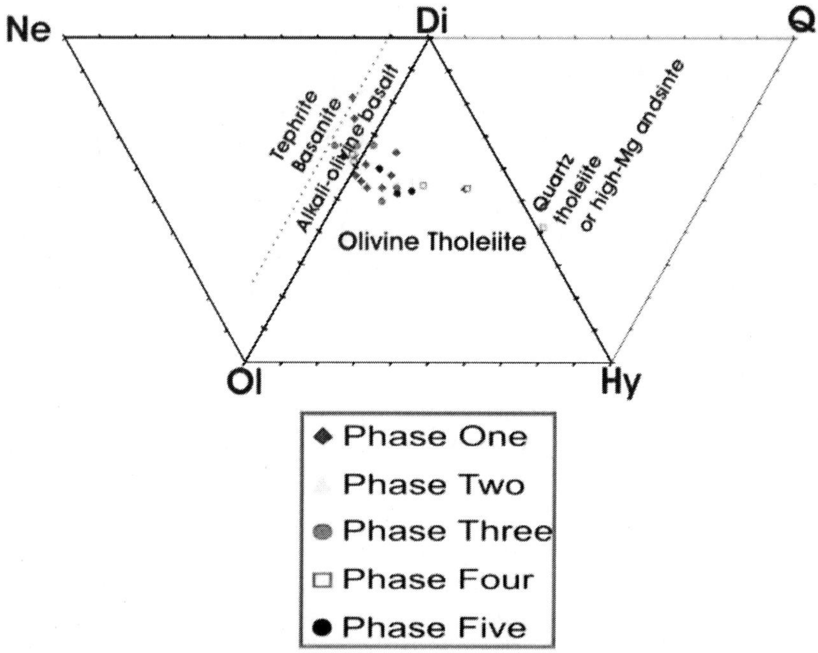

Figure 9. The classification of the studied samples according to their CIPW normative composition expressed as Ne-Ol-Di, Ol- Di-Hy, Di-Hy-Q (basalt tetrahedron, after Thompson, [26]).

Vague correlations between MgO and most other oxides in the studied samples probably suggest that these are related by different degrees of partial melting rather than by fractional crystallization. The relatively low MgO of less than 7.5 wt% and Ni (81 - 193 ppm) and Cr (238 - 361 ppm) contents indicates that the studied basaltic rocks exhibit slight to moderate fractionation of olivine and/or spinel. Consequently, the basaltic rocks are not primary magmas.

On average in the studied samples, the lava flows of Phase 5 and two samples from Phase 3 are more primitive. MgO is an important component of the solid phases in equilibrium with mafic melts and shows a great deal of variation, either as a consequence of the breakdown of magnesium phases during partial melting or as a result of their removal during fractional crystallization. Below 7% MgO-CaO correlates positively with MgO, indicating the removal of CaO-MgO from the liquid due to the crystallization of plagioclase and clinopyroxene. However, above 7% MgO and CaO-MgO correlate negatively, because this part of the trend is controlled entirely by the olivine. Inflections are most obvious where the number of fractionating minerals is small, Rollinson, [24] .

Trace element data provide a clearer view than the major and minor elements of distinct sources for the various magma types. In the current study the focus is on the trace elements, which due to enrichment and depletion are revealing about the nature of the basalt plateau. The trace element data can be used to evaluate the source composition, degree of partial melting and subsequent processes. The ratios of incompatible elements have been employed to distinguish between source reservoirs. The concentrations of the compatible trace elements (Ni, Cr, Co, Cu) are low in the studied samples, which supports our contention that they are not primary magma.

A characteristics feature of the studied volcanic rocks is their relatively constant ratios of certain incompatible trace elements (Nb/Zr and Rb/Zr) (Table 4). Relative constancy of ratios of incompatible elements provides strong evidence of a common source (homogenous). Also, the (Zr/Nb) ratios of studied basaltic rocks are less than 10, which indicates an ocean island basalt (OIB) source, Winter [22] .

Crustal contamination would certainly alter the ratios of Zr/Nb and Hf/Ta. The studied mafic samples have similar ratios, so the effect of crustal contamination on the studied basaltic rocks is probably minor, Baker [29] .

Several variants of the spider diagram have been used in the literature, with different elements and different normalization schemes. The order of the elements along the abscissa may vary slightly between different authors. Spider diagrams are flexible, and a variety of elements and normalizations have been displayed using them, Winter, [22] . In the current study, the comparison of the geochemical composition of the different rock samples, representative of different phases, with the MORB-normalized trace element patterns was used by Pearce [30] .

Figure 10 compares the incompatible elements of each sample with the most abundant igneous rocks on the planet. The large ionic lithophile (LIL) elements are on the left-hand side of the diagram and the high field strength (HFS) elements are on the right. Both are arranged in order of increasing incompatibility away from the margins, so that the most incompatible elements are just left of the centre of the diagram, Winter [22] .

The MORB-normalized trace element patterns of representative samples of different volcanic phases are given in Figure 10. Based on these diagrams, the patterns show a fairly close similarity in the shape and concentration range. The diagrams show a slightly different concentration of some incompatible trace elements within samples of the same phase, and also between samples representative of different phases, but they showed nearly identical behaviour for the less incompatible trace elements (Hf, Sm, Ti, Yb).

The studied basaltic rocks showed more peaks and troughs, reflecting the variable behaviour of the elements involved. Basalts from the study area are clearly enriched by the whole spectrum of more incompatible elements relative to MORB samples (normalized values greater than one). They show a broad central hump in which both LIL (K, Ba), and HFS (Ta) elements also reveal slight negative anomalies in Rb, Nb and Zr. The rocks display similar patterns of the peaks and troughs, which strongly suggests they share a common parent and subsequent processes, Winter [22] . Thus, based on this, the basalts could have a common genesis.

The basaltic samples from different volcanic phases show a fairly similar character to each other in their geochemical behaviour. Consequently, one phase cannot be distinguished from another based on their trace element contents.

From the consistent negative slopes, it can be deduced that the slope of the studied basaltic rocks are distinct from MORB slope and appear to originate in the lower enriched mantle reservoir, although a very low degree of partial melting may also produce LREE-enriched melts from primordial or slightly depleted sources, Cox et al. [19] . The pattern is regarded as typical of melts generated from relatively undepleted mantle material.

Spikes at LIL (Rb, Ba, K) may reflect enrichment, perhaps due to contamination of magma by continental crust. Negative Nb anomalies are also characteristic of the continental crust and may be an indicator of crustal involvement in magma processes, Rollinson [24] . Enriched LIL and the slight alkaline of the studied basaltic rocks suggest derivation from an enriched mantle source, Aboazom [31] .

Table 4. Ratios between some trace elements of the studied mafic rocks that could be used for their petrogenesis.

	Phase One											Phase Two		
Sample No.	97	19	12	13	17	5	49	50	52	53	54	64	23	80
Nb/Zr	0.135	0.116	0.135	0.134	0.106	0.144	0.136	0.137	0.139	0.121	0.117	0.120	0.138	0.143
Rb/Zr	0.078	0.056	0.063	0.074	0.063	0.075	0.070	0.095	0.069	0.084	0.062	0.064	0.067	0.065

	Phase Two			Phase Three			Phase Four					Phase Five		
Sample No.	94	95	35	43	51	55	30	33	69	92	108	86	90	87
Nb/Zr	0.129	0.120	0.125	0.126	0.128	0.122	0.127	0.120	0.145	0.129	0.142	0.127	0.128	0.151
Rb/Zr	0.065	0.056	0.066	0.048	0.065	0.051	0.092	0.086	0.081	0.083	0.062	0.081	0.085	0.075

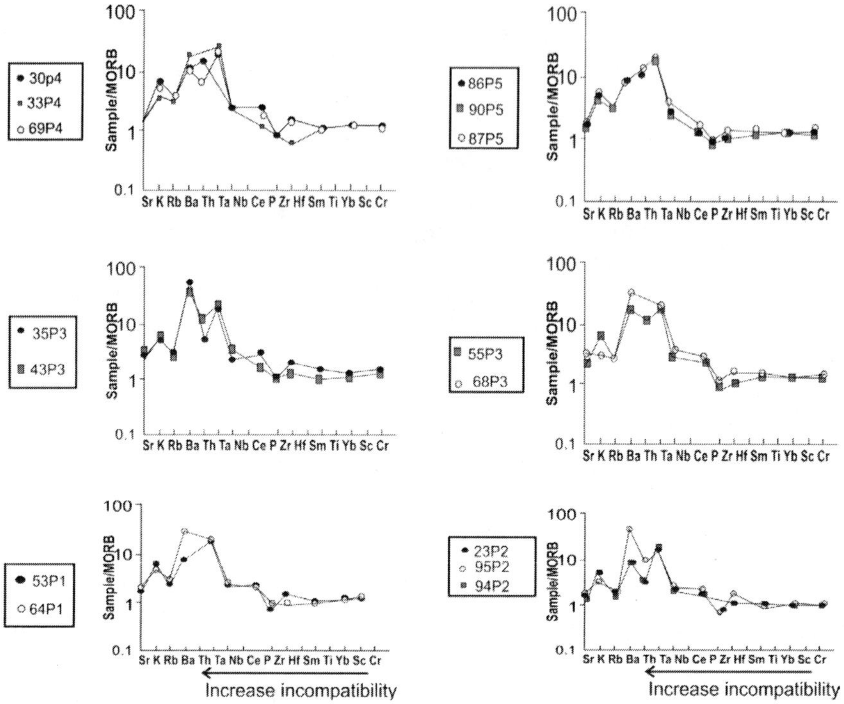

Figure 10. MORB-normalized patterns for selected samples from the study area, representative of different phases from Phases 1 to 5 (normalized data is greater than one). The Ba concentration of the studied samples shows extreme variation in some samples due to very high sensitivity to secondary processes.

The high LIL/HFS pattern and negative Nb anomaly is a feature of many intracontinental volcanics, suggesting this pattern may be also a feature of the subcontinental lithospheric mantle (Fitton et al. [32] ; Hawesworth [33]).

As the degree of partial melting increases, the proportion of a more incompatible element with respect to a less incompatible element decreases as it is progressively diluted, so that (as illustrated by Figure 11) the La/Nb ratio should decrease and the La/Zr ratio should increase. The extent of partial melting should thus increase from the upper left to the lower part of the mantle partial melting curve, Winter [22] .

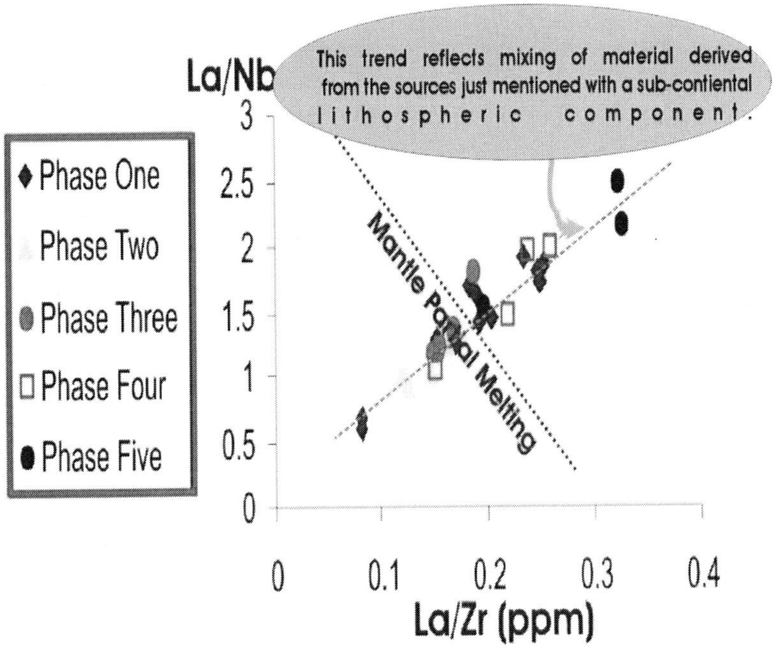

Figure 11. La/Zr vs La/Nb for the studied basalt samples, after Hopper et al. [34] .

The distinct phases suggest a single source and probably similar degrees of partial melting. The positive trend in Figure 12 reflects the mixing of material derived from the source, Hopper et al. [34] .

The elements of the first transition series (Ti, V, Cr, Mn, Fe, Co, Ni, Cu, Zn) vary in geochemical behaviour, and most of them are considered to be compatible elements. Transitional element plots have been mainly used with basalts as a means of exploring the geochemical behaviour of the first transitional elements, Rollinson [24] .

Figure 12 shows consistent trends for most studied samples representative of different phases on chondrite- normalized plots. They show progressive depletion from Ti to Ni and have a positive Ti anomaly and negative Cr anomaly. Cu and Zn are more variable and their concentrations may not reflect the original igneous values, (Longmuir et al. [35] . Anomalies in Ni and Cr concentrations may reflect the role of olivine (Ni) and clinopyroxene or spinel (Cr); Ti anomalies indicate the role of Fe-Ti oxides as opaque minerals.

Despite differences in the major element chemistry, the pronounced and very close patterns of similarity of trace elements of all the samples representative of the different phases certainly suggest there is a common source for the mantle peridotite beneath Al-Haruj province.

3.1. Comparison of the Studied Mafic Rocks with the Mid-Oceanic Ridge (OIB)

A broad spectrum of trace elements for the studied basalts rock can be compared using composite (average) OIB, as proposed by Sun [37] (Figure 13). The patterns of the studied basaltic rocks are similar in the shape to OIB, but vary in the concentration range of some incompatible trace elements. OIB enrichment in LIL (Sr, K, Rb) thus suggests a somewhat depleted mantle source relative to OIB, Winter [22] .

3.2. The Results of Correspondence Analysis

There is a large body of literature on the statistical treatment of geochemical data. Correspondence analysis is a descriptive and exploratory technique designed to analyse multi-way tables containing some measures of correspondence between rows (observations) and columns (variables). The results provide information that is similar in concept to those produced by correlation techniques. The term "inertia" in correspondence analysis can be thought of as a measurement of variation in data.

Obviously, the final goal of correspondence analysis is to find a graphical interpretation (meaning) from extracted dimensions (dimension 1 on the x-axis and dimension 2 on the y-axis). Thus, one can make the interpretation that the great similarity between the average of the elements of each phase with regard to the correlation between most HFS, LIL and transitional metal elements is more in Phase 4. However, as has been indicated, it is appropriate to make general statements about the nature of the dimensions (Figure 14).

The variance is explained by coordinates (dimensions) of more than 98% (96.03% of inertia on dimension 1 and 2.192% of inertia on dimension 2), which implies the relation between the average of the different phases and relation of elements between them and the relation between the elements and average of the five phases are highly significant, and so it can be stated with great confidence.

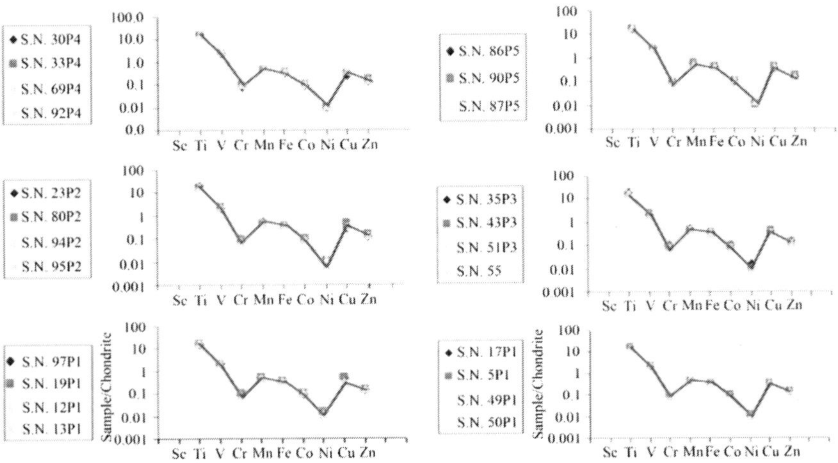

Figure 12. Transitional metal concentrations in most studied samples representative of the five phases, normalized to chondritic values, showing the same behaviour of transitional metal of different phases. The normalized values are taken from Kay and Hubbard [36] .

Judging the quality of the average of five phases expresses the value and importance of each phase in the current study, so in Table 5 it can be clearly observed that all phases are of highly significant importance. The correspondence analysis is useful for drawing attention to the likelihood that all phases were generated by sharing a common source.

3.3. Tectonic Environments

The approach of using geochemical analyses to try to discriminate magmas into different tectonic settings is described by several authors, and is probably best attributed to Pearce ([38] [39]), Pearce and Cann ([40] [41]), Pearce et al. ([42] [43]), Pearce and Norry [44] and Wood et al. [45] . They accepted that the major, minor and trace elements of basalts should reflect their geotectonic position, because the geochemistry of basalts is dependent on the crustal nature of the area in which they were formed.

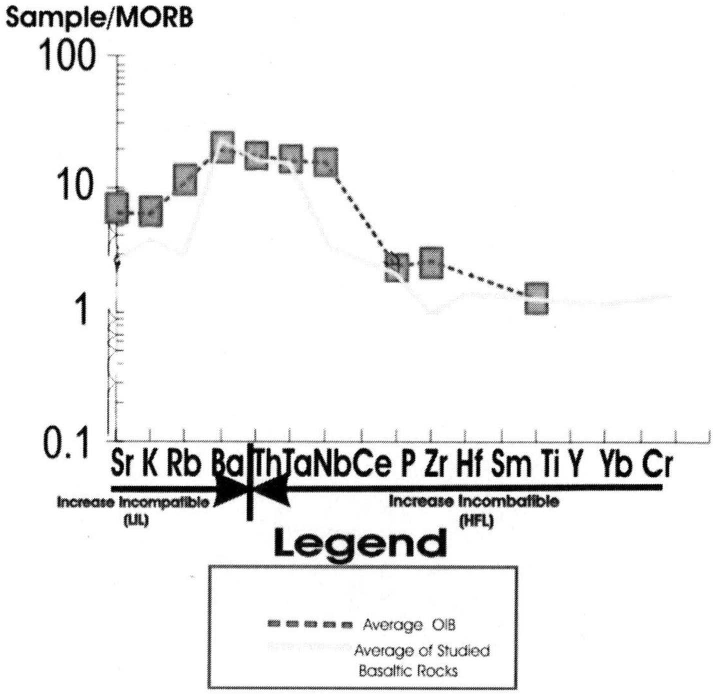

Figure 13. MORB-normalized spider diagram for average OIB, data from Rollinson, [24] and average representative analyses of the studied samples.

Floyd and Winchester [46] found that basalts with similar chemistry may be produced in different tectonic regimes, since their compositions reflect the nature of the particular mantle and the melting processes operating rather than the regime in which they are found. Probably, this often makes the chemical designation of ancient basalt less certain.

Wood et al. [47] considered the effects of crustal contamination which were not considered by Pearce and Cann [41] and yet which are likely to have a significant influence in the appearance of continental flood basalts and may be a contributory factor in their persistence in classification, Rollinson [24] . There are a large number of discrimination diagrams applicable to basalts and basaltic andesites, which use trace elements and minor and major elements to distinguish the different environments of basalt. Of course, the investigator should not rely wholly on any single technique or plot, Winter [22] .

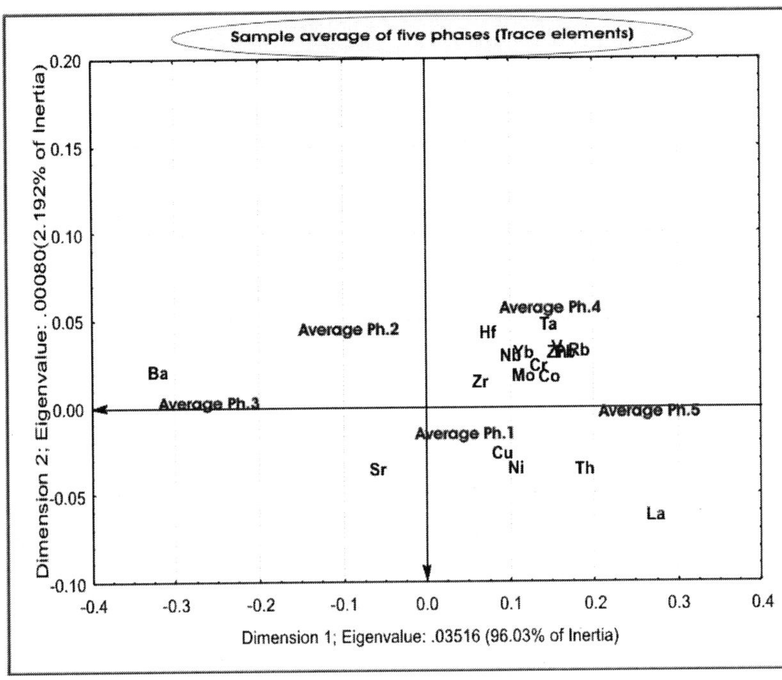

Figure 14. Correspondence analysis for trace elements of the average five phases presented in the area investigated, plotted on coordinates (on dimension 1 and dimension 2).

Pearce [38] developed a method that classified basalts in term of their tectonic setting on the basis of their major element patterns. He calculated the discriminant function based upon the eight major oxides and presented discriminant diagrams to identify mid-oceanic ridge basalt (MORB), within plate basalt, oceanic island basalt (OIB), continental flood basalt (CFB), calc-alkaline basalts (CAB), island arc tholeiites (IAT) and shoshonites (SB).

Table 5. The results of correspondence analysis for the average of the five phases.

Aver. Phases	Colum	Coordinate	Coordinate	Quality
Average Ph. 1	1	0.052836	−0.032650	0.818930
Average Ph. 2	2	−0.093243	0.028226	0.904941
Average Ph. 3	3	−0.262192	−0.0125540	0.995548
Average Ph. 4	4	0.135643	0.042508	0.970742
Average Ph. 5	4	0.261218	−0.020566	0.991675

Two discriminant functions, F1 and F2, were identified as follows:

$$F1 = +0.0088\mathrm{SiO_2} - 0.0774\mathrm{TiO_2} + 0.0102\mathrm{Al_2O_3} + 0.0066\mathrm{FeO} - 0.0017\mathrm{MgO}$$
$$- 0.0143\mathrm{CaO} - 0.0155\mathrm{Na_2O} + 0.0007\mathrm{K_2O}$$

$$F2 = -0.0130\mathrm{SiO_2} - 0.0185\mathrm{TiO_2} - 0.0129\mathrm{Al_2O_3} - 0.0134\mathrm{FeO} - 0.0300\mathrm{MgO}$$
$$- 0.00204\mathrm{CaO} - 0.0481\mathrm{Na_2O} - 0.071\mathrm{K_2O}$$

It is clear that the majority of the samples investigated are plotted in the field of within-plate basalts (Figure 15).

Wood [47] proposed the Th-Hf-Ta ternary discrimination diagram based upon the immobile (HFS) elements. Using this classification (Figure 16(a)), the studied basaltic rocks plot closer to the fields of within-plate basalts. The elements Th, Hf, and Ta existed in very low concentrations in basalts and cannot accurately determined by XRF analysis.

The advantage of using minor elements as discriminants between basalts is that they are readily detected and more accurately determined than trace elements, Rollinson [24] .

Pearce et al. [42] found in an empirical study that it was possible to discriminate between oceanic and continental basalts with a single straight boundary line on a TiO_2-K_2O-P_2O triangular diagram (Figure 16(b)). The field oceanic basalts included MORB and OIB. The studied basaltic rocks are plotted around the line separating the two fields.

4. SUMMARY AND DISCUSSIONS

The tertiary volcanic rocks of Jabal Al Haruj represent the largest volcanic province in Libya covering an area of ≈ 45,000 km^2, which originated during several phases of eruptions that were closely associated in space and time.

The study area is located in north-eastern Jabal Al Haruj al Aswad, an area which has not been studied in a thorough manner so far. Nevertheless, the work of Klitzsch [13] gave research details showing the distribution of six major phases and one minor phase. The most important study in this respect was done by Busrewil and Suwesi [2] on the Al Haruj al Aswad. The authors gave a detailed volcanological study of the area and distinguished six volcanic phases.

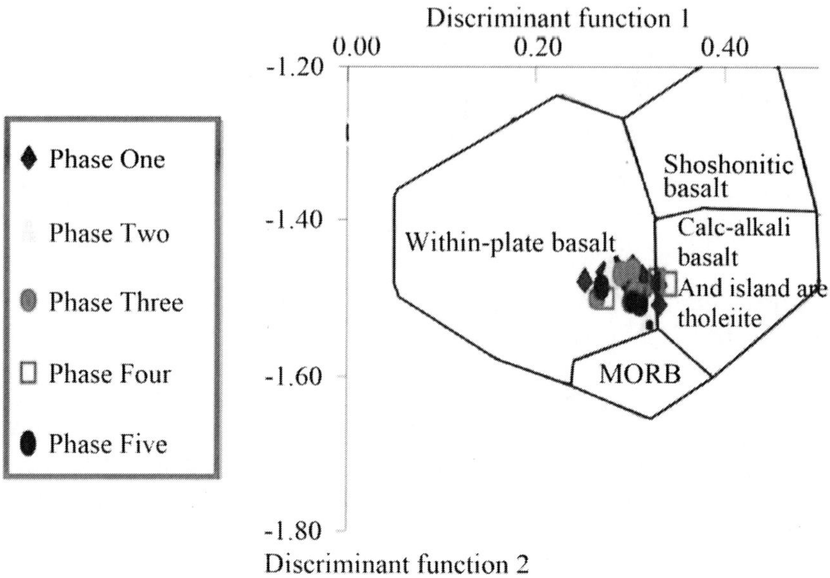

Figure 15. Major element discrimination diagram for basalts, after Pearce [38] , showing the most studied samples plot in the field of within-plate basalts.

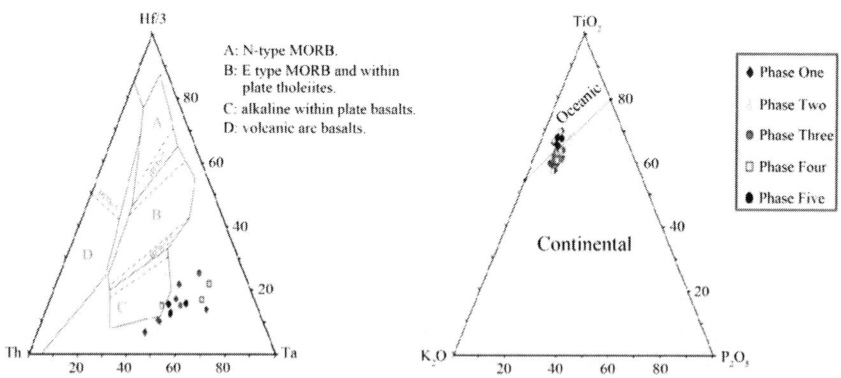

Figure 16. (a) The Th-Hf/3-Ta ternary discrimination diagram for basalts (after Wood, [47]); the broken lines indicate transitional zones between basalts types. (b) Triangular discrimination diagram between oceanic and continental fields, after Pearce et al. [42] .

The division of the volcanic phases used in the present work is based on similar principles to those developed by the earlier authors, which are as follows:

Colour variation of the lava flows of different phases on Landsat TM images.

Degree of weathering of the rocks.

Accessibility of lava fields.

Morphological and lithological characteristics observable in the field.

Superposition of the lava flows and drainage pattern.

Field evidence has revealed that each of these criteria are complementary and cannot be used separately, Busrewil and Suwesi [2] . Field studies included geologic mapping and the identification of mode of occurrence of the volcanics. According to modern age dating, the volcanic activity in the area started in the Late Miocene and lasted until the Late Pleistocene. The first five of the six volcanic phases of the Al Haruj volcanic province are present in the study area. The geological setting, the petrography, the geochemistry and tectonic setting of these occurrences are examined in this study. The following is a summary of the above-mentioned aspects and the main conclusions obtained during the present work.

Field geological mapping was carried out using Landsat TM multispectral satellite imagery. This imagery formed the base of the geological map, which was completed by the addition of topographic contours, locality names and drainage pattern of wadis, as well as other field criteria. The study area is remote, rather inaccessible and extremely difficult terrain. Due to this, the field work was carried out in parallel with the Industrial Research Centre (IRC), Tripoli during their regional geological mapping of SE Libya in March 2006.

The topographic form of the old volcanic phases is highly modified, while the forms of more recent phases are usually fresh and very well preserved. In general, each basalt flow unit begins at the bottom with the most massive part, characterized by a fine to micro-vesicular texture, and it evolves upwards into an obviously vesicular texture. The basalts are dark to medium and pale gray with a porphyritic appearance. A fresh sample was taken from the massive part of each flow unit.

Characteristic features of Phase 1 are the blocky, rough, subangular to subrounded rocky structure and have smooth surfaces, massive and less vesicular. On the satellite image it is distinguished by a bluish colour.

Phase 2 covers a limited geographical extent and outcrops in small scattered localities in parts of the study area. Lava flows from this phase tend to disintegrate into small blocks and slabs, less than 50 cm in diameter. The vesicles are often elongated in the direction of flow. The presence of lamination is pronounced by the alteration of the variable concentration of vesicles. The accessibility of the area covered by Phase 2 is particularly bad. On the satellite image it is discriminated by a greyish to reddish blue colour, and an almost homogenous surface without any readily recognizable features.

The lava field of Phase 3 covers a large area with subangular blocks that form the rough type of morphology. The blocks are coated by thin dark brown crust providing a rather characteristic field appearance. The fresh broken surface of the rock is black or dark grey. The rocks are readily recognizable on the satellite image by their brown tone or hue, and the hilly morphology dotted by number of large balta (kipukas). The kipuka, or according to their Arabic name balta, are closed shallow depressions within the lava flows surrounded by lava tongues. In Phase 3 the geomorphological dimension of baltas may vary.

The exposures of Phase 4 are widespread in the central Al Haruj al Aswad and in the study area. From the central part of Al Haruj al Aswad large volumes of Phase 4 lava flow poured downward in a more or less radial pattern. The flows are distinguished by their morphology and the preservation of surface features. Lava flowing down pre-existing wadis acquired an elongated form as they filled the wadis.

Phase 4 overlies the Maradah Formation in some places, but the contact is not always clear due to talus covers, but in other places the contact is quite sharp. This phase is represented by dark grey to black vesicular lavas in the upper part with polygonal columnar joints and dense masses in the lower part. This flow is characterized by hilly morphology, pressure ridges and steep distinctly fresher lava scarps. On the satellite image, it has a black colour and features a large number of baltas.

Phase 5 is represented by a well-preserved volcanic form with the freshest of rocks, a vesicular form, and the conspicuous features of the pahohoe structure. The rocks of this phase cover a relatively small area confined to the south-west part of the study area. The excellent preservation and appearance of this phase reflects the younger age of these basaltic flows. On the satellite image the rocks are dark black with dots, representing small baltas of 5 to 20 m in size.

The drainage pattern developed on the surface of basaltic lava flows shows systematic alteration with the age of the basaltic lava flows, Busrewil and Suwesi [2] . From the Landsat TM image, the drainage frequently decreases as the flows of basaltic rock become progressively younger, as Phase 5 has a less dense drainage pattern.

The Libyan volcanic provinces are spatially related to two structural trends: early Palaeozoic structural elements of a prevalent NNW-SSE direction and by late Palaeozoic to Mesozoic structures with a general ENE- WSW trend. These elements represent the regional structures that have played an important role in the geological evolution of Libya, Peregi et al. [5] . Klitzsch [8] suggested that the basalts of Al Haruj al Aswad erupted at the crossing zone of NW trending early Palaeozoic elements (south Al Haruj uplift) and the NE trending late Paleozoic elements (Tibesti-Sirt uplift), and they probably also used some of the upper Cretaceou-Tertiary faults as conduits.

According to Busrewil and Suwesi [2] , the major structural trend of the central part of Al Haruj al Aswad volcanic province is controlled by the alignment of spatter cones and the distribution of the more recent lava flows (Phases 5 and 6) along NW-SE and NE-SW directions. Herein, the present area is located to the north- eastern part of Al Haruj al Aswad, where the data taken from the investigated area reveal that Phases 4 and 5 are controlled by the NE-SW direction, which is coincident with the pattern of the most distinct structures in the region (Figure 17).

From the direction of flow of the more recent lava (Phases 4, 5, 6) and from the alignment of cones, it is suggested that the situation of the Al Haruj lava field is related to tensional lines, one parallel to the Dur al Qussah fault system and the other running along the prolongation of the Hun graben, Busrewil and Suwesi [2] . These lines probably represent extensional stresses reactived by Proterozoic basement faults in the direction NW-SE, NNW-SSE, NE, NNE and E-W during the late Cenozoic, reflecting the evolution of the Red Sea, EL-Akhal [48] . The rising magma erupted along the fault planes cutting across the lower most resistant areas at the cross points of these tensional lines. The lava field and its associated volcanoes are thus considered to have resulted from a rather continuous release of pressure in the main central feeder. After the activity waned in Al Haruj volcanic province (after the Pleistocene), and the area was cut by a conspicuous rejuvenated NW-SE set of faults, Busrewil and Suwesi [2] .

The different rock varieties of the study area are petrographically described in 42 thin sections showing certain petrographic features signifying the different varieties. Mineralogically, most of the studied basaltic rocks contain olivine as an essential mineral together with clinopyroxenes (augite), plagioclase (labradorite to bytownite) feldspars and opaque accessory minerals (probably ilmenite). The recorded olivines are unzoned and either fresh or altered.

Phases 1, 3 and 4 contain a higher modal olivine content and olivine phenocrysts that are larger in size than those from Phases 2 and 5.

The volcanic trends (Phases 4 and 5) are well documented in the area investigated by the continuation of the NE-SW structural trend.

Therefore, the studied rocks can be considered as olivine bearing or olivine basaltic rocks, since they contain a significant amount of modal olivine.

The degree of alteration of the olivines in the studied rocks varies greatly; alteration is manifested around the edges and along the cracks of Phases 3 and 4. On the other hand, highly altered basaltic rocks are found in Phases 1 and 2. The alteration products are either partial or complete to deep reddish iddingsite. Clinopyroxene (augite) is detected in the investigated rocks: it has a maximum extinction angle of 44° to 51°, and is characterized by a pale brown colour, and is subhedral to anhedral and prismatic in shape. Pleochrism is not observed in most samples.

The rocks from Phase 5 are characterized by low plagioclase modes (labradorite to bytownite) and clinopy- roxene (augite) and high contents of vesicles, amygdales and cryptocrystalline to glassy groundmass and microphenocrysts of olivine, which are fresher when compared to rocks from other phases. Plagioclase laths are the most abundant, representing 37% to 48% of the matrix in most of the studied samples. The composition of the plagioclase varied between An 58% to An 78% (labradorite to bytownite). They are both pilotaxitically and randomly arranged, and sometimes they exhibit a trachytic arrangement around olivine phenocrysts. Brown glass and cryptocrystals fill the interstices between laths of plagioclase. The rocks exhibit a porphyritic texture with phenocrysts embedded within the groundmass, whereas the groundmass shows intergranular, intersertal, ophitic and subophitic textures. Olivine phenocrysts occur sometimes as clots displaying a glomerporphyritic texture.

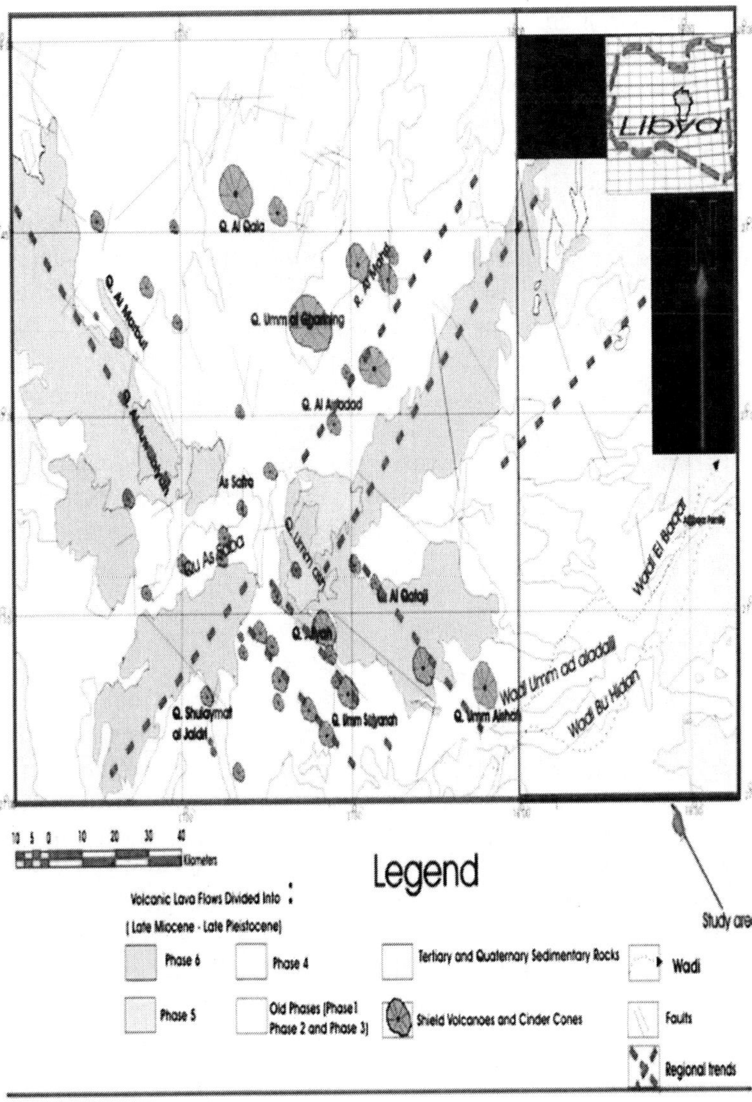

Figure 17. The structural alignment of the central part of al Haruj al Aswad volcanic province (Phases 5 and 6) and a scatter of cones following NE-SW and NW-SE trends, Busrewill and Suwesi [2] .

The order of crystallization of the minerals in the rocks from the various phases is similar. Phenocrysts of olivine result from a slower cooling beneath the surface in the magma chamber prior to eruption. Therefore, they indicate the presence of enough time to equilibrate with the melt at mantle depth. Upon eruption, the remaining liquid crystallizes to form fine laths and tabular or equant crysts comprising the groundmass and resulting in a dense network of plagioclase microlites and pyroxene granules, with opaque oxides of a small size. Glass may solidify at late stage, filling spaces between minerals and forming an intersertal texture, as in Phase 5.

Inclusions of opaque minerals should have formed at an earlier stage than the plagioclase (labradorite to bytownite) that enveloped them. In the ophitic texture, laths of plagioclase (labradorite to bytownite) enveloped by larger pyroxene (augite) is commonly interpreted as indicating that the clinopyroxene (augite) formed later, Winter, [22] . An anhedral or skeletal form of olivine indicates that their formation is in the later or final stages of rock crystallization, Winter [22] . From petrographical and mineralogical criteria alone, it is evident that the investigated basaltic rocks have more olivine phenocrysts than the pyroxene (augite). The groundmass typically consists of interlocking plagioclase (Labradorite to Bytwonite) laths with stubby prisms of clinopyroxene (augite) granules of olivine and opaque minerals.

This petrography study concluded that the five phases are typical alkaline basalts, which supports the findings of the study of Busrewil and Wadsworth [1] . Busrewil and Suwesi [2] suggested that they are a "subalkaline basaltic series which show some similarities except that they are more clearly subophitic with the presence of more opaque minerals". Consequently, the studied mafic volcanic rocks representative of the five phases are probably transitional between alkaline and subalkaline series. But, it must be emphasized that it is necessary to establish their geochemical character to define their petrogenetic affinity within the framework that will establish the magma type. The difference between the various phases is due to the circumstances of magma solidification, not a manifestation of major variances in mineralogical and chemical composition.

The findings of the present study are a positive supplement to previous works such as that by Peregi et al. [5] , which suggested that "the petrographic similarity of the mafic rocks that should be noticed, cannot be distinguished frequently in terms of volcanic phases and flow units based on their petrographic characteristics. The difference sometimes may be more significant in samples taken from the surfaces and deeper part of the same lava flow, than between the different volcanic rock phases."

Again, the above study agrees with that by Farahat et al. [14] who stated that "differences in groundmass textures may be related to differential degrees of cooling and crystallization at various levels of the outpouring of magma."

The geochemistry of the basaltic rock under consideration has been revealed through the analysis of 29 samples for their major and trace elements determined quantitatively by Philips (2404) X-ray fluorescence (XRF). Several geochemical parameters and various diagrams were adopted to recognize the chemical characteristics of these rocks, to evaluate their chemistry within the framework of established magma types and to determine the tectonic environment of their phases.

The major elements of the investigated rocks and the content of their trace elements demonstrates that all the phases are similar in chemical composition. In terms of different methods of recent chemical classification based on major, minor and trace elements, the investigated basaltic rocks have a subalkaline to alkaline affinity. If the term "transitional basalts" is accepted, the investigated rocks are best named as "transitional".

The most studied samples are hypersthene-normative. Wilkinson [49] considered hypersthene-normative rocks to be transitional or mild alkaline, and other samples were classified as nepheline-normative or alkaline basalts, and one sample was located at the boundary between olivine tholeiite and quartz tholeiite, which could represent a distinctive feature of tholeiitic basaltic rock.

In the current study, the focus has been on trace elements, which are generated as enriched or depleted and reveal more about the nature of basaltic rocks. Concentrations of compatible trace elements are low in the studied samples, which supports the contention that they are not primary magma. A characteristic feature of the studied volcanic rocks is their relatively constant ratios of certain incompatible trace elements such as Nb/Zr and Rb/Zr. The constancy of incompatible element ratios provides strong evidence that they were generated from a common source. The Zr/Nb ratios of the basaltic rocks are less than 10, which also indicates an OIB source. The studied samples have similar ratios of Zr/Nb and Hf/Ta, so the effect of crustal contamination on the studied basaltic rocks is probably minor.

Based on MORB normalized trace element patterns of representative samples of different volcanic phases, the patterns show fairly close similarities in their shape and concentration range, whereas they are characterized by very slightly different concentrations of some incompatible trace elements within samples of the same phase and also between samples representative of different phases, but those showed a nearly similar pattern in the less incompatible trace elements.

The studied samples of the different phases, which were enriched in all incompatible trace elements over MORB (normalized values greater than one) in proportion to the degree of incompatibility, also showed slightly positive anomalies in K, Ba and Ta and slightly negative anomalies in Rb, Nb and Zr. The rocks display similar patterns of peaks and troughs, strongly suggesting that they share a common parent and subsequent processes, Winter [22] .

The basaltic samples from the different volcanic phases showed remarkable similarities with each other, but with slight geochemical variations in some incompatible elements. Consequently, the phases cannot be distinguished from one another on the basis of their trace element contents.

Spikes at Pb, Ba and K may reflect an enrichment component, perhaps due to their contamination by continental crust, Winter, [22] . Enriched LIL and the slight alkaline of the studied basaltic rocks could suggest a derivation from an enriched mantle source. The high LIL/HFS pattern and negative Nb anomaly is a feature of much intracontinental volcanism, suggesting a distinctive feature of the subcontinental lithospheric mantle. The studied samples tend to be depleted in Ni and Cr, along with Mg# (Mg/(Mg + Fet)) (52 - 62), suggesting the fractionation of olivine and spinel prior to eruption.

Despite differences in the major element chemistry, the pronounced similarities in the pattern of trace element between different phases certainly suggest that there is a common source of the mantle peridotite beneath the Al Haruj volcanic province.

The results of correspondence analysis are explained by the variance on the coordinates (dimensions) of more than 98% (96.03% of inertia on dimension 1 and 2.192% of inertia on dimension 2), which implies that the relation between the average of the different phases is highly significant. The quality of the average of the five phases expresses the value and importance of each phase, so it can clearly be noticed that all phases are of very significant importance.

The basalts of the area representing all phases were formed in an intracontinental (within plate) tectonic setting and they represent alkali to tholeiitic basalts (transitional basalt), most of them having undergone slight to moderate fractionation. The studied basaltic rocks were plotted around the separating line between the two fields (oceanic and continental), Pearce et al., [42] ; Figures 5-14(b)). Probably the volcanic activity responsible for these basalts occurred in a zone of initial continental rifting.

The trace element distribution patterns of the studied rocks is similar to OIB, which probably represents basaltic melts from the enriched asthenospheric mantle source, often associated with an uprising mantle plume, Peregi et al. [5] .

4.1 Comparison of the Tertiary Volcanics from Different Parts of Libya

Generally, Libya is situated on the Mediterranean foreland of the African shield, which extends over a platform of cratonic basins. The major diastrophic disturbances include the Caledonian and Hercynian disturbances during the Cretaceous, middle Tertiary (Oligocene through to Miocene) and Holocene times, causing uplifts, subsidence, tilting, faulting and intrusions. However, the effects of major diastrophic events were generally broad, and there are very few compressional folds, Goudarzi [50].

Apart from the different ages of the four principal volcanics in Libya, much similarity can be detected among them, particularly in the mechanism and mode of eruption. The common criterion for Tertiary volcanics is that volcanism occurs mainly along well-defined old structural elements. Those structural elements are basically developed from three structural periods characteristic of the middle part of North Africa, Klitzsch [51] .

The basement underneath the Tertiary volcanism areas is bound to the overlap of older structural elements forming the boundaries of the major current basins in Libya. The overburden sequence of those areas are of the minimum thickness, and occasionally the Precambrian crystalline rocks are exposed in limited and comparatively small areas such as in Jabal al Hasawinah, west of Jabal Nuqay, and the Tibesti area (Busrewil and Oun

[28] ; El-Makhrouf [52] ; Goudarzi [50]). The volcanic fields of the four principal provinces in Libya are strongly correlated to the current basement elevated areas. Consequently, it seems to be that the attenuation of the crust within specific areas during Paleozoic and Mesozoic times is one of the major contributory factors that facilitated the outpouring in the volcanic provinces in Libya during the Tertiary Period.

The Gharyan volcanic province (GVP) is the smallest of the principal Tertiary volcanic areas in Libya, and it forms a generally continuous sheet in the central part of Jabal Nafusah. The others are Jabal as Swada, Jabal Al Haruj and Jabal Nuqy. These volcanic fields represent the Tertiary-Quaternary outpouring of volcanic activity and define a NNW-SSE linear trend with a decrease in age from NNW to SSE. The GVP lies at the intersection of two major structures, the NNW trending Tripoli-Tibesti uplift formed in Caledonian times, and the E-W trending "Nafusah uplift" formed during the Hercynian orogeny (Anketell and Ghellali [53] ; Goudarzi [50] ; Mikbel [54]).

The volcanic field of Jabal al Hasawinah is related to the intersection of the early Paleozoic NNW-SSE Tripoli-Tibesti uplift and the Hercynian ENE-WSW Al Qarqaf uplift, (Busrewil and Oun, [28] ; Jurak, [55]). Further NE of the Jabal al Hasawinah volcanic field, there is the Al Qarqaf uplift crossing the early Paleozoic NNW-SSE structural element of the Al Haruj uplift; Jabal as Swada is spatially related to this area (Woller and Fediuk, [9]).

The situation of the Al Haruj lava field, contrary to the other volcanic provinces, is related to three structural elements: the early Paleozoic Al Haruj uplift of a NNW-SSE direction, the late Paleozoic to Mesozoic NE fault system that is associated with Tibesti-Sirt uplift, and the prolongation of the Cretaceous-Tertiary Hun graben, Busrewil and Suwesi [2] ; Woller and Fediuk [9]). This may account for the greater apparent size of the Al Haruj volcanic field compared to the other volcanic provinces, Farahat et al. [14] .

Chemical analyses of major, trace and rare earth elements (REE) from Al Haruj volcanics are presented in Appendix III, El-Shaafi [6] , together with analyses of representative rocks from different parts of Libya for comparison purposes (Tables B.1-B.4). The analysed samples from representative rocks from different parts of Libya are depicted on variation diagrams. The consistent patterns of the plotted data suggest that apart from oxidation, there has been little alteration.

The subalkaline basalts from the different volcanics in Libya tend to be relatively evolved with a differentiation index (DI) extending from 29.9 in the most primitive basalts to 39.9 in the most evolved hawaiites, Busrewil and Suwesi, [2] . There are no associated less-evolved rock types exposed that might be representative of the parent magma. All the chemically analysed rocks have relatively high silica and low TiO_2 and total alkalis, and although there is some partial overlap with the alkali basalts of the main lava series, they are clearly subalkali. They could be related to other lavas generated at lower pressure and/or higher degrees of partial melting of the parent magma. Fractional crystallization at shallow levels, involving major minerals (ol, cpx, ±opaques) occurring in the rock, can account for the observed series, Busrewil and Suwesi, [2] .

For classifying the rocks on a geochemical basis, the variation of silica and total alkalis is one of the most important discrimination techniques. The IUGS total alkalis (TAS) classification of volcanic rocks of Le Bas et al. [56] is adopted and shown in Figure 18.

The Gharyan volcanics, the present study volcanics and the study made by Busrewil and Suwesi, [2] of the basaltic rocks of the central part of Al Haruj al Aswad, illustrate the suite of subalkaline basalts and alkali basalts, and most of them fall within the basalt field of the TAS diagram. The more differentiated types are simply referred to as hawaiites and mugearites (Figure 19(a)). Twenty-seven samples of the studied basaltic rocks from GVP were chemically analysed, Gafeer [57] for major and trace elements (Tables B.1-B.4, Appendix III). Fresh samples were selected as far as possible and attention was paid to the homogeneity of the samples.

The volcanics in the GVP comprise four petrologic and chronologic phases: the Old Lava Series (OLS); its subordinate, the Young Lava Series (YLS; Early-Eocene to Quaternary); phonolite-trachyite intrusions (Late Eocene-Early Oligocene); and the late volcanic centres (LVC) of the Middle-Late Miocene (see Figure 19(a); Aboazoam et al. [58] ; Almond et al. [59] ; Busrewil and Wadsworth [1] ; Busrewil and Wadsworth [3] ; Hegazy [60] ; Piccoli [61]).

The Gharyan lava series demonstrates a subalkaline affinity, as shown by its basalt and basaltic andesite presented in Cox et al. ([19] ; Figure 19(a), Figure 19(b)), which typically contain normative hypersthene (although three of the analysed samples contain minor amounts of normative nepheline). Generally, the term “transitional” applies to basaltic rocks that exhibit a nepheline-free norm, Le Bas et al. [56] .

Wilkinson [49] considered hypersthene-normative rocks to be transitional or mildly alkaline, a characterization which Almond et al. [59] had used to define the suite of the lava series within the Gharyan volcanic province. Phonolites were not observed in Jabal Al Haruj al Aswad by previous investigators or in this study, but the emplacement of phonolites in Jabal Awayant, the Gharyan volcanic province and Jabal Hassawnah, Jurak [55] suggests that a genetic relationship between the basement highs, broad domal uplifts and the emplacement of phonolites exists Busrewil and Suwesi [2] .

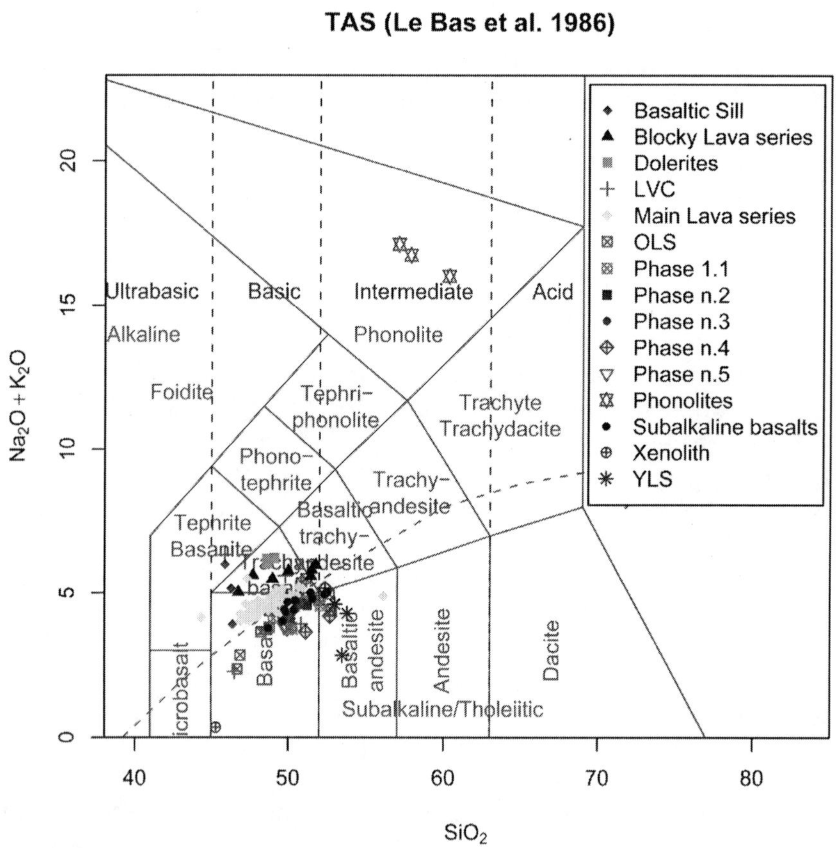

Figure 18. The chemical classification and nomenclature of volcanic rocks studied in the different volcanic provinces in Libya, using the total alkalis versus silica (TAS) diagram of Le Bas et al. [56] .

The divisions of Al Haruj volcanics into plateau lavas, blocky lava, shield volcanoes and doleritic dyke-like masses, based on field appearance have already been established by Busrewil and Suwesi [2] . These data together with chemical observations are used to identify the various rock types present and to assess their petrogenesis. Variations of silica and total alkalis demonstrate a subalkaline affinity trending towards alkaline field (Figure 19(b); in Cox et al. [19]).

Eighty-eight samples of the studied basaltic rocks from Al Haruj al Aswad were chemically analysed by Busrewil and Suwesi [2] for major and trace elements (Tables B.1-B.4, Appendix III). Fresh samples were selected as far as possible and attention was paid to the homogeneity of the samples. Although there is similarity between the lava series of the current study data (Figure 19(c)) to counterparts in the study by Busrewil and Suwesi [2] , the above suite shows a wider compositional range as it attains a wider differentiation index (DI) than the present study.

Twenty-nine samples of the studied basaltic rocks from Al Haruj al Aswad were chemically analysed (the current study) for major and trace elements (Tables B.1-B.4, Appendix III; El-Shaafi [6]). Fresh samples were selected as far as possible and attention was paid to the homogeneity of the samples, (Figure 19(c); Cox et al. [19]).

Based on geochemical data, alkali-olivine basalt and transitional basalt of the study area appear to have been emplaced continentally within plate. The use of immobile HFS elements Th-Hf-Ta is proposed by Wood [47] for basalt plots in the field of “within plate basalts” (Figure 20), whereas the most differentiated members of subalkaline transitional basalts lie in the range of “continental basalts” field on FeOt, MgO, Al_2O_3, (Figure 21; Pearce et al. [43]), which encompasses both subalkaline basalts and basaltic andesites.

The within plate tendency for the analysed basaltic rocks from various regions are pronounced in the Zr/Y-Zr diagram (Figure 22).

The chemical data for different volcanic provinces are normalized to the primitive mantle (Figure 23) in which the observed Nb anomaly is a characteristic feature of continental and oceanic alkali basalts. The contamination by crustal rocks is the main factor controlling the behaviour of K, Ba and Rb (Najafi et al., 1981, as cited in Aboazom, [31]).

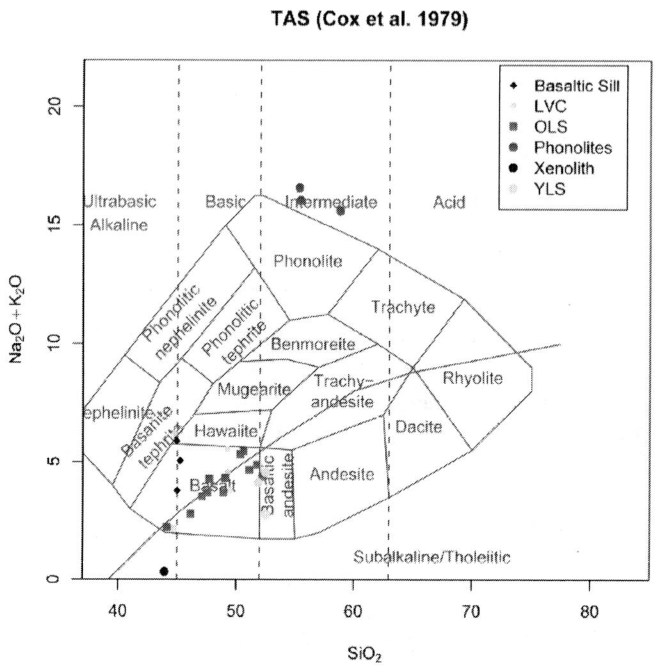
TAS (Cox et al. 1979)
Basaltic Sill
LVC
OLS
Phonolites
Xenolith
YLS
Ultrabasic
Alkaline
Basic
Intermediate
Acid
Phonolite
Phonolitic nephelinite
Phonolitic tephrite
Trachyte
Benmoreite
Mugearite
Trachy-andesite
Rhyolite
ephelinite
Basanite tephrite
Hawaiite
Dacite
Basalt
Basaltic andesite
Andesite
Subalkaline/Tholeiitic
$Na_2O + K_2O$
SiO_2
0
5
10
15
20
40
50
60
70
80

(a)

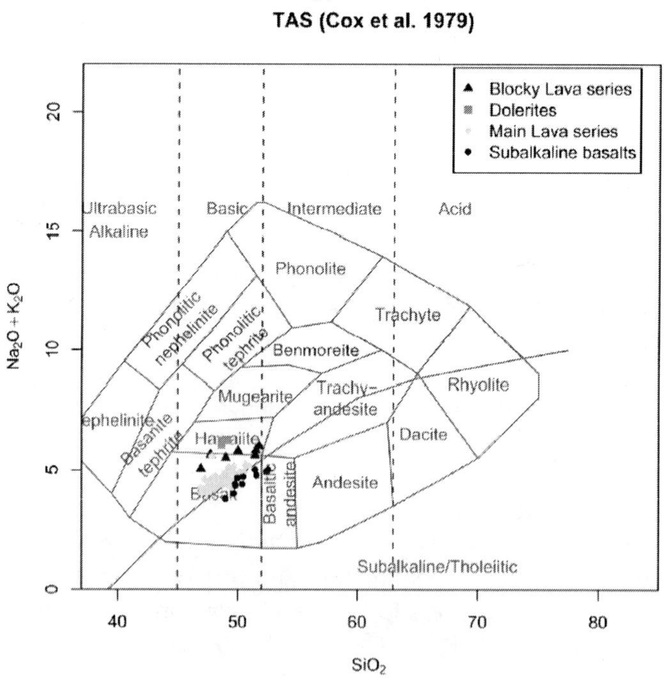
TAS (Cox et al. 1979)
Blocky Lava series
Dolerites
Main Lava series
Subalkaline basalts
Ultrabasic
Alkaline
Basic
Intermediate
Acid
Phonolite
Phonolitic nephelinite
Phonolitic tephrite
Trachyte
Benmoreite
Mugearite
Trachy-andesite
Rhyolite
ephelinite
Basanite tephrite
Hawaiite
Dacite
Basalt
Basaltic andesite
Andesite
Subalkaline/Tholeiitic
$Na_2O + K_2O$
SiO_2
0
5
10
15
20
40
50
60
70
80

(b)

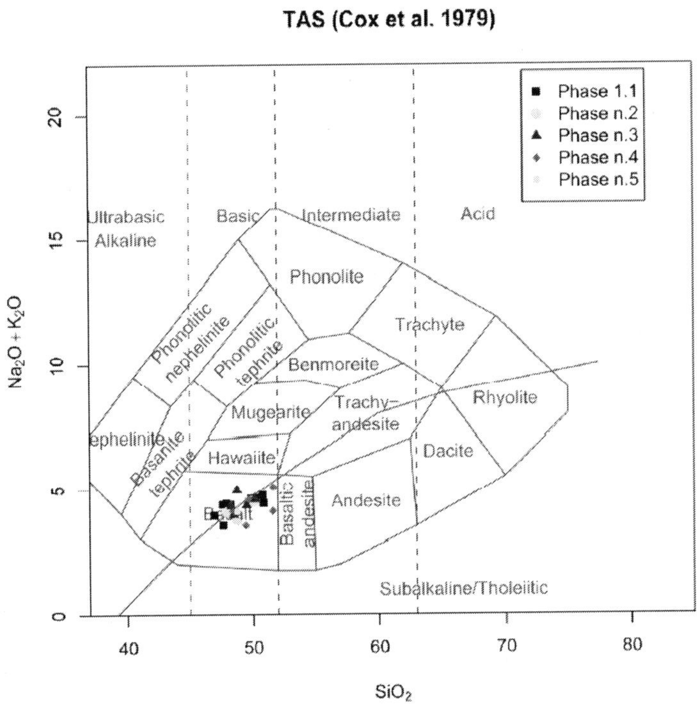

(c)

Figure 19. Alkali versus silica diagrams of Cox et al. [19] showing (a) Plot of total alkalis vs silica for Gharyan province after Gafeer [57] ; (b) Plot of total alkalis vs silica from Busrewil and Suwesi [2] ; (c) Plot of total alkalis vs silica for the area investigated (current study).

The basalts lie within the field of ocean island basalts (Figure 24; Mullen, [62]), which represents basaltic melts from an enriched asthenospheric mantle source. This characteristic has been confirmed for the major outpourings in volcanic provinces in Libya, such as Jabal as Swada, Al Haruj Al-Abiyad (Aboazoam et al., [58] ; Hegazzy, [60] ; Peregi et al., [5]).

Finally, it is interesting to ask whether the clustering in each and every plot shown in this summary illustrates that the current study represents an evolved portion or a different degree of partial melting from a common igneous source. Further research may give the answer.

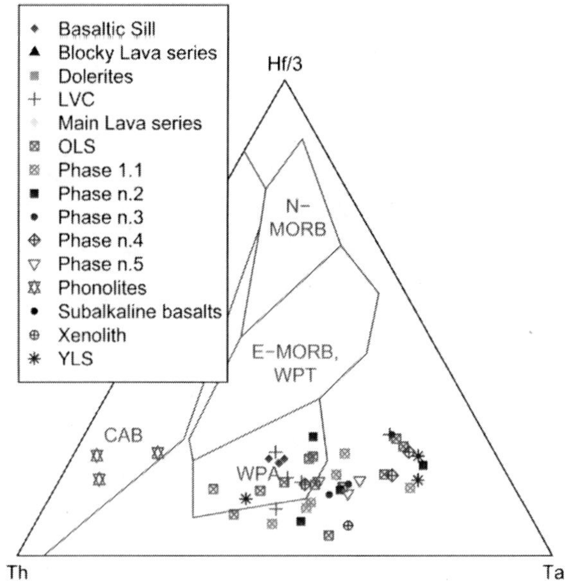

Figure 20. The Th-Hf-Ta discrimination diagram for alkali and tholeiite basalts, after Wood [47] .

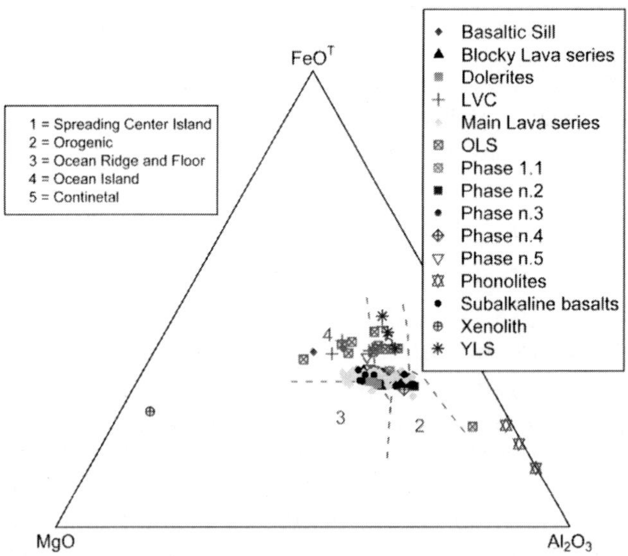

Figure 21. The MgO-FeOt-Al_2O_3 diagram, after Pearce et al. [43] , showing the discriminant boundaries between tectonic settings based on the compositional range of recent volcanics. The diagram can only used with subalkaline rocks in the silica range of 45% - 56%.

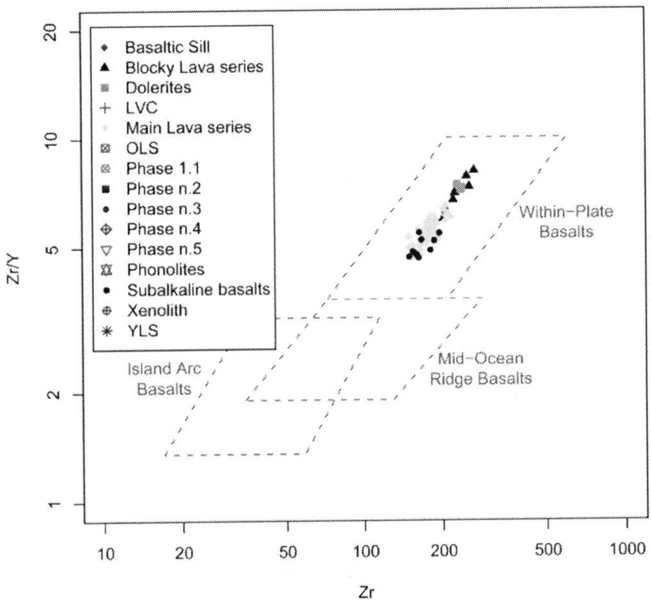

Figure 22. Discrimination diagram for basalts based upon the Zr-Zr/Y variation. The diagram has a logarithmic scale. The analysed samples of Busrewil and Suwesi [2] from the central part of Al Haruj al Aswad in the field within plate basalts (WPB; after Pearce and Norry, [44]).

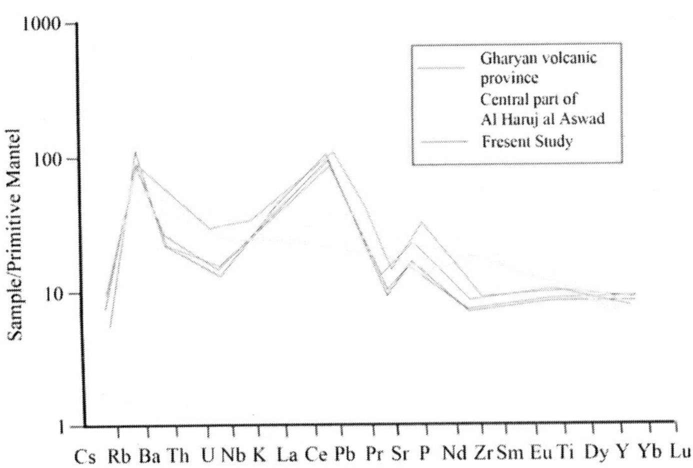

Figure 23. Trace element concentrations of some representative samples from the Gharyan volcanic province by Gaffer [57] and Al Haruj al Aswad by Busrewil and Suwesi [2] and the current study, normalized to the composition of primitive mantle.

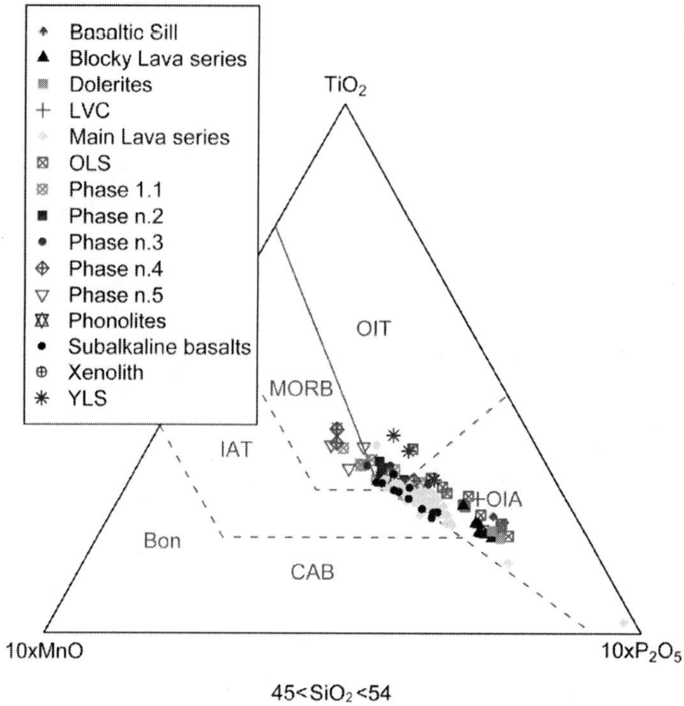

Figure 24. The MnO-TiO_2-P_2O_5 discrimination diagram for basalts and basaltic andesites of 45% - 54% SiO_2, after Mullen, [62] . The fields are: MORB; OIT?oceanic island tholeiite; OIA?oceanic island alkali basalts; CAB?is- land arc calc-alkaline basalt; IAT?island arc tholeiite; Bon?boninite.

ACKNOWLEDGEMENTS

The authors would like to express their appreciation to A. El-Arnauti and A. El-Makhrouf for their valuable remarks. We feel obliged to express gratitude to M. J. Aburkheis, the former Head of the Popular Committee for the Industrial Research Centre (IRC), and are also indebted to I. Al. Abeid, the Director of the Geological Research and Mining Department and to S. Turki, the Director of the Mapping Department for their encouragement and kind support to the present work while in the field. Thanks and appreciation must go to Rami Gebreel for his valuable help in statistics. Appreciations is expressed to the geological staff of the (IRC): A. Buagrab, A. Garbee, N. Buziad, K. Etrish and Chtronbe from Belgrade University , for their kind assistance during the field work. Thanks are also due to the crew from the IRC mapping project, SE Libya for their help and kind assistance on the camp.

N. M. Al-Hafdh, Abd El-Salam S. El-Shaafi Thanks also extend to M. El-Amawi from the Department of Earth Sciences, University of Benghazi for the coordination and communication with the Central Laboratories of the Geological Survey of Egypt to access the analyses of the samples in question.

REFERENCES

1. Busrewil, M.T. and Wadsworth, W.J. (1980) Preliminary Chemical Data on the Volcanic Rocks of Al Haruj Area, Central Libya. In: Salem, M.J. and Busrewil, M.T., Eds., The Geology of Libya, Vol. 3, Academic Press, London, 1077-1080.

2. Busrewil, M.T. and Suwesi, S.Kh. (1993) Geological Map of Libya 1:250,000, Sheet. Al Haruj Al Aswad NG 33-4. Explanatory Booklet. Industrial Research Centre, Tripoli, 95.

3. Nemeth, K., Suwesi, Kh., Peregi, Z., Gulacsi, Z. and Ujszaszi, J. (2003) Plio/Pleistocene Flood Basalt Related Scoria and Spatter Cones, Rootless Lava Flows and Pit Craters, Al Haruj Al Abiyad, Libya. Geolines, 15, 107-112.

4. Busrewil, M.T. and Wadsworth, W.J. (1996) Tertiary-Quaternary Alkaline Subalkaline Magmatism in Gharayan Area: Field Aspects and Petrography. PRJ, 8, 13-23.

5. Peregi, Zs., Less, G.Y., Konrad, Gy., Fodor, L., Gulacsi, Z., Gyalog, L., Turki, S.M., Suwesi, S.Kh., Sherif, Kh. and Dalub, H. (2003) Explanatory Booklet. Geological Map of Libya 1:250,000. Sheet: Al Haruj Al Abyad NG 33-8. Industrial Research Centre, Tripoli, 248.

6. El-Shaafi Abd El-Salam, S. (2008) Geochemistry and Petrology of Basic Volcanic Rocks of the NE. Part of Jabal AL-Haruj AL-Aswad, Central Libya. Unpublished MSc Thesis, University of Benghazi, Libya.

7. Hallett, D. (2002) Petroleum Geology of Libya. Elsevier, Amsterdam.

8. Klitzsch, E. (2000) The Structural Development of the Murzuq and Kufra Basins: Significance for Oil and Mineral Exploration. In: Sola, M.A. and Worsley, D., Eds., Geological Exploration of the Murzuq Basin, Elsevier, Amsterdam, 143-150.

9. Woller, F. and Fediuk, F. (1980) Volcanic Rocks of Gabal as Sawda. In: Salem, M.J. and Busrewil, M.T., Eds., The Geology of Libya, Vol. 3, Academic Press, London, 1081-1093.

10. Pesce, A. (1966) Uau en Nammus: South Central Libya and Northern Chad. Excursions fùhrer der petroleum Exploration Society of Libya, Tripoli, 47-51.

11. Ade Hall, F.M., Reynolds, P.H., Dagley, P., Musset, A.G., Hubbard, T.B. and Klitzsch, E. (1974) Geophysical Studies of North African Cenozoic Volcanic Areas Al Haruj Al-Assuad, Libya. Canadian Journal of Earth Science, 11, 998- 1006.

12. Less, Gy., Turki, S.M., Suwesi, S.Kh., Peregi, L.F., Koloszar, L., Kalmar, J., Sherif, Kh., Csaszar, G., Gulasci, Z., Dalum, H. and Al Tajuri, A. (2006) Explanatory Booklet. Geological Map of Libya 1:250,000. Sheet: Waw Al Kabir NG 33-12. Industrial Research Centre, Tripoli, 295.

13. Klitzsch, E. (1968) Der Basalt Vulkanismus des Djebel Haroudj, Ostfezzan, Libya. Geologische Rundschau, 57, 585- 609.

14. Farahat, E.S., Abdel Ghani, M.S., Aboazom, A.S. and Asran, A.M.H. (2006) Mineral Chemistry of Al Haruj Low- Volcanicity Rift Basalts, Libya: Implications for Petrogenetic and Geotectonic Evolution. Journal of African Earth Sciences, 45, 198-212.

15. Vesely, J. (1985) Geological Map of Libya 1:250,000 Sheet: Zallah NH 33-16. Explanatory Booklet. Industrial Research Centre, Tripoli, 125.

16. Woller, F. (1984) Geological Map of Libya 1:250,000, Sheet: Al Fuquaha NG 33-3. Explanatory Booklet. Industrial Research Centre, Tripoli, 123.

17. Mackenzie, W.S. and Adams, A.E. (1995) A Color Atlas of Rocks and Minerals in Thin Section. Manson, London .

18. Hatch, F.H., Wells, A.K. and Wells, M.K. (1972) Petrology of the Igneous Rocks, Fifth Impression. Thomas Murby, London .

19. Cox, K.G., Bell, J.D. and Pankhurst, R.J. (1979) The Interpretation of Igneous Rocks. Allen & Unwin, London.

20. Middlemost, E.A.K. (1980) Iron Oxidation Ratios, Norms and the Classification of Volcanic Rocks. Chemical Geology, 77, 19-26.

21. Middlemost, E.A.K. (1975) The Basalt Clan. Earth Science Reviews, 11, 337-364.

22. Winter, J.W. (2001) An Introduction to Igneous and Metamorphic Petrology. Prentice Hall.

23. Winchester, J.A. and Floyd, P.A. (1976) Geochemical Discrimination of Different Magma Series and Their Differentiation Products Using Immobile Elements. Chemical Geology, 20, 325-343.

24. Rollinson, H.R. (1993) Using Geochemical Data, Evolution, Presentation, Interpretation. Longman Scientific & Technical, Oxford/John Wiley, New York, 1-206.

25. Yoder, H.S. and Tilley, C.E. (1962) Origin of Basalt Magmas: An Experimental Study of Natural and Synthetic Rock Systems. Journal of Petrology, 3, 342-532.

26. Thompson, R.N. (1984) Dispatches from the Basalts Front. 1. Experiments. Proceedings of the Geologists' Association, 95, 249-262.

27. Wilkinson, J.F.G. (1986) Classification and Average Chemical Composition of Common Basalts and Andesites. Jour- nal of Petrology, 27, 181-194.

28. Busrewil, M.T. and Oun, K.M. (1991) Geochemistry of the Tertiary Alkaline Rocks of Jabal al Hasawnah, West Libya. In: Salem, M.J., Busrewil, M.T. and Ben Ashour, A.M., Eds., Geology of Libya, Vol. 7, Elsevier, Amsterdam, 2587- 2598.

29. Baker, B.H. (1987) Outline of the Petrology of the Kenya Rift Alkaline Province. In: Fitton, J.G. and Upton, B.J., Eds., Alkaline Igneous Rocks, Geological Society, Special Publications, London, 293-311.

30. Pearce, J.A. (1983) Lithosphere in Magma Genesis at Active Continental Margins. In: Hawesworth, C.J. and Norry, M.J., Eds., Continental Basalts and Mantle Xenoliths, Shiva.

31. Aboazom, A.S. (2005) Petrogenesis of Phase 6 of Al Haruj Al Aswad Volcanic Province, Central Libya. Sebha University Journal, 4, 96-113.

32. Fitton, J.G., James, D., Kempton, P.D., Ormerod, D.S. and Leeman, W.P. (1988) The Role of the Lithospheric Mantle in the Generation of Late Cenozoic Basic Magmas in the Western United States. J. Petrol. Special Lithosphere Issue, 331-349.

33. Hawkesworth, C.J., Hempton, P.D., Ellam, R.M. and Calsteren, P. (1990) Continental Mantle Lithosphere and Shallow Level Enrichment Processes in the Earth's Mantle. Earth and Planetary Science Letters, 96, 256-268.

34. Hopper, P.R., Bailey, D.G. and Hawkeworth, C.J. (1993) Isotopic and Geochemical Constraints on the Origin and Evolution of the Columbia River Basalt. Journal of Petrology, 34, 1203-1246.

35. Longmuir, C.H., Beneder, J.F., Bence, A.E., Hanson, G.N. and Taylor, S.R. (1977) Petrogenesis of Basalts from the Famous Area: Mid-Oceanic Ridge. Earth and Planetary Science Letters, 36, 133-156.

36. Kay, R.W. and Hubbard, N.J. (1978) Trace Elements in Ocean Ridge Basalts. Earth and Planetary Science Letters, 38, 95-116.

37. Sun, S.S. (1980) Lead Isotopic Study Volcanic Rocks from Mid-Ocean Ridges, Ocean Islands and Islands Arcs. Philosophical Transactions of the Royal Society A, 297, 409-445.

38. Pearce, J.A. (1976) Statistical Analysis of Major Element Patterns in Basalts. Journal of Petrology, 17, 15-43.

39. Pearce, J.A. (1980) Geochemical Evidence for the Genesis and Eruptive Settings of Lavas from Tethyan Ophiolites. Proceedings of the International Ophiolite Symposium, Cyprus 1979, Institute of Mining and Metallurgy, 261-272.

40. Pearce, J.A. and Cann, J.R. (1971) Ophiolite Origin Investigated by Discriminants Analysis Using Ti, Zr and Y. Earth and Planetary Science Letters, 12, 339-349.

41. Pearce, J.A. and Cann, J.R. (1973) Tectonic Setting of Basic Volcanic Rocks Determined Using Trace Element Analyses. Earth and Planetary Science Letters, 19, 290-300.

42. Pearce, T.H., Gorman, B.E. and Birkett, T.C. (1975) The TiO_2-K_2O-P_2O_5 Diagram: A Method of Discriminating between Oceanic and Non-Oceanic Basalts. Earth and Planetary Science Letters, 24, 419-426.

43. Pearce, T.H., Gorman, B.E. and Birkett, T.C. (1977) The Relation between Major Element Chemistry and Tectonic Environment of Basic and Intermediate Volcanic Rocks. Earth and Planetary Science Letters, 36, 121-132.

44. Pearce, J.A. and Norry, M.J. (1979) Petrogenetic Implications of Ti, Zr, Y and Nb Variations in Volcanic Rocks. Contributions to Mineralogy and Petrology, 69, 33-47.

45. Wood, D.A., Tarney, J., Varet, J., Saunders, A.D., Bougault, H., Joron, J.L., Treuil, M. and Cann, J.R. (1979) Geochemistry of Basalts Drilled in the North Atlantic by IPOD Leg 49: Implications of Mantle Heterogeneity. Earth and Planetary Science Letters, 42, 77-79.

46. Floyd, P.A. and Winchester, J.A. (1975) Magma Type and Tectonic Setting Discrimination Using Immobile Elements. Earth and Planetary Science Letters, 27, 211-218.

47. Wood, D.A. (1980) The Application of a Th-Hf-Ta Diagram to Problems of Tectonomagmatic Classification and to Establishing the Nature of Crustal Contamination of Basaltic Lavas of the British Tertiary Volcanic Province. Earth and Planetary Science Letters, 50, 11-30.

48. El-Akhal, H. (2004) Contribution to the Petrology, Geochemistry and Tectonic Setting of the Basalt Flows of the Umm-Qais Plateau, North Jordan.

49. Wilkison, J.F.G. (1986) Classification and Average Chemical Composition of Common Basalts and Andesites. Journal of Petrology, 27, 181-194.

50. Goudarzi, G.H. (1980) Structure: Libya. In: Salem, M.J. and Busrewil, M.T., Eds., The Geology of Libya, Vol. 3, Academic Press, London, 879-892.

51. Klitzsch, E. (1971) The Structural Development of Parts of North Africa since Cambrian Times. In: Gray, C., Ed., First Symposium on the Geology of Libya, Faculty of Science, University of Libya, Tripoli, 253-262.

52. El-Makhrouf, A.A. (1988) Tectonic Interpretation of Jabal Eghei Area and Its Regional Application to Tibesti Orogenic Belt, South Central Libya. Journal of African Earth Science, 7, 945-967.

53. Anketell, J.M. and Ghellali, S.M. (1991) A Palaeogeologic Map of the Pre-Tertiary Surface in the Region of the Jifarah Plain and Its Implication to the Structural History of Northern Libya. In: Salem, M.J., Busrewil, M.T. and Ben Ashour, A.M., Eds., Geology of Libya, Vol. 7, Elsevier, Amsterdam, 2681-2687.

54. Mikbel, S.R. (1977) Basement Configuration and Structure of West Libya. Libyan Journal of Science, 7A, 19-33.

55. Jurak, L. (1978) Geologic Map of Libya 1:250,000, Sheet: Jabal Al Hasawanah NH 33-14. Explanatory Booklet. Industrial Research Centre, Tripoli, 74-100.

56. Le Bas, M.J., LeMaitre, R.W., Strecheisen, A.L. and Zanettin, B. (1986) A Chemical Classification of Volcanic Rocks Based on the Total Alkali-Silica Diagram. Journal of Petrology, 27, 745-750.

57. Gafeer, A.S. (2007) The Petrology and Geochemistry of Gharyan Volcanic Province, NW Libya. Unpublished MSc Thesis, The University of Garyounis, Benghazi.

58. Aboazom, A.S., Asran, A.S.H., Abdel Ghani, M.S. and Farhat, E.S. (2006) Geologic and Geochemical Constraints on the Origin of Some Tertiary Alkaline Rift Volcanics, Gharyan Area, Northwestern Libya. Assiuit University Journal of Geology, 35, 25-47.

59. Almond, D.C., Busrewil, M.T. and Wadsworth, W.J. (1974) The Gharian Tertiary Volcanic Province of Tripolatania, Libya. Geological Journal, 9, 17-28.

60. Hegazzy, A.H. (1999) Tertiary Volcanics in Libya: Evidence for the Direction and Rate of the African Plate Motion. Proceedings of the Fourth International Conference on Geochemistry, Alexandria University, Egypt, 401-419.

61. Piccoli, C. (1971) Outlines of Volcanism in Northern Tripolitania. In: Gray, C., Ed., Symposium on the Geology of Libya, Fac. Sci., Univ. Padova.

62. Mullen, E.D. (1983) $MnO/TiO_2/P_2O_5$: A Minor Element Discriminant for Basaltic Rocks of Oceanic Environments and Its Implications for Petrogenesis. Earth and Planetary Science Letters, 62, 53-62.

Index